FORESTRY + WATER CONSERVATION IN

South Africa

History, Science + Policy

FORESTRY + WATER CONSERVATION IN

South Africa

History, Science + Policy

Brett Bennett + Fred Kruger

Australian
National
University

PRESS

WORLD FOREST HISTORY SERIES

ANU PRESS

Published by ANU Press
The Australian National University
Acton ACT 2601, Australia
Email: anupress@anu.edu.au
This title is also available online at press.anu.edu.au

The *World Forest History Series* aims to produce rigorous histories of forestry that inform contemporary environmental policy debates and provide enduring scholarly landmarks for future generations of historians and environmental researchers. Each book, published in hardcopy and available as a free electronic copy for download, is available to scholars and the public around the world. The series is affiliated with the Centre for Environmental History at The Australian National University.

Series Editors: Gregory A. Barton, Professor of History, Western Sydney University, Affiliate, Centre for Environmental History, The Australian National University; Brett M Bennett, Senior Lecturer in History, Western Sydney University, Affiliate, Centre for Environmental History, The Australian National University.

National Library of Australia Cataloguing-in-Publication entry

Creator: Bennett, Brett M., 1983- author.

Title: Forestry and water conservation in South Africa : history,
                  science and policy / Brett M. Bennett,
                  Frederick J. Kruger.

ISBN: 9781925022834 (paperback) 9781925022841 (ebook)

Series: World forest history series.

Subjects: Afforestation--South Africa.
                  Forests and forestry--South Africa.
                  Forest policy--South Africa.
                  Water conservation--South Africa.

Other Creators/Contributors:
                  Kruger, F. J. (Frederick J.), 1944- author.

Dewey Number: 634.9560968

Cover design and layout by ANU Press. Cover image: Ethel Edwards, *Jonkershoek Valley with exotic trees and indigenous flowers*, date unknown (circa 1905-1920).

Contents

List of Illustrations

Acknowledgements

The authors contributed equally to this work and are given in alphabetical order on the title.

I owe a great debt of gratitude to all those who infused in me the capacity to learn about my surroundings, forests included, beginning with my parents Barbara May and Cecil Eric, and including many from the Lebenya peoples of the Mount Fletcher district, the peoples of Kavango, forest workers from the Transkei, and elsewhere. For a time, I had as my neighbours the white labourers in the settlement at Jonkershoek, and from them too I learnt. Christiaan Wicht, as one of the professors in the Forestry Faculty at the University of Stellenbosch, taught me not only forestry and ecology but also to see the Cape landscapes through many lenses, and he was my promoter for the Master's degree. J. D. M. Keet was my grandfather, and over time he shared much of his professional and tacit knowledge of South African forestry with me. This included a memorable excursion in the early 1960s, during his full-time engagement to provide technical support to the Westfalia forestry program. At an age of over 80 years, he took me on a strenuous excursion to the indigenous forests on the heights adjoining Woodbush, where, at times scrambling on our hands and knees, we found the sawpits of the woodcutters who supplied the mines 70 years before, 500 km away.

I thank the Hans Merensky Foundation for the financial support that allowed visits to Stellenbosch, and travel to and from the United States in 2010 to access the libraries at Cornell University (whose librarians provided a marvellous service). While in the US, the Forest History Society bore my costs for a brief but very fruitful visit to their library in Durham, NC. The University of Western Sydney (now Western Sydney University) brought me to Australia in 2013 for a rewarding period of work in the libraries there. I thank the Council for Scientific and Industrial Research (CSIR) for permission to access its forestry collections.

My gratitude is also due to Susan Wicht Clarke for a correspondence that threw important light on her father Christiaan's life and personality. Greg Forsyth of the CSIR in Stellenbosch has been unfailingly helpful and has kept a watchful eye on the Jonkershoek collection there, while Karen Eatwell has helped me to access the Pretoria collections of the Wicht Papers and the files of the South African Forestry Research Institute (collections that she has resolutely monitored as they were shunted from one location to another and another). Marius Bakkes of the Mpumalanga History Interest Group helped me with materials on the old Eastern Transvaal; Richard Green of the Department of Agriculture, Forestry and Fisheries in Nelspruit has provided key documents which otherwise would have been overlooked; Roger Godsmark of Forestry South Africa provided key data on the recent history of the industry; Rodney Constantine, at the Anglo-Boer War Museum, Bloemfontein, uncovered key facts on the internment camp at Tokai and drew my attention to accounts of woodcutters in the eastern Cape frontier territory; Desiree Lamoral of the Institute for Commercial Forestry Research in Pietermaritzburg provided me with key historical publications; Corine Viljoen of the Southern African Institute of Forestry helped me access important forestry literature; Tom Vorster of the Department of Agriculture, Forestry and Fisheries in Pretoria showed me the treasure trove of old forestry photographs and maps that he has carefully protected.

To my son Laurence, my thanks for hospitality and support during a crucial period, and to his siblings, Alexandra, Oliver and Elizabeth, my gratitude for their constant support. My thanks go also to Caroline Stiebler and Stefan Mangos for their hospitality in Pretoria, and to Brett Bennett and Greg Barton for their hospitality while in Australia. And to Luzelle Naudé, for faith.

Fred Kruger
Skukuza, South Africa, February 2015

This book would not exist if I had not received a chance email in February 2010 from my co-author, Fred. I had just presented a draft chapter of my dissertation at Stellenbosch University when I received an email from him informing me that he had received the paper from a friend who attended the meeting. Would I care to speak with him because he had ideas that may be useful for my ongoing research? As it turned out, I was soon to fly to Pretoria, where he lived. Fred generously invited me to stay at his house during my visit; I agreed. The rest is history, or more accurately, it is a history book.

A second meeting with Fred in the Kruger National Park in 2011 solidified our mutual interests and raised a few possibilities for collaboration. The decision to co-author a book was made during a transcountry archival research trip in 2012 across Southern Africa with Fred Kruger and Gregory Barton. We drove from Pretoria, South Africa, to Maun, Botswana, and Windhoek, Namibia, before returning to South Africa. Fred was taking a creative writing unit and decided to write about Christiaan Wicht. Perhaps inspired by Greg's vivid dreaming as a result of a fever in the Kalahari (when the air conditioning broke down in Fred's Land Rover), we decided to toy with the idea of doing a book. Fred and I agreed to write the book together while eating Kalahari truffles and hartebeest for dinner in Windhoek. Discussions continued—apologies to Greg who had to endure them for around 20 hours in the car as we drove all the way back to Pretoria. The conversation has never really ended.

Though much of the book focuses on the Cape, we drafted most of it in the Kruger National Park. Many of the ideas we had for the book emerged while driving in the evening throughout the Kruger National Park after we finished writing in the late afternoon. Thanks go to Laurence Kruger for housing me in Skukuza, and for introducing me to the Kruger and wider region, especially the Escarpment and Sabie River catchment, which are the focus of one of the next projects Fred and I are undertaking.

The project would not have been possible without funding and support from my colleagues at Western Sydney University (WSU). I owe a great deal to my Dean, Peter Hutchings, who has supported me unfailingly throughout my time at WSU. Peter allowed me to use funds to bring Fred out to WSU to finish the book, a promise that is now fulfilled. Funding from the School of Humanities and Communication Arts allowed this book to happen; without such support I would not have had the ability to research and write it with Fred. The interdisciplinary nature of WSU has helped me to conceive and write the book. My friend and colleague Jason Shaw helped me to conceive of the value of history to other disciplines. Corridor colleagues from various backgrounds— Drew Cottle (political economy), Chris Fleming (sociology), Robert Lee (history), Robert Mailhammer (historical linguistics), John Hadley (philosophy)—always gave me food for thought. My history colleagues at WSU were constant friends and intellectual companions. The Institute for Culture and Society provided financial support for various trips to South Africa. Thanks go to Tim Winter (now at Deakin University), Denis Byrne, Paul James, and Ien Ang for supporting my research. I cannot state how much I appreciate the support of everyone at Western Sydney University.

Colleagues around the world supported me in various ways. I was an Honorary Research Associate at the University of Cape Town from 2012 to 2014. Thanks go to Lance van Sittert for helping me to establish myself as a visiting fixture at the

University of Cape Town and for always being an honest critic of my work. Sandra Swart at Stellenbosch University has always been a mentor who has brought me into her amazing environmental history program there. Thanks go to David Richardson at Stellenbosch for his support. Jane Carruthers has been a constant help in Australia, South Africa, and throughout the world (via email). She has at times challenged me in a good way and I thank her greatly. Gerald Groenewald encouraged me to join the University of Johannesburg (UJ) as a Senior Research Associate in 2015. Thanks also go Natasha Erlank at UJ. In Australia, Glenda Sluga, Libby Robin, Tom Griffiths, Ian Tyrell, Ian McCalman, Andrea Gaynor, John Dargavel, Adam Lucas, James Beattie, and others who I probably forgot to name (force me to shout a round to make good), have encouraged and helped me with my work on South Africa. Both Fred and I have received advice and criticism from our friend in the UK, Simon Pooley. Simon's work dovetails with ours and we hope that together we can reveal the importance and uniqueness of the history of South African environmental policies and science.

Some of the archival research was done with dissertation support funds. I received financial support from the Social Science Research Council (SSRC) for the International Dissertation Research Fellowship, a National Science Foundation Improvement Grant, the Forest History Society for a Bell Fellowship (twice), the J. B. Harley Map Fellowship, and the University of Texas at Austin. Once again I must thank my graduate supervisors, Bruce Hunt, Wm. Roger Louis, Gail Minault, James Vaughn, and my numerous graduate school friends.

Lastly, I must thank my mum, Wendy, my dad, Mike, and my brother, Jared, for giving me the foundations that allowed me to develop an interest in subjects far removed from the Great Plains where I spent most of my youth. I feel an affinity towards forests as a result of spending summers with my grandfather Bob, his wife Joy, and my lovely cousins in Washington State, a place that I consider to be my true home in the world despite all of my travels and work overseas suggesting to the contrary.

Brett Bennett
Sydney, Australia, August 2015

Author Biographies

Brett Bennett is Senior Lecturer in History in the School of Humanities and Communication Arts at Western Sydney University. He is also a Senior Research Associate in the Faculty of Humanities at the University of Johannesburg and an Associate in the Centre for Environmental History at the Australian National University.

Bennett is author of *Plantations and Protected Areas: A Global History of Forest Management* (Cambridge, MA: MIT Press, 2015), editor with Joseph Hodge of *Science and Empire: Knowledge and Networks Across the British Empire, 1800–1970* (Basingstoke: Palgrave Macmillan, 2011), and author of many chapters and articles published in peer-reviewed journals.

Fred Kruger is a Research Associate in the Centre for Environmental Management at the University of the Free State. During his career, Kruger has served in a variety of research and executive positions, including the Officer in Charge at the Jonkershoek Forestry Research Station, Director of the South African Forestry Research Institute, Director of Forestek (Council for Scientific and Industrial Research), and as a consultant and educator. Kruger played a leading role in creating policies in South Africa to manage fire and catchments, jointly developed and then helped direct the first global survey on exotic species through the Scientific Committee on Problems of the Environment in the 1980s, and served as a leading contributor to the new National Forests Act and National Veld and Forest Fire Act in the 1990s. Currently, he pursues ecological research, teaches on the South Africa program of the Organisation for Tropical Studies, and writes about the history of science in environmental management in South Africa.

Preface

How does tree planting for economic purposes help or hinder wider imperatives to conserve water and preserve biodiversity? The question is one of considerable importance in parts of the world where tree planting for economic purposes has had a significant ecological or hydrological impact. Reconciling diverse economic and environmental demands associated with tree planting is a serious challenge in countries that lack commercially viable indigenous forests or tree species and therefore must either import large volumes of wood or grow exotic trees for domestic consumption. In many countries, there is considerable conflict between the forest sector, on the one hand, and environmentalists and scientists who are concerned about the impact of tree planting, on the other. The tension is likely to intensify in the coming decades as a result of predicted climatic changes, population growth, and the rise of living standards, all of which will put pressure on the world's finite resources. New insights are necessary to resolve the demands of diverse sectors of society in the face of these pressures. We need to move beyond polarised debates between those who view the issue from either economic or ecological perspectives, which have grown increasingly divergent in the past three decades.[1]

Against the backdrop of global concerns, this book examines how scientific debates about the hydrological impact of afforestation in South Africa shaped the development of modern scientific ideas and state policies relating to timber plantations, water conservation, invasive species control, and biodiversity management. The history of forest hydrology in South Africa is of significance because insights gained from over 60 years of long-term research have directly shaped South African policies and legislation. South African researchers played a central role in creating knowledge that is fundamental to the contemporary disciplines of hydrology, forestry, and invasion biology. The interpretation put forward in this book helps to explain why programs relating to invasive trees,

---

1   B. M. Bennett, *Plantations and Protected Areas: A Global History of Forest Management* (Cambridge, MA: MIT Press, 2015), 3.

water conservation, and biodiversity in South Africa are held up as being world-leading 'models'. It explains for the first time how existing hydrological and forestry policies and ideas evolved out of an older system that has not been properly situated within its historical, scientific, or policy contexts.

Modern South African history is a story of humans attempting to overcome resource constraints in the face of pressures unleashed by migration, colonisation, and globalisation. In terms of forests, South Africa's economy has, since the origins of the European colonial period, been constrained by a lack of sizeable indigenous forests and abundant, accessible water supplies. These resource constraints encouraged, from the earliest period of colonisation, a program of afforestation to augment the domestic timber supply. Tree planting later became an imagined means to improve water supply and climate. The desire to plant trees for economic and climatic purposes intensified in the last quarter of the nineteenth century in response to the origins of state forestry programs, first in the Cape Colony and then in the early twentieth century in neighbouring South African colonies.

Tree planting received wide public and scientific support among white South Africans, but dissent arose with regard to the claimed climatic and hydrological benefits of exotic afforestation as a result of large-scale state-sponsored afforestation after Union in 1910 that accompanied a rapid expansion of private plantings. Afforestation significantly changed the ecology and economy of many regions of South Africa from the 1910s onward. In response to this large-scale planting, critics blamed exotic afforestation for drying up streams, destroying indigenous vegetation, and disfiguring the country's landscape. This scientific and public controversy led the then Division of Forestry to create a Forest Influences Research Station in 1935 in the Jonkershoek Valley, outside of Stellenbosch in the south-western Cape, to investigate the relationship between forests, water, and climate.

In the six decades following its establishment, findings from Jonkershoek reshaped South African scientific understandings of how indigenous and exotic vegetation each influenced water supply in South African conditions. Its first director, Christiaan L. Wicht, used evidence from Jonkershoek to help design a comprehensive national catchment management strategy that aimed to harmonise the afforestation of exotic trees, the preservation of indigenous vegetation, and the rights of downstream water users. This framework—embodied in various pieces of legislation and policies—determined national catchment management strategies from the late 1960s to the 1990s. At its heart was the concept that all vegetation types used water, and although exotic trees often used more water than other types of indigenous vegetation, in many circumstances exotic trees were the most cost effective and beneficial land use. Foresters would manage all of the nation's mountain catchments. The policies aimed at an integrated system to produce timber, protect downstream water users, conserve biodiversity, and control invasive plants in the nation's catchments. Other research from

Jonkershoek led to the recognition that invasive exotic trees could lower streamflow to downstream users. This led to the development of a Forestry Department program to control catchments using fire to improve streamflow, control invasive species, and maintain biodiversity.[2]

The national program of catchment and plantation management fell into decline from the early 1990s as the result of policy changes. Current legislation and policy frameworks—the centrepiece being the *National Water Act* of 1998—have reoriented key tenets of forest policy established from 1935 onward. Catchment management was handed to the provinces in the late 1980s, and then, ambiguously, to catchment management authorities in the mid-1990s. Today, laws such as the *Mountain Catchment Areas Act* are largely in abeyance, and there are only two catchment management agencies for the country's 19 water management areas called for through the *National Water Act*. Exotic timber plantations have to date been the only land use classified in the *National Water Act* as a 'Stream Flow Reduction Activity'. Anti-exotic tree attitudes run deep among a variety of groups. There is strong antagonism to the planting of exotic trees based on fears about how much water they transpire and their economic and ecological impacts. Exotic trees are sometimes conflated (wrongly) with invasive trees despite the fact that only some exotic trees, in certain conditions, become invaders, and even fewer become problematic invaders.

South Africa's highly developed forest sector, one of the fastest growing parts of the South African economy during the second half of the twentieth century, stagnated and then contracted after the passing of the *National Water Act* and the onset of regulations. Whether or not the forest sector in South Africa has reached its 'natural' limits is a question that is contested by foresters, hydrologists, environmental planners, government officials, and members of the public right now. There is considerable fear and anxiety on the part of private timber growers because of questions of land tenure and the inability to expand plantings. At the same time, government-owned forest corporations have seen the loss of expertise, transient leadership, and declining productivity. The forest industry has made investments to improve efficiency, but the question about the sustainability of existing timber supplies is driving investment abroad and could diminish the country's economic growth, needed to pull millions of people from poverty. Forestry provides unique flow-on benefits at each stage—from growth, transport, production, to retail—in regional areas that often have limited economic options. Even with the digital revolution, per person use of paper and board in sub-Saharan Africa will almost certainly need to increase to improve living standards, from its present level of around one-twentieth or less of that of the developed world—to which solid timber for construction and other uses must be added. If South Africa does not produce this timber, it

---

2   See S. Pooley, *Burning Table Mountain: An Environmental History of Fire on the Cape Peninsula* (Basingstoke: Palgrave Macmillan, 2014).

will need to be imported from somewhere else, which creates current account deficits and can merely outsource environmental problems to countries with less developed regulatory systems. Yet one cannot just plant trees recklessly or promote one sector of the economy over another.

A recent meeting at Stellenbosch University between stakeholders has expressed the need for robust information to move forward on new policies to deal with plantations.[3] In the spirit of this ongoing process, this study provides needed historical background to enable a full discussion on the future directions of forestry and catchment management in South Africa. As will be argued, the policy framework and institutional arrangement prior to the 1990s utilised an integrated approach that brought together researchers, industry leaders, and government agencies in order to harmonise economic development and nature protection at national and provincial scales. This model attempted to integrate diverse scientific perspectives into policy and planning in order to direct afforestation to suitable sites, while using knowledge gained from forestry research to protect the nation's catchments and ecosystems. The origins of contemporary environmental policies and scientific perspectives are often to be found in this earlier program, although the present-day regime is lagging in establishing an equivalent integrated approach. We hope that this historical account will usefully inform the progressive improvement of the present-day regime.

Note

Some of the chapters in this book include material from previously published articles that have been modified significantly prior to this publication. Chapter One includes material from B. M. Bennett, 'Naturalising Australian Trees in South Africa: Climate, Exotics and Experimentation', *Journal of Southern African Studies*, 37 (2011): 265–80; Chapter Two is republished from B. M. Bennett and F. J. Kruger, 'Forestry in Reconstruction South Africa: Imperial Visions, Colonial Realities', *Britain and the World*, 8 (2015): 225–45; Chapter Three draws on text from B. M. Bennett, 'The Rise and Demise of South Africa's First School of Forestry', *Environment and History*, 19 (2013): 63–85; Chapter Six has sections taken from B. M. Bennett and F. J. Kruger, 'Ecology, Forestry and the Debate over Exotic Trees in South Africa', *Journal of Historical Geography*, 42 (2013): 100–9; Chapter Nine includes sections from F. J. Kruger and B. M. Bennett, 'Wood and Water: An Historical Assessment of South Africa's Past and Present Forestry Policies as they Relate to Water Conservation', *Transactions of the Royal Society of South Africa*, 68 (2013): 163–74.

---

3   B. van Wilgen, 'Plantation Forestry and Invasive Pines in the Cape Floristic Region: Towards Conflict Resolution', *South African Journal of Science*, 111 (2015): 1–2.

Introduction

The importance of forests and trees in South African history

How exotic trees are managed in South Africa is a contested subject. One can quickly ascertain the contested nature of this issue by asking people their opinion about exotic trees or by reading articles in the opinion section of newspapers advocating for and against the removal of exotic trees.[1] Some South Africans want to maintain existing exotic trees or even plant more to expand the economy, maintain heritage aesthetics, and provide shade in treeless environments. Environmentalists, government policy, and international advocates are more critical of exotic trees, which are seen as a threat to the integrity of indigenous ecosystems, the survival of threatened species, and the conservation of water. Advocates of tree planting worry that criticisms of invasive trees and the existing water legislation is a threat to the viability of the timber sector, an important contributor to the national and regional economy. Yet critics argue that biological invasions already cost billions of Rand in possible revenue and use South Africa's limited water supplies. Climate change projections suggest that the south-western part of the country, which is the most invaded, will see declining rainfall. The issue is complex and requires a nuanced understanding that is evidence-based and takes into account social, economic, and environmental considerations.

Whatever the perspective, it is undeniable that exotic trees are a significant part of South Africa's ecosystems, economy, and modern history. One cannot drive throughout the upland areas of the KwaZulu-Natal, Mpumalanga, or Limpopo provinces without seeing large plantations of introduced species of *Acacia*,

---

1   I. Dickie et al., 'Conflicting Values: Ecosystem Services and Invasive Tree Management', *Biological Invasions*, 16 (2014): 705–19.

Eucalyptus and *Pinus* stretching across the countryside. A small percentage of the country's plantations can be seen in select locations in the Western Cape Province. There are an estimated 650,000 hectares of *Pinus*, 500,000 hectares of planted *Eucalyptus*, and 100,000 hectares of Australian *Acacia* (called wattle) in South Africa. Other populations of exotic trees, some of them self-reproducing and others the remnants of past planting efforts, dot millions of hectares of land.

Despite negative to neutral public attitudes towards planted trees, there are significant benefits that flow from planted exotic trees in South Africa. Exotic trees support a multi-billion Rand timber industry concentrated in the country's higher rainfall regions in its northern and eastern provinces. Plantation-grown timber is used to make products as diverse as sawn timber boards, mine support packs, cellulose pulp for rayon, pressed board, the poles that are used for the safari architectural style seen at national parks and game lodges, and in indigenous fences and housing. South Africa is home to two of the world's largest multinational forest products companies, Sappi and Mondi, companies that originated from private afforestation programs during the first half of the twentieth century. The South African company Steinhoff International started operating in South Africa using domestic timber resources; from this base it became the second-largest furniture company in the world. No other country with comparable environmental resources has a forest product sector that is so productive. The question of how South Africa, a country that is naturally deficient in forests, created such a dynamic forest sector is one that has not been subjected to proper investigation. Though the main contributions are economic, planted trees also contribute positively to ecosystem services in many regions, the most obvious being north-east South Africa where the afforestation along the escarpment improves the quality of water flowing through the Kruger National Park.

Yet there are powerful reasons to be wary of planting exotic trees in South Africa that are known invaders or planting trees in catchments with water shortages. The ecological impact of planted and invasive exotic trees is substantial. Tree planting significantly changed the landscapes and ecosystems in South Africa. Invading alien trees are one of the greatest environmental threats in South Africa. Ecologists and environmental managers now seek to remove and manage invasive trees that form dense groves along river and stream banks, and in water catchment areas, and that outcompete indigenous vegetation. Over half a billion dollars (USD) has been spent by the Working for Water program to remove invasive plants, mostly exotic trees such as *Pinus pinaster* and *Acacia mearnsii*. These removals are based on the recognition that exotic trees use more water than do indigenous vegetation types, such as grassland or fynbos (a heath and shrub-dominated vegetation type). In a country with limited water supplies, invasive trees can create economic as well as conservation problems. The removal of invasive species is part of a wider effort to preserve the country's unique biological diversity, especially in the species-rich fynbos in the Western Cape Province, a biome that is significantly invaded. Without radical interventions, large swaths of the world-famous fynbos will become highly invaded 'novel ecosystems' with little chance of them returning to their previous condition.

Academic discussions about exotic trees usually focus on ecological, economic, heritage, or aesthetic considerations. The most well-developed scholarly literature focuses on the dynamics of invasive exotic plants, water conservation, and biodiversity management in South African and global contexts. South Africa is seen as a type of a global 'model'. South African researchers have since at least the 1980s been recognised as global leaders in invasion biology and ecology, with a particular emphasis on exotic trees. At the same time, South Africa's efforts to control the spread and impact of trees through Working for Water have often been praised by members of the international scientific community for creating a world-leading program that targets invasive species and provides employment for disadvantaged South Africans.

Though many scientists around the world know about South Africa's current research and policies, much less is known about how these evolved out of a long-term research program on forest hydrology. That program—its origins, development, and legacy—is the focus of this book. It is difficult to properly understand South African environmental history or the country's current research agendas and state legislation and policies without knowing about this program. The historical analysis offered in this book provides knowledge that can be used to assess and continue to improve South Africa's national policies and legislation. The history of tree planting and forest hydrology discussed in the book should be of interest to scholars in China, Australia, and India, among other places, because the issues addressed—how to balance ecological and economic considerations when planting trees—are salient. It is also important for hydrologists and foresters because South African research in forest hydrology helped to produce knowledge that is foundational within each discipline today.

While the book uses history to contribute to science and policy, it concurrently seeks to add to research on South African environmental history. There are many new developments within the field of South African environmental history, a sub-field that originated in the 1980s and has progressively developed since then. In the past decade, historians have turned their attention to how valuations of ecosystems and indigenous and exotic species changed throughout the twentieth century.[2] This work demonstrates that contemporary attitudes celebrating indigenous species and criticising exotic species grew stronger during the second half of the twentieth century as a result of environmental, political, and cultural trends.[3] Another important development within the field is the growth of scholarship that seeks to engage with ongoing policy issues, and to use history to inform policy and science.[4] Historians are increasingly seeking to engage with scientists on key issues, while there is an equally growing interest among scientists to learn from history in order to assess and solve environmental problems, such as biological invasions.[5]

This book adds to the fields of history and science by focusing on the wider history of forestry, especially timber plantations and hydrology. Forestry played a pivotal yet hitherto under-recognised role in shaping the history of South Africa's environmental sciences, economy, and its ecosystems. Unlike other countries with more developed forest histories, South Africa lacks an authoritative history of forestry or its forests.[6] Foresters have written most histories of forestry, and these histories are rarely based on archival research and tend not to situate themselves properly within the wider historiography

---

2   B. M. Bennett and F. J. Kruger, 'Ecology, Forestry and the Debate Over Exotic Trees in South Africa', *Journal of Historical Geography*, 42 (2013): 100–9; S. Pooley, 'Recovering the Lost History of Fire in South Africa's Fynbos', *Environmental History*, 17 (2012): 55–83; S. Pooley, 'Pressed Flowers: Notions of Indigenous and Alien Vegetation in South Africa's Western Cape, c. 1902–1945', *Journal of Southern African Studies*, 36 (2010): 559–618; F. Sundnes, 'Scrubs and Squatters: The Coming of the Dukuduku Forest, an Indigenous Forest in KwaZulu-Natal, South Africa', *Environmental History*, 18 (2013): 277–308; J. Comaroff and J. Comaroff, 'Naturing the Nation: Aliens, Apocalypse and the Postcolonial State', *Journal of Southern African Studies*, 27 (2001): 627–51; J. Carruthers and L. Robin, 'Taxonomic Imperialism in the Battles for *Acacia*: Identity and Science in South Africa and Australia', *Transactions of the Royal Society of South Africa*, 65 (2010): 48–64; L. van Sittert, 'Making the Cape Floral Kingdom: The Discovery and Defense of Indigenous Flora at the Cape, c.1890–1939', *Landscape Research*, 28 (2003): 113–29; L. van Sittert, 'From Mere Weeds and Bosjes to a Cape Floral Kingdom: The Re-Imagining of Indigenous Flora at the Cape, c.1890–1939', *Kronos*, 28 (2002): 102–26; L. van Sittert, '"Our Irrepressible Fellow-Colonist": The Biological Invasion of Prickly Pear (*Opuntia ficus-indica*) in the Eastern Cape c.1890–c.1910', *Journal of Historical Geography*, 28 (2002): 397–419.
3   B. M. Bennett, 'Model Invasions and the Development of National Concerns Over Invasive Introduced Trees: Insights from South African History', *Biological Invasions*, 16 (2014): 499–512; B. M. Bennett, 'Margaret Levyns and the Decline of Ecological Liberalism in the Southwest Cape, 1890–1975', *South African Historical Journal*, 67 (2015): 64–84.
4   F. J. Kruger and B. M. Bennett, 'Wood and Water: An Historical Assessment of South Africa's Past and Present Forestry Policies as they Relate to Water Conservation', *Transactions of the Royal Society of South Africa*, 68 (2013): 163–74; S. Pooley, *Burning Table Mountain*; W. Beinart and L. Wotshela, *Prickly Pear: The Social History of a Plant in the Eastern Cape* (Johannesburg: Wits University Press, 2011); Pooley, *Burning Table Mountain*, 178–83.
5   Much of this work is being driven by David Richardson and the Stellenbosch Centre for Excellence in Invasion Biology. Timm Hoffman at the University of Cape Town has demonstrated the value of repeat photography in assessing landscape changes. Scientists and managers in South Africa express a growing a interest for historical information.
6   There have been some publications covering specific aspects. See M. Lawes, H. A. C. Ealey, C. M. Shackleton, and B. G. S. Geach (eds), *Indigenous Forests and Woodlands in Southern Africa: Policy, People, and Practice* (Pietermaritzburg: University of KwaZulu-Natal Press, 2004).

of South Africa and the world.[7] For their part, professional historians have focused their attentions on larger problems (e.g. white supremacy, colonisation, African resistance and agency) within social, economic, and political history. The history of forestry, with a few important exceptions, has been relatively neglected as a subject.[8]

The history of South African forestry is significant for many reasons, but this book focuses specifically on three areas where it is of particular importance for understanding South Africa's environments, past and present. First, we argue that the environmental constraints in Southern Africa led the first generation of state foresters to focus their efforts on establishing exotic timber plantations. Foresters in Southern Africa devoted more efforts to plantations—their establishment, management, and utilisation—than any other issue. Only by understanding plantations is it possible to situate historical debates about water conservation or scientific concerns about invasive plants. Our focus on plantations builds on the existing historiography, which has usually been divided between studies that focus on the 'external' origins of forestry or pursue detailed regional histories without situating them in national and global contexts.[9] This emphasises the importance of developments and impacts, and through this we seek to understand how South Africa's forestry and water policy framework evolved across the twentieth century.

Second, we propose a theoretical explanation for how human interaction with environmental constraints led to the creation of new ecosystems, scientific ideas, and conservation policies in South Africa. The theoretical framework attempts to describe how key interactions between humans, non-humans, and the wider environment changed in Southern Africa during the periods from the onset of European colonialism to post-apartheid. The model accounts for the history of tree species introductions by explaining why some tree introductions 'succeeded' and 'failed' from biological, ecological, and social perspectives. The key question is to explain why the success rate of species introductions improved during the twentieth century. To explain this, we focus on foresters who created a methodology for selecting exotic trees that sought, in the words of the Cape forester David E. Hutchins, to 'fit the tree to the climate'.

7　　See N. L. King, 'Historical Sketch of the Development of Forestry in South Africa', *Journal of the South African Forestry Association*, 1 (1938): 4–16; M. Grut, *Forestry and Forest Industry in South Africa* (Rotterdam: A. Balkema, 1965); W. F. E. Immelman, C. L. Wicht, and D. P. Ackerman (eds), *Our Green Heritage: A Book About Indigenous and Exotic Trees in South Africa, about Trees and Timber in our Cultural History and About Our Extensive Silvicultural, Forestry and Timber Industries* (Cape Town: Tafelberg, 1973); W. D. Reekie, 'The Wood from the Trees: Ex Libri ad Historiam Pertinentes Cognoscere', *South African Journal of Economic History* 19 (2004): 67–99; W. Olivier, *There Is Honey in the Forest: The History of South African Forestry*, 2nd ed. (Pretoria: Southern African Institute of Forestry, 2010).

8　　See J. Tropp, *Natures of Colonial Change in the Making of the Transkei* (Athens, OH: Ohio University Press, 2006); Pooley, *Burning Table Mountain*.

9　　Pooley, *Burning Table Mountain*, is an exception.

Cape foresters' emphasis on bio-climatic comparison and matching, and their interest in experimenting with exotic trees, helps explains the success of exotic tree plantations in South Africa in the twentieth century. Successful bioclimatic matchings is one reason why South Africa has one of the highest percentages of trees as invasive plants of any region in the world.[10]

Third, the book traces the history of hydrological research and policy in South Africa by focusing on the evolution of ideas, science and institutions relating to forests and water. The history of hydrology in South Africa is significant because it shaped modern understandings of catchment hydrology and invasive species, as well as created a coherent national framework for afforestation. Current management programs, such as Working for Water, developed out of the sustained, national effort to determine whether tree planting had a negative or positive influence on the water balance of South Africa's catchments.

This effort began in earnest with the establishment of a research station in 1935 in the Jonkershoek Valley, just outside of the university town of Stellenbosch. Researchers from Jonkershoek played a disproportionate role in creating the scientific concepts and national legislation used to manage plantations, catchments, and invasive species at regional and national scales. The interdisciplinary effort in the program led to the creation of a national framework that sought to reconcile competing interest groups (e.g. foresters, downstream users, and ecosystems) by directing afforestation toward regions in the country where it was environmentally and economically appropriate. This policy began to break down in the late 1980s to late 1990s during the transition towards democracy. The new regulatory regime that came into existence in the 1990s included the *National Water Act* of 1998, a piece of legislation that defined forestry as the country's only streamflow reduction activity.

Reorienting South Africa's forest historiography

South Africa has a unique forestry history that deviated from European and Indian traditions soon after its inception as a state program in the Cape Colony in the early 1880s. The first generation of professional foresters in the Cape Colony faced the tasks of protecting and managing scarce and dwindling resources of indigenous, mixed-species forests; conserving water in an arid climate; and creating plantations from the ground up in unknown environments with

---

10   Scientists have recognised since the 1980s that forestry has left a legacy of invasive alien species in different regions of the world. See D. M. Richardson, 'Forestry Trees as Invasive Aliens', *Conservation Biology*, 12 (1998): 18–26. It has only been recognised recently that the history of bioclimatic matching in South Africa may go some way towards explaining the proportionally high number of of naturalised and invasive forest tree species. See J. E. Donaldson, et al., 'Invasion Trajectory of Alien Trees: The Role of Introduction Pathway and Planting History', *Global Change Biology*, 20 (2014): 1527–37.

almost no suitable indigenous species from which to select. Foresters turned their attentions to tree planting, an action they believed would encourage soil and water conservation as well as alleviate the pressure on the country's finite indigenous forests.

Not only did environmental constraints limit foresters, the economic demands caused by the mineral revolution—based on the mining of diamonds from the early 1870s and gold from the late 1880s—generated massive need for timber for mining. Neither the Cape nor Transvaal had adequate supplies of indigenous forest to provide for these demands. Private producers soon entered into the market, but with little success because they lacked the knowledge of which species to plant and how to manage them. Foresters from the Cape who moved to the Transvaal after the South African War were tasked with making sure the mines were provided with timber. Foresters had a difficult task: there were few indigenous forests and fewer plantations.

As a result of their emphasis on tree planting, foresters in the Cape, and eventually foresters throughout the rest of South Africa, developed their own identity in contradistinction to their European or British Empire counterparts. South African foresters saw themselves as innovative creators of new forest systems rather than the more traditional manager's role as a *conservator* of existing forests. The rapid expansion of plantations led critics outside and inside of South Africa to worry that plantations had negative conservation consequences on indigenous vegetation and hydrological regimes. Fears about the desiccating influence of certain exotics, such as eucalypts, were expressed in the mid-nineteenth century, but these concerns become more pronounced in droughts during the early twentieth century.

This argument amends historical interpretations that imply that the forestry regime that developed in South Africa was 'European' or 'Indian' in theory *and* practice.[11] It is undeniable that forestry science originated first in Europe and spread throughout the world through globalisation. India, too, acted as a central hub in the development of an 'empire forestry' movement, which spread the gospel of forestry conservation throughout the British Empire. For the past decade these diffusionist studies offered a useful framework that helped to understand how professional foresters spread state forestry globally. Yet there is now a growing recognition of the limitations of global diffusionist perspectives, which often obscure cross-imperial networks, national histories and the importance of

---

11   For India, see G. Barton, *Empire Forestry and the Origins of Environmentalism* (Cambridge: Cambridge University Press, 2002), 98–104; for Europe, see R. Rajan, *Modernizing Nature: Forestry and Imperial Eco-Development 1800–1950* (Oxford: Oxford University Press, 2006); K. Brown, 'The Conservation and Utilisation of the Natural World: Silviculture in the Cape Colony, c. 1902–1910', *Environment and History*, 7 (2001): 427–47, 433; K. Brown, '"Trees, Forests and Communities": Some Historiographical Approaches to Environmental History on Africa', *Area*, 35 (2003): 343–56, 346.

place.[12] Unlike Australia and New Zealand, where forest historians have amended these frameworks, South Africa's forest historiography, with a few exceptions mentioned below, has remained anchored to these origins stories.

A series of studies by Bennett have argued that South African forestry deviated from European practice and principles soon after the arrival of professional foresters in the 1880s. This book draws on and advances this argument by documenting how foresters from the Cape developed unique ideas about the importance of bioclimatical modelling. By the mid-1900s, foresters opened up the first African forestry school in order to train students about the unique problems associated with forestry in Southern Africa. South African foresters developed distinct practices for selecting species and creating plantations that created considerable international interest, and even controversy, that lasted until the 1950s. Forest hydrology research in the 1950s and 1960s also generated considerable controversy among leading hydrologists. Interestingly, in each case—climatic modelling, plantation methods, and hydrological research—South African methods eventually gained international acceptance, and even became standard procedures elsewhere in the world.

Plantation-grown timber output increased steadily throughout the twentieth century, following from steady afforestation in the period from the late 1920s to the 1930s, culminating in the 1960s to the 1970s.[13] This expansion reflected sustained government investments to expand timber plantations in grassland or shrubland around or between patches of indigenous forest. Government plantations were composed primarily of pines (*P. radiata* in the Mediterranean climate of the Cape and *P. patula* and *P. elliottii* in the summer-rainfall eastern provinces). Private plantations expanded at an even faster rate throughout the twentieth century in response to domestic demand for eucalypt mine-support timbers and global demands for wattle bark, and later, pine and eucalypt paper and cellulose pulp.

---

12   This trend is particularly well developed in New Zealand and, to a lesser extent, Australia. See J. Beattie, *Empire and Environmental Anxiety: Health, Science, Art and Conservation in South Asia and Australasia, 1800–1920* (Basingstoke: Palgrave Macmillan, 2011); G. A. Barton and B. M. Bennett, 'Edward Harold Fulcher Swain's Vision of Forest Modernity', *Intellectual History Review*, 21 (2011): 135–50; M. Roche, 'Forestry as Imperial Careering: New Zealand as the End and Edge of Empire in the 1920s–40s', *New Zealand Geographer*, 68 (2012): 201–10; M. Roche, 'Colonial Forestry at its Limits: The Latter Day Career of Sir David Hutchins in New Zealand 1915–1920', *Environment and History*, 16 (2010): 431–54; M. Roche, 'The New Zealand Timber Economy 1840 to 1935', *Journal of Historical Geography*, 16 (1990): 295–313; M. Roche, 'The State as Conservationist, 1920–60: "Wise use" of forests, lands and water', in T. Brooking and E. Pawson (eds), *Environmental Histories of New Zealand* (Melbourne: Oxford University Press, 2002): 183–99. Much of this work is based on research on 'networks' and the historical geography of science. See D. N. Livingstone, *Putting Science in its Place: Geographies of Scientific Knowledge* (Chicago: University of Chicago Press, 2003); A. Lester and D. Lambert (eds), *Colonial Lives across the British Empire: Imperial Careering in the Long Nineteenth Century* (Cambridge: Cambridge University Press, 2006); B. M. Bennett and J. M. Hodge (eds), *Science and Empire: Knowledge and Networks of Science across the British Empire, 1800–1970* (Basingstoke: Palgrave Macmillan, 2011).
13   K. B. Showers, 'Prehistory of Southern African Forestry: From Vegetable Garden to Tree Plantation', *Environment and History*, 16 (2010): 295–322.

Exotic trees reshaped ecosystems and created new economies and social communities.[14] Plantations drew praise and criticism. Examining the history of plantations offers a way to tie together hitherto fragmented strains in fields as diverse as environmental history, social history, economic history, the history of science, global history, and political history.

At the most basic level, plantations required land and human labour to plant trees. Finding labour and sufficient land proved to be an early problem for South Africa's foresters. In many districts, labour was initially scarce, and attempts by government foresters to hire locally agitated the white farming community, who feared competition for workers and increased wage burdens. After Union, foresters found themselves embroiled in political interventions to reverse urbanisation and address the 'poor White problem'.[15] This led to a national program for buying and appropriating land suitable for afforestation and establishing white forestry settlements. Alongside these settlements, government instituted a discriminatory system that subsidised wages for whites and paid African and coloured labourers less for similar work.

The increased output of timber grown in domestic plantations partially explains how South Africa's small, indigenous closed-canopy forests stayed largely intact throughout the twentieth century. By the early twentieth century, most of South Africa's remaining indigenous forests around Knysna, the Transkei, Natal, and Transvaal had been placed under the control of state foresters.[16] Foresters negotiated with and often struggled against white landowners, politicians, diverse groups of woodcutters, and rural African communities in order to demarcate and control South Africa's finite and scattered forest resources. Once demarcated and policed, the Forestry Department had to figure out how to regulate the harvest from indigenous forests. They established a research program in the early 1920s to investigate how to regenerate and sustain native tree species. The first ecologist employed under this scheme, John Phillips, who worked in Knysna from 1922 to 1927, went on to develop the first ecological critique of South African forestry.[17]

---

14   H. Witt, 'The Emergence of Privately Grown Industrial Tree Plantations', in S. Dovers, R. Edgecombe, and B. Guest (eds), *South Africa's Environmental History: Cases and Comparisons* (Athens, OH: Ohio University Press, 2002): 90–112.

15   A. Grundlingh, '"God het ons Arm Mense die Houtjies Gegee": Poor White Woodcutters in the Southern Cape Forest Area, c. 1900–1939', in H. Wolpe (ed.), *White but Poor: Essays on the History of Poor Whites in Southern Africa 1880–1940* (Pretoria: University of South Africa, 1992): 40–56. The paper was first presented at the History Workshop conference 'The Making of Class', held in Johannesburg on 9–14 February 1987; T.R. Roach, 'The White Labour Resettlement Program in South Africa 1917–1938' (MA Thesis, University of the Witwatersrand, 1989).

16   J. Tropp, 'Displaced People, Replaced Narratives: Forest Conflicts and Historical Perspectives in the Tsolo District', *Journal of Southern African Studies*, 29 (2003): 207–33.

17   See Bennett and Kruger, 'Ecology, Forestry and the Debate over Exotic Trees in South Africa'.

Afforestation played a central role in state policies for conserving water. Orthodox forestry theories inherited from the mid-1800s posited that trees encouraged rain, stopped erosion, and regulated a more even flow in the streams.[18] In the twentieth century, heterodox forestry researchers, such as Phillips, and prominent scientific and public critics challenged this view, culminating in the debate at the fourth British Empire Forestry Conference held in South Africa in 1935. Despite dissent, the view that trees had a positive influence on the water balance dominated forest policy from the 1880s (the foundation of state forestry) until the late 1940s, when research initiated at Jonkershoek Forest Research Station began to provide evidence that trees did in fact use more water than indigenous grasses and fynbos. Findings from this research led to the development of a national system for regulating afforestation. This new framework sought to balance the competing interests of different groups in order to achieve sustainable and equitable economic, environmental and social outcomes.

Environmental constraints, biological introductions and ecological change

South Africa is poor in forest resources compared with many other places in the world.[19] Less than 0.3 per cent of its land area is covered with closed-canopy forest and only a small fraction of the country receives more than about 750 mm of rain per year, the amount necessary to profitably produce timber: given other physiographic and economic constraints, only about 2–3 million ha—no more than about 2.5 per cent of the land area—is suitable for forestry.

---

18   Barton, *Empire Forestry*, 99–104; W. Beinart, *The Rise of Conservation in South Africa: Settlers, Livestock, and the Environment 1770–1950* (Oxford: Oxford University Press, 2003), 95–8; R. Grove, 'Early Themes in African Conservation: The Cape in the Nineteenth Century', in D. Anderson and R. Grove (eds), *Conservation in Africa: Peoples, Policies and Practice* (Cambridge: Cambridge University Press, 1987): 21–38; R. Grove, 'Scotland in South Africa: John Croumbie Brown and the Roots of Settler Environmentalism', in T. Griffiths and L. Robin (eds), *Ecology and Empire: Environmental History of Settler Societies* (Seattle: University of Washington Press, 1997): 139–53.

19   The question of whether Africa had or has adequate resources is inherently tied to our understanding of the history and potential productivity of African ecosystems; these views were also shaped by state and scientific attitudes towards indigenous land use in the recent past. For much of colonial history, foresters believed that African ecosystems, such as savanna, were denuded forests destroyed by indigenous people and early European settlers. Observers in South Africa began to challenge this view vigorously from about 1900 onward. More recently, social scientists have drawn upon science and history to critique foresters for 'misreading' landscapes. See J. Fairhead and M. Leach, *Misreading the African Landscape: Society and Ecology in a Forest-Savanna Mosaic* (Cambridge: Cambridge University Press, 1996); for southern Africa, see K. B. Showers, *Imperial Gullies: Soil Erosion and Conservation in Lesotho* (Athens, OH: Ohio University Press, 2005). Though foresters indeed often misread landscapes and ecosystems, the fact remains that given population growth and the ecological capacity of Southern African forests, resources scarcity would have become more pronounced. See F. J. Kruger's book review of J. Tropp, *Natures of Colonial Change* in *Britain and the World*, 2 (2010): 263–70. This view is echoed in recent work on factor endowments in Africa. See G. Austin, 'Resources, Techniques and Strategies South of the Sahara: Revising the Factor Endowments Perspective on African Economic Development, 1500–2000', *Economic History Review*, 61 (2008): 587–624.

European colonisation, first in the Cape (1652 – early 1800s), and then later in the interior (post-1830s), provided the key catalyst that reshaped the ecology and economy of wood usage in Southern Africa, markedly accelerated by the mineral revolution from 1867 onward.

Within the Southern African context, scarcity created what James Beattie called 'environmental anxieties' that led settlers and scientists to use science to try to understand and control nature.[20] A desire to overcome the inadequate forest resources drove Europeans to introduce hundreds of varieties of trees from the mid-seventeenth century to the mid-twentieth century. Anxieties about environmental constraints also explain the intense hope that tree planting could change southern Africa's climate for the 'better'. Government attempts to create a domestic sawmilling and timber plantation industry in South Africa sought to directly overcome the region's biological (i.e. not enough valuable species) and ecological (i.e. not enough forests) deficiencies by importing and planting exotic species of pines, eucalypts and wattles.

Today, there are 750 species of exotic plants that are regarded as invasive in South Africa, the result of hundreds of years of human-mediated species introductions, which has had significant long-term impacts on the country's economy, terrestrial ecosystems and scenery.[21] We define 'naturalisation', which follows an initial species 'introduction', as a process that 'starts when abiotic and biotic barriers to survival are surmounted and when various barriers to regular reproduction are overcome'.[22] In short, when a species no longer needs human intervention to maintain, and potentially expand, a population. The next step in this process is 'invasion', when self-propagating populations spread to distant sites with or without human aid.

Historians have tended to view the broader process of species naturalisation and invasion as being a mix of accident, purposefulness, and innate ecological 'fit'.[23] This historiography lies in the long shadow of Alfred Crosby, who posited that Old World flora and fauna (what he called a 'portmanteau biota') succeeded in New World conditions because of evolutionary and historical reasons. Historians subsequent to Crosby have gone out of their way to eschew coherent interpretations of species movement, and have instead traced out geographic

20   Beattie, *Empire and Environmental Anxiety*.
21   B. W. van Wilgen, 'The Economic Consequences of Alien Plant Invasions: Examples of Impacts and Approaches to Sustainable Management in South Africa', *Environment, Development and Sustainability*, 3 (2001): 145–68, 146.
22   D. M. Richardson, et al., 'Naturalization and Invasion of Alien Plants: Concepts and Definitions', *Diversity and Distributions*, 6 (2000): 93–107.
23   For a review of literature see W. Beinart and K. Middleton, 'Plant Transfers in Historical Perspective: A Review Article', *Environment and History*, 10 (2004): 3–29.

randomness of exchange.[24] Others argue against Crosby by pointing out that the expansion of Australian *Acacia* and *Eucalyptus* globally negates his emphasis on the one-way expansion of Eurasian biota versus reverse flows from the New World to Eurasia, specifically Europe.[25]

We seek to create a model for tree-species naturalisation and invasion in Southern Africa that accounts for multiple timescales and geographies without rejecting the importance of a theoretical explanation or falling into geographic determinism based on a single location. Introduction, naturalisation, and invasion required a broader 'package' of processes, including human disturbance of pre-existing ecosystems, exotic species introductions, and the purposeful and accidental creation of novel ecosystems.[26] Successful naturalisations and invasions were not entirely random, nor did they occur at the same rate throughout the periods being discussed. Purposeful human actions served as the means for creating conditions necessary for most naturalisations and invasions to occur.[27] It took, for instance, over half a century for people to find *Eucalyptus* species that were adapted to tropical climes in Africa and Asia. The seeming 'one-off' success, such as *Acacia mearnsii* in the midlands of Natal, was not random; trees only became invasive threats *after* foresters and plantation owners planted them widely once they had recognised their superior performance in comparison with other *Acacia* species propagated in South Africa (e.g. *Acacia decurrens*, and *A. dealbata*); even then, populations required human intervention to create ideal growing conditions by controlling competing indigenous vegetation and managing pests in plantations.

From 1652 to the 1880s, residents in Southern Africa did not develop a specific methodology for selecting species of exotic trees to plant; instead they relied on word of mouth and examples of successful trees that could be found about the landscape. Prior to the 1880s, would-be tree planters had very little knowledge about the climatic requirements of exotic trees. They had no methodology for determining what tree species would succeed except through the process of trial and error based on extensive trials—in effect a random process. During this period, human agency helped to establish almost all populations of exotic trees.

---

24   L. van Sittert, '"The Seed Blows About in Every Breeze": Noxious Weed Eradication in the Cape Colony, 1860–1909', *Journal of Southern African Studies*, 26 (2000): 655–74; C. Kull and H. Rangan, 'Acacia Exchanges: Wattles, Thorn Trees, and the Study of Plant Movements', *Geoforum*, 39 (2008): 1258–72.

25   See Beinart and Middleton, 'Plant Transfers in Historical Perspective', 6, 10; J. Radkau, *Nature and Power: A Global History of the Environment* (New York: Cambridge University Press, 2008), 21–2, 159; B. R. Tomlinson, 'Empire of the Dandelion: Ecological Imperialism and Economic Expansion 1860–1914', *Journal of Imperial and Commonwealth History*, 26 (1988), 89. For 'wattles', the common name for Australian *Acacia*, see Kull and Rangan, 'Acacia Exchanges', 1258–72; Carruthers and Robin, 'Taxonomic Imperialism', 48–9.

26   This was Crosby's bigger point, which has been overlooked by most critiques of his work which cast him as arguing for a supposed European dominance of species exchange.

27   See B. M. Bennett, 'A Global History of Australian Trees', *Journal of the History of Biology*, 44 (2011): 125–45.

Humans propagated many of the most aesthetically valued or otherwise useful species (e.g. European Oak, *Quercus robur*, or the Grey Poplar, *Populus canescens*) through purposeful planting and management. Only a handful of introduced species formed self-reproducing populations that spread without human aid into undisturbed ecosystems (e.g. *Pinus pinaster* in the south-western Cape; *Salix* spp. throughout the river systems). In the instance of *P. pinaster*, invasion followed extensive plantings and broadcast sowings.

A massive boom in tree planting occurred in the second half of the nineteenth century because of the need for wood, geographic dispersal of settlers, and increased immigration of white settlers who were buoyed by the belief that trees could increase rain and change the climate. They had access to new species, many from Australia, that they believed would grow quickly to produce valuable, useful trees. Yet this enthusiasm faded as the exotic seeds and saplings failed to grow into forests; those that did seemed to have little influence on the climate; and some people began to even fear that the trees that did grow used more water than indigenous vegetation.

The onset of state forestry in the Cape Colony in the early 1880s occurred at the peak of this speculation. Foresters began to note that the species that proved successful were planted in Southern African climates similar to those in which they grew in their native habitats. This insight led to explorations that sought to find regions of the world with similar climates to those in Southern Africa, and to select trees from those regions to test in Africa. David Ernest Hutchins, a leader of this movement, told foresters to 'fit the tree to the climate', a mantra that guided foresters in the region for the entire twentieth century. By the 1940s, new insights into genetics allowed foresters to pinpoint the provenance of the most 'superior' trees growing in Southern Africa; botanists and foresters then went to foreign localities to find elite breeds to bring back to South Africa for trial and selection. As a result, foresters introduced species with traits and evolutionary histories most likely to naturalise in South Africa.

The expansion of plantations themselves caused new environmental constraints as well as afforded conservation opportunities. Timber plantations increasingly became implicated in a debate about water conservation and later, invasion biology. Since the second half of the nineteenth century some white settlers and scientists argued that exotic trees used more water than indigenous vegetation types. Eucalypts and wattles, in particular, were perceived to be water-demanding species that could dry *vleis* (swamps), small streams, and springs. Over time these concerns grew into a national debate about the effects of tree planting that peaked in the mid-1930s. At the same time, a small but engaged and powerful group of botanical enthusiasts in the south-western Cape began in the 1930s and 1940s to call for the removal of exotic trees from unique, diverse indigenous vegetation types, first the 'Cape flora' (now known as fynbos) and later from other types. Many foresters sympathised with these concerns, and in 1945, Christiaan Wicht chaired the country's first major report on the conservation of the fynbos, which emphasised the problem of invasive trees.[28]

Foresters believed that the positive benefits of plantations outweighed the negative, at the same time that they agreed that exotic trees had some negative side effects. They argued that timber produced in plantations allowed foresters to protect indigenous forests. That South Africa's indigenous forest cover stayed essentially the same, or grew, throughout the twentieth century is chiefly attributable to state and private plantation timber output. Whether or not exotic trees used more water than indigenous plants and vegetation types was a question that would not be settled until the late 1940s. By the late 1960s, foresters began devising methods for managing invasive species as well as streamflow in catchments using controlled fires.[29] This activity fit with the active ethos of foresters, inherited from the first generation of state foresters who saw South Africa's environmental constraints and conditions as something to be understood, manipulated, and controlled for human benefit.

The origins, history and development of hydrological research in South Africa

South Africa is by no means the only region where people debated the hydrological and climatic influence of forests, but it is undoubtedly one of the most important places, if one judges in terms of the longevity and intensity of the debate, and the knowledge produced solving it.[30] Given South Africa's limitations in terms of wood and water it is unsurprising that a conflict arose, yet what makes the South African experience significant is the sustained national efforts that went into defining and attempting to solve this question.

South Africa's post-1998 forest and water policies differ significantly from their pre-1998 antecedents, yet cannot be understood without reference to the period before 1998. Prior to 1998, legislation and policy frameworks did not discriminate against planted forests as a land use at the national level, something the current policy framework does. Rather, South African policies from the 1960s to the mid-1990s were predicated on the assumption that the permitting of afforestation and hence, indirectly, the allocation of water rights

---

28   C. L. Wicht, *Preservation of the Vegetation of the South Western Cape, Report for the Royal Society of South Africa* (Cape Town: The Society, 1945).

29   Pooley, *Burning Table Mountain*, 90.

30   For instance, see D. Davis, *Resurrecting the Granary of Rome: Environmental History and French Colonial Expansion in North Africa* (Athens, OH: Ohio University Press, 2007); Fairhead and Leach, *Misreading the African Landscape*; Showers, *Imperial Gullies*.

should be determined regionally, as well as in relation to geographical patterns of water supply and demand and the economic efficiency of water users. This framework was situated within a broader catchment conservation strategy, which stipulated that the vast majority of state 'forest land' remain un-forested and the protection of catchments on privately owned land would be managed by the Department of Forestry through the *Mountain Catchment Areas Act* of 1970, while an amended *Forest Act* allowed for the regulation of afforestation through a permit system. This legislation and framework guided national forest and water policies in catchment until new national legislation and regulatory frameworks were established from the late 1980s to the late 1990s during the decline of the apartheid government and the emergence of post-1994 democracy in South Africa.

The transition from apartheid to democracy in South Africa significantly changed the structure of institutions, policies, and legislation for managing the environment. The Department of Forestry experienced some of the greatest changes, which began in the 1980s and early 1990s in response to economic and political changes within South Africa that encouraged the central government to corporatise plantations, merge, and shed responsibilities to provincial governments.[31] The Department of Forestry went from controlling the country's catchments and managing the biodiversity and invasive species on this formerly extensive state forest land—which totalled over 2 million hectares—to having little role in these areas today. Recent studies have shown how Department of Forestry researchers from the late 1960s to the late 1980s used research to create a coherent policy for using prescribed burns to control invasive species, maintain biodiversity and conserve water.[32] The financial decline of the apartheid government in the late 1980s led to the fragmentation of this national policy, leaving an institutional and policy vacuum that is still largely unfilled, even after the establishment of Working for Water in 1995. In a recent article, the historian Simon Pooley concludes, 'What remains tantalizing is what the longer term environmental outcomes might have been if the collapse of the apartheid state had not truncated the state conservation forestry research and management program in South Africa in the early 1990s'.[33]

One of the key missing pieces from recent historical research and critical assessments of South Africa's current environmental policies is an understanding of how South Africa's forest and water conservation policies developed across

---

31   For the economics of forestry see W. J. A. Louw, 'General History of the South African Forestry Industry: 1975 to 1990', *South African Forestry Journal*, 200 (2004): 77–86; W. J. A. Louw, 'General History of the South African Forest Industry: 1991 to 2002', *South African Forestry Journal*, 201 (2004): 65–76.

32   S. Pooley, 'Recovering the Lost History of Fire'; in South Africa's Fynbos', *Environmental History*, 17 (2012): 55–83; B. W. van Wilgen, 'The Evolution of Fire and Invasive Plant Management Practice in Fynbos', *South African Journal of Science*, 105 (2009): 335–6.

33   Pooley, 'Recovering the Lost History of Fire', 76.

the twentieth century. To fill this gap, this book offers an historical examination of the origin and evolution of South Africa's forest and water policies from the 1920s to the present. We pay particular attention to the history of South Africa's first hydrological research station located in the Jonkershoek Valley near Stellenbosch. The Jonkershoek Forest Influences Research Station, as it was originally called,[34] was the centrepiece of South Africa's forestry research in relation to water conservation from the mid-1930s to the mid-1990s. The Forestry Division of the Union of South Africa founded the station in 1935 in order to resolve a deep-seated national debate about the hydrological impact of exotic trees. Research from the Jonkershoek station, especially that of its first director, Christiaan Wicht (hereafter Wicht), played a critical role in shaping South African policies surrounding forestry and water conservation from the mid-1930s until today.

Debates about the hydrological influence of afforestation in Southern Africa began in the second half of the nineteenth century, but only came to the forefront of national attention after 1902 as a result of different rural development drives following the South African War (1899–1902), the First World War (1914–1918) and the extensive rural resettlement of poor whites pursued first by Jan Smuts and later by J. B. M. Hertzog.[35] These schemes all emphasised the importance of irrigation farming. The establishment of new irrigation schemes and farms, especially in the eastern Transvaal, created new claimants on water supplies who saw large-scale afforestation as a threat, rather than a means to an improved water economy. Criticisms of afforestation grew proportional to the expansion of plantations, which increased in size in the late 1920 and early 1930s just as Hertzog's policy of resettling whites on forest settlements in the Cape, Natal, and Transvaal provinces began to take real effect.

Initial steps to acquire a site for the purpose of 'forest influences' research began in 1932, with the purchase of land in the Jonkershoek Valley, eight kilometres south-east of the university town of Stellenbosch. In 1935, the Forestry Department established a 'forest influences' research station there and work on the research program began after the resolution of the fourth British Empire Forestry Conference, held in South Africa that same year. The conference provided the imprimatur to the idea of such a program, which itself was the culmination of a series of attempts to gather evidence about afforestation effects, beginning with the catchment experiment set up in the eastern Transvaal at Jessievale in 1910. The conference affirmed the political and scientific urgency of the research, paving the way for foresters to commit proper resources to the

---

34   It was later called the Jonkershoek Forest Research Station, and then the Jonkershoek Forestry Research Center.

35   Concerns prior to the war focused on particular colonies or metropolitan areas. See Pooley, *Burning Table Mountain*, Chapter 3.

program. Jan Smuts and the ecologist John Phillips articulated a comprehensive list of science and policy objectives to be satisfied through research, and again, these views gave the justification that allowed researchers to address questions beyond the narrow concern of the water issue, as was the case in the Jessievale design, opening the door to an inclusive ecosystems approach. As a result, the vision for research at Jonkershoek was broad and inclusive.

Jonkershoek was a central piece of what Saul Dubow describes as the 'South Africanisation' of science after the South African War (1899–1902).[36] The two decades following Union in 1910 saw the progressive, intentional assembly of a body of scientifically trained foresters with advanced degrees from leading European and American universities, meant to augment and diversify the small though tenacious cadre of colonial foresters. Foresters shared a common purpose of creating a distinctly 'South African' science based on the construction of a unique white South African identity that stressed the importance of understanding environmental problems from a distinctly local, South African perspective. This belief was grounded in the desire to create a distinct national identity, and also reflected the economic imperative to build a modern, industrial South African economy. Foresters believed that the hydrological question of afforestation required a broader understanding of other problems within the fields of botany, ecology, physiology, meteorology, forest growth dynamics, timber mechanics, and economics. The Forestry Department sent forestry trainees to Dresden, Oxford, and Yale to receive training that was not available in South Africa following the closing of the forestry school at Tokai, as discussed further in Chapter 3.

Jonkershoek acted as a key locale in the national and international history of hydrology.[37] Its location close to Stellenbosch University and the University of Cape Town enabled the ready exchange of ideas that Wicht needed as he worked to solve technical methodological problems, which required the help of experts from diverse disciplines to assist him, as he confronted the problems of running large experiments in difficult terrain and unpredictable climate. He designed a multiple-catchment experiment that sought to accommodate physiographic complexity and to unravel how secular climate change influenced catchment hydrology. This multiple-catchment design faced criticism from world-leading experts in hydrology, but ultimately became the basis for South Africa's national hydrological research agenda.

---

36   S. Dubow, *Scientific Racism in South Africa* (Cambridge: Cambridge University Press, 1995); S. Dubow (ed.), *Science and Society in Southern Africa* (Manchester: Manchester University Press, 2000).
37   R. Kohler, 'Practice and Place in Twentieth-Century Field Biology: A Comment', *Journal of the History of Biology*, 45 (2012): 579–86.

Research produced by scientists at Jonkershoek and satellite stations informed a coherent national policy of catchment management and informed ecosystems management more generally. Wicht and his successors used findings from Jonkershoek to guide the expansion of afforestation while trying to balance the needs of downstream water users, including farmers, urban areas, and indigenous ecosystems. The hallmark of this system was a vision of consensus and nation-building that brought together disparate, often competing groups to direct the national economy by recognising the diverse needs of constituent regions and stakeholders.

Jonkershoek was the seed of a bigger program. The work had hardly begun before a reconnaissance party had scouted the Drakensberg Mountains on foot and horseback, and selected Cathedral Peak as the next site to complement Jonkershoek. But Jonkershoek was much more that just the origin, it was the methodological test-bed for the program as a whole that eventually included experiments in all the forestry regions that tested the main plantation species and diverse catchment management regimes.

We contextualise the wider significance of the research and policies that flowed from Jonkershoek. There has been a tendency in current policies to reduce Jonkershoek's findings to the key concept that exotic trees use more water than many indigenous vegetation types. The policy recommendations based on research findings from Jonkershoek must be understood as having been implemented within a national water and forest management strategy that sought to account for a variety of forms of land usage, including forestry, agriculture, and indigenous ecosystem conservation. National policy regarding forests and water from the late 1940s onwards, directed by findings from the network of catchment experiments that began at Jonkershoek, sought to direct afforestation to areas with higher rainfall and profitability, where there was little competition for water use, while encouraging catchment management on public and private lands through various Acts and policies. These policies led to the development of a coherent and effective national policy for managing catchments that was deconstructed from the late 1980s to the late 1990s. Current policy and public discourse on water resources have, by contrast, reduced the issue to one of governing the distribution of water-use rights among competitors and overlooking the sustainable management of catchments as ecosystems.

We conclude by analysing the vicissitudes of the hydrology program during a period of political uncertainty lasting from the late 1980s to the late 2000s. The research station was transferred, along with the South African Forestry Research Institute, to Forestek within South Africa's Council for Scientific and Industrial Research, CSIR, where it resided from 1990 to 1995. Since 1995–1996, funding has come mainly from the Water Research Commission on the basis of competitive bids. The Jonkershoek program remained alive after direct government funding ended in 1995 owing to the efforts of a small, dedicated group of scientists who kept the research station going, and now shows the signs of revival, under new auspices. The infrastructure of the network, a large body of knowledge, and the core of an intellectual capital are a legacy now being rebuilt under the current rubric of global climate change: the South African Environmental Observation Network, a new initiative, is resuming elements of the monitoring.

Chapter 1

'Fit the Tree to the Climate':
The Cape Model of Forestry

David Ernest Hutchins remarked amusingly in a letter from 1890 that 'there is a twinkle in old [Dietrich] Brandis's eye when he talks of the forest officers at the Cape who have never seen a regular forest!'[1] Dietrich Brandis's description of Cape forestry, though somewhat exaggerated, nevertheless reflected an environmental reality. Southern Africa, and much of the former Cape Colony, was 'wonderfully devoid of trees', in the words of the Scottish-born botanist T. R. Sim.[2] Today, closed-canopy forests cover less than 0.3 per cent of South Africa's land surface, a figure that probably approximates the size of these forests in the late nineteenth century.[3] Nor did the Cape's indigenous species prove easy to propagate. Valuable Yellowwood (*Afrocarpus falcatus* and *Podocarpus latifolius*), Black Stinkwood (*Ocotea bullata*), Black Ironwood (*Olea capensis*) and Sneezewood (*Ptaeroxylon obliquum*) are slow-growing species difficult to raise in the nursery and establish in the field, and regenerate only under ideal conditions, as was well known at the time.[4]

Foresters in the Cape responded to these ecological conditions, and to their inherent economic consequences, by embarking on the then unprecedented course of using introduced species of tree for domestic forest production.

---

1   Dietrich Brandis was the former Inspector General of Forests of India. He is considered the 'father' of forestry in India by many. Hutchins to Fourcade, 10 July 1890, Fourcade Bequest BC246, C5, University of Cape Town Archives (UCT).

2   T. R. Sim, *Treeplanting in South Africa, Including the Union of South Africa, Southern Rhodesia, and Portuguese East Africa* (Pietermaritzburg: Natal Witness Limited, 1927), 1.

3   L. Mucina and M. C. Rutherford (eds), *The Vegetation of South Africa, Lesotho and Swaziland* (Pretoria: South African National Biodiversity Institute, 2006), 32.

4   See T. R. Sim, *Tree Planting in Natal* (Pietermaritzburg: P. David and Son, 1905), xviii, 48, 278.

The inevitable landscape change, speculation and dispute about the consequence to water supplies and climate, and the price of this change, caused the emergence of the scientific search for evidence to resolve this conflict.

To support its population and economy, the colony had to import far more timber than was produced at home, at great cost. The imperative to plant forests forced them to seek new ways of growing exotic trees successfully and cost-effectively. Many of the first plantations failed. Yet failure, together with the motivating force of some early successes, compelled Cape foresters to devise methods for selecting and experimenting with exotic trees. This chapter argues that during the 1880s to 1900s, Cape foresters initiated a coherent and globally unique research program that sought to select and then grow climatically suitable genera and species in timber plantations. Hutchins, a leader of this movement, summarised this view succinctly in his favourite phrase: 'fit the tree to the climate'. Research that began in the Cape later spread to other South African colonies after the conclusion of the South African War in 1902. South Africa's first forestry school focused on training foresters in this unique Cape method of forest development (Chapter 3). Foresters who trained and worked within this tradition reshaped South Africa's landscapes, economies, and cultures by planting trees throughout suitable and less suitable areas of the country.

Foresters working in the Cape Colony during the 1880s and 1890s faced some of the most difficult conditions in the world, similar to conditions in South Australia, the Punjab, Algeria, Tunisia, and Morocco. Attempts to grow exotic trees suffered many setbacks throughout the colony, most especially in the vast arid Karoo. Attempts at propagating trees of the indigenous forests to expand timber resources proved futile. The Cape's Parliament was penurious and the Minister for Agriculture worried about unrestrained spending. Many rural residents looked with scepticism on the propagandistic claims of how trees could improve the climate and push back the arid Great Karoo. White settlers and indigenous Africans contested the boundaries and meanings of Crown land.[5] There was a shortage of labour during the first decades of the Forestry Department's existence. Yet with a belief in the rightness of their duty, colonial foresters set about their tasks with diligence and perseverance.

In its tenets and practices, the 'Cape model' of forestry that evolved in the colony diverged knowingly from core northern European forestry. Foresters in the Cape chose to rely on planting exotic trees from all over the world, and had

---

5   Throughout this work, we use the term 'Crown land' to refer to land held by the state when writing about the period of Crown colonies, but 'state land' for the period after Union in 1910, even though the term 'Crown land' continued as a statutory category until South Africa left the Commonwealth in 1961. The same applies to 'Crown forest' and 'state forest'. The situation with state land and state forests is complicated after the 1936 *Native Trust and Lands Act*, when a category of land generally termed 'Trust land' emerged (also 'Trust forests'). Where necessary, we distinguish this category of state asset.

little use for European forestry theories, which failed to explain how to select and manage exotic trees for extra-tropical climates. They had to learn how to domesticate and cultivate species of tree that were entirely novel to their discipline. Once they established a coherent plantation program that clearly deviated from European orthodoxy, foresters began to draw criticism from imperial foresters and botanists in Britain. Not for the last time would South African forestry methods draw disapproval from foresters overseas.

The unique set of conditions that shaped foresters during the late nineteenth century must be understood within the wider history of the Cape Colony, dating back to the foundations of Cape Town in 1652. European colonists introduced a new biotic regime to the south-western Cape region, bringing with them what Crosby described as a 'portmanteau biota', including livestock, plants, pests, and diseases. Environmental constraints, including aridity, recurrent wildfire, and a highly diverse but largely treeless shrub-dominated indigenous Cape flora, meant that settlers were constantly seeking new sources of timber, first from sites near Cape Town, and later in forests further away as well as in plantations. Yet numerous problems hindered the success of plantations and forest conservation during the eighteenth century and the first three quarters of the nineteenth century. It was these historical and contemporary constraints that compelled the first generation of professional foresters to find innovative ways of selecting and acclimatising exotic trees.

The environmental and colonial context of post-1652 botanical introductions

Southern Africa has for millions of years been a relatively forest-less landscape.[6] Paradoxically, the subcontinent also has climatic and soil conditions that could theoretically have allowed for the growth of trees. A variety of evolutionary, edaphic, and climatic conditions led to the development of a number of forest-free biomes. South Africa's two largest biomes—grassland (27.9 per cent of South Africa's surface area) and savanna (32.5 per cent)—are 'fire-driven' ecosystems where trees are kept in check by recurring grass fires that kill trees before they can crowd out grasses.[7] Fynbos (6.6 per cent), the dominant biome of the south-western Cape, where the Dutch first settled in 1652, characterised by the heath,

---

6   We use the term 'forest' in this text to mean the indigenous, evergreen, closed-canopy forest, mostly of the Afrotemperate Mistbelt, Coastal and Scarp types, as in Mucina and Rutherford, *The Vegetation of South Africa, Lesotho and Swaziland*, 587.

7   Ibid., 37.

proteoid, and restoid growth forms, lacks trees, is also fire-dependent and deficient in mineral nutrients.[8] Trees are scarce also in the neighbouring Nama-Karoo and Succulent Karoo biomes, where aridity is a limiting factor.

The introduction of trees was part of the onset of a new biotic regime that began in 1652 with the Dutch settlement in Cape Town. Chosen for its strategic location, the natural environment of the Cape presented many difficulties to early settlers, among which was the scantiness of wood. The south-western and southern Cape is relatively devoid of forests except for occasional small patches on south-facing mountain slopes, and in sheltered valleys. From its initial settlement, the Vereenigde Oost-Indische Compagnie—the VOC—passed laws regulating the cutting of indigenous timber at the Cape. Resource scarcity, especially timber shortages, remained a pressing issue in the Cape for the next three centuries.

To overcome the Cape's natural timber deficiency, Jan van Riebeeck (Commander, 1652–1662) and Simon van der Stel (Commander, 1679–1691; Governor, 1691–1699) required settlers to plant trees. Though settlers used some indigenous species, such as the Keurboom (*Virgilia oroboides*) and Wild Almond (*Brabejum stellatifolium*), most species planted were imported from Europe (such as oak, *Quercus robur*) and Asia (camphor, *Cinnamomum camphora*).[9] Introduction of the stone pine (*Pinus pinea*) and cluster pine (*Pinus pinaster*) probably accompanied the arrival of Huguenots in the 1680s.[10] Settlers planted trees in cities, alongside roads and around homesteads (van der Stel required this by law). Thus, exotic trees always followed close behind (and in some cases advanced ahead of) the steady expansion of European settlement and Christian missions into the interior of Southern Africa.

In the eighteenth century, the VOC established some timber plantations across the Cape Flats and at Silvermine, Tokai, and Cecilia, but the extent and effectiveness of these plantations is unknown.[11] The well-watered locations at Tokai and Cecilia (located on the south side of Table Mountain, which receives up to 2,270 mm of precipitation annually) later became sites of larger plantations in the last half of the nineteenth century. It is probable that plantations suffered

---

8   Ibid., 37, 79.

9   G. A. Zahn and E. J. Neethling, 'Notes on the Exotic Trees in the Cape Peninsula', *South African Journal of Science*, 26 (1929): 211–34; S. Pooley, 'Jan van Riebeeck as Pioneering Explorer and Conservator of Natural Resources at the Cape of Good Hope (1652–62)', *Environment and History*, 15, 1 (2009), 3–8; R. Grove, *Green Imperialism: Colonial Expansion, Tropical Island Edens and the Origins of Environmentalism, 1600–1860* (Cambridge: Cambridge University Press, 1996), 140–4.

10   This is widely referenced in historical literature, with no attributable source. One of the early descriptions of this introduction date comes from *Edinburgh Philosophical Society Journal*, 9 (1830), 401.

11   N. Visser, 'Wood Production and Environmental Legislation at the Cape during VOC Rule, c. 1652–1795', draft. Also see, A. Appel, 'Die Geskiedenis van Houtvoorsiening aan die Kaap, 1652–1795' (MA Thesis, University of Stellenbosch, 1966).

greatly from illegal extraction of timber by burghers because of the high costs of imported timber.[12] The plantations were not enough to compensate for the continued destruction of indigenous forests in the south-western Cape given the expanding frontier of timber harvesting, which by the late eighteenth century stretched all the way east to Knysna.

The geopolitical upheaval during the Napoleonic Wars led to the Cape finally being ceded to the British in 1814, a formal recognition of Britain's *de facto* rule of the Cape dating back to 1806. This official transfer encouraged new flows of people, plants, and ideas between the Cape and British colonies. British rule led to Lord Grey's 1820s settlement of British and, later, German immigrants in what is now the Eastern Cape. The banning of slavery in 1833 by the British caused more conservative farmers to trek into the interior in 1835–1836 in search of new lands where they could farm, live in self-ruled Boer republics, and follow the rules of Calvinism. Frontier wars, initially between Boers and Xhosa in the last 20 years of the eighteenth century, intensified with British settlements as settlers and British military forces came into conflict with African groups east of the Great Fish River, a colonial boundary that up to the 1820s and 1830s had demarcated the end of European settlement and the beginning of numerous Bantu African settlements.[13] Trekkers skirmished to the north and east with the Matabele and Zulu, as well as Sotho-speaking polities. To counter the Boer expansion, the British annexed the Boer Natalia Republic, established in 1839, and re-christened it Natal Colony in 1843. This led most trekkers to move further into the interior where they settled in what became the Orange Free State and South African Republic (ZAR).

The new areas that European immigrants settled after 1820, as a general rule, lacked closed-canopy forests. The extensive highveld grasslands of the Free State and ZAR had few trees. The trees that dotted South Africa's different types of savanna in ZAR and Natal could be used as firewood and some for house construction and furnishing, but the varied quality and low quantity of timber made it unsuitable for large-scale industry. Exceptions included forests of the Eastern Cape Frontier and Transkei, along the kloofs of Natal, and some

12   Visser, 'Wood Production and Environmental Legislation at the Cape during VOC Rule, c. 1652–1795'.
13   There is a historiographical debate about the nature of the lands that *voortrekkers* came across. Prior to the 1980s, scholars often assumed that much of the land was 'empty' because of Mfecane, a period of unrest and destruction unleashed on Nguni groups neighbouring the Zulu Kingdom led by King Shaka. See N. Etherington, *The Great Treks: The Transformation of Southern Africa, 1815–1854* (London and New York: Pearson Education, 2001); for a disputed critique of orthodox views before the 1980s see J. Cobbing, 'The Mfecane as Alibi: Thoughts on Dithakong and Mbolompo', *The Journal of African History*, 29 (1988): 487–519.

parts of eastern and north-eastern Transvaal, such as the Woodbush forest.[14] Given the finite forests and their slow growth, the expanding population of Europeans, coupled with current and prior wood use by indigenous Africans, led to the depletion of their stocks of large trees and the loss of some forests. This destruction increased rapidly in all forests after the discovery of diamonds in Kimberley in the late 1860s, and gold in the east of the ZAR and on the Witwatersrand in 1886, which sparked a gold rush in Johannesburg. As a result, woodcutters harvested indigenous timber at unsustainable levels given the slow growth rate of indigenous trees, the small size of the forests and the great demands for timber for mines and industry.

Introductions of new tree species in nineteenth-century Southern Africa

The paucity of tree cover in Southern Africa provided the environmental context for a comprehensive tree-planting campaign during the second half of the nineteenth century. One of the peculiar facets of nineteenth-century colonialism, especially in settler colonies, was that economic, cultural, and scientific views merged together into a powerful impulse to plant trees. European settlers, in Southern Africa as elsewhere, put their faith in the belief that trees would 'improve' climates, beautify landscapes, increase productivity, mark boundaries, and aid colonisation.[15] Settlers planted exotic trees for timber, for shelter from the sun and wind, and because they supposed that trees increased rainfall in dry environments.[16] Little was known about the climate and meteorology of the subcontinent of Africa. As a result, South Africa's aridity, heat, strong seasonal contrasts, and tropical diseases were all attributed at one time or another to the country's lack of forest cover. This was predicated on eighteenth-century ideas about the morality of climates (e.g. that tropics led to degradation) and built on nineteenth-century ideas of desiccation popularised by Alexander von Humboldt and a cadre of prominent German and French foresters and botanists, as well as George Perkins Marsh in the USA.[17]

---

14   See Mucina and Rutherford, *The Vegetation of South Africa, Lesotho and Swaziland*. There is a debate about the historic size of the forests in the Transkei in relation to the development of state forestry. See Tropp, *Natures of Colonial Change in the Making of the Transkei* and F. J. Kruger, 'Natures of Colonial Change: Environmental Relations in the Making of the Transkei', *British Scholar Journal*, 3 (2010): 263–71.

15   The first professionally trained forester in the Cape, the Comte de Vasselot de Regné, made this clear when he argued that with forestry 'the development of colonization will be powerfully helped forward'. Cape of Good Hope, *Report of the Superintendent of Woods and Forests* (Cape Town, 1882), 45. Hutchins discussed using eucalypt trees with wire fencing in the Eastern Cape to denote boundaries. Cape of Good Hope, *Report of the Superintendent of Woods and Forests for the Year 1886* (Cape Town, 1887), 32.

16   See Barton, *Empire Forestry*, 98–105; Grove, 'Early Themes in African Conservation: The Cape in the Nineteenth Century', 21–39.

17   Grove, *Green Imperialism*, 366; Barton, *Empire Forestry*, 15–7.

Tree planting served different purposes and employed various species. Before the second half of the nineteenth century, settlers planted trees for homestead utility using a few proven species such as Grey Poplar (*Populus canescens*), Common Oak (*Quercus robur*), and to the north-east, syringa (*Melia azedarach*), although amateur and professional botanists, in certain instances, did facilitate certain exchanges of new species into the region. For instance, the first known introduction of Australian *Acacia longifolia* probably came in 1827 when James Bowie, a plant collector for Kew Gardens, arrived in the Cape with seeds from England.[18] Settlers planted whatever seed they could get their hands on, or, in the case of the Grey Poplar, possibly the most popular homestead tree, vegetative truncheons. A series of complex familial, commercial, and state networks developed that exchanged seeds from different parts of the world. Many exchanges occurred when people brought seeds from one colony to another. *Eucalyptus globulus* (Blue Gum) was probably introduced into the Cape in 1828 when Sir Galbraith Lowry Cole, the new governor, brought the species with him from Mauritius.[19] Settlers in remote regions worked through word of mouth and read print materials to find out what seeds were working in Southern Africa and elsewhere in the world.

The successes and failure of Australian trees, especially *Eucalyptus,* in South Africa illustrate the blend of myth, pragmatic and widespread trial, and later, systematic scientific enquiry, which finally led to successful acclimatisation of exotic trees. Initially, a potent brew of uncertainty and hope led to the rise of much hearsay regarding the properties of certain species and genera. For a period, eucalypts and other Australian trees become the wonder of the tropical and subtropical settler world. The popularity of Australian trees soared among Cape colonists in the 1860s after botanical enthusiasts in Australia and France peddled grandiose claims about their properties, especially those of the genus *Eucalyptus.* The Australian botanist Ferdinand von Mueller boasted that the timber of *E. globulus* rivalled the world's most valuable timbers. Mueller and other botanists also argued that eucalypts helped to prevent malaria and other tropical diseases, both by draining swamps due to their vigorous growth and through the secretion of their scented, powerful oils, which subscribers to the miasmic theory of disease believed would kill malaria.[20]

---

18   G. L. Shaughnessy, 'Historical Ecology of Alien Woody Plants in the Vicinity of Cape Town, South Africa' (PhD Thesis, University of Cape Town, 1980), 104–5.
19   See Zahn and Neethling, 'Notes on the Exotic Trees in the Cape Peninsula'; J. Noble, *History, Productions, and Resources of the Cape of Good Hope* (Cape Town: W. A. Richards and Sons, 1886), 150.
20   B. M. Bennett, 'The El Dorado of Forestry: The Eucalyptus in India, South Africa, and Thailand', *International Review of Social History*, 55, S18 (2010): 27–50, 30–2.

The Victorian belief that the location of Australia and Southern Africa in the southern hemisphere made them geographically and botanically related further justified interest in planting Australian trees. The botanist Joseph Hooker, who noted similarities amongst the floras on different continents and islands across the entire southern hemisphere, hypothesised that Australia and Southern Africa might have been linked as part of a large ancient southern continent.[21] The popular and prolific naturalist, Alfred Russel Wallace, argued that the southern hemisphere was characterised by 'detached areas, in which rich floras have developed … but [which are] comparatively impotent and inferior beyond their own domain'.[22] The only exception to this rule was Australia's forest flora, which naturalised in the southern hemisphere on sites outside of its original geographic range, such as the Cape.[23] Scientists in the Cape Colony expressed similar beliefs. At least one Cape Colony forester, Hutchins, noted the dominance of the Australian flora over the Cape's when arguing for the importation of Australian trees into Southern Africa.[24] This survival-of-the-fittest mentality fitted well with the popular fallacies of Darwinian natural selection when they were applied to explain why some floras and species seemed more 'aggressive' and others 'impotent'.

Formal botanical exchanges between Australia and Southern Africa increased in intensity from the 1850s to the 1890s as a result of the creation of new botanical gardens in Australia and Southern Africa.[25] Gardens opened in Cape Town, Melbourne, Adelaide, Durban, Pietermaritzburg, King William's Town, Graaff-Reinet, Perth, and Grahamstown in the 1850s to 1870s. Botanical gardens in Natal and the Cape Colony prominently featured Australian trees. Founded in 1881, the Cape Colony's Department of Agriculture pursued the largest institutional program of tree planting in Southern Africa in the 1880s and 1890s. One estimate suggested that its Forestry Division sent out 300 million wattle (Australian *Acacia*) seeds alone to Cape colonists from 1882 to 1893.[26] It is doubtful that

---

21   See the introductory essays in J. D. Hooker, *Flora Novae Zeelandiae* (London: Lovell Reeve, 1853) and *Flora Tasmaniae* (London: Lovell Reeve, 1860).

22   A. R. Wallace, *Island Life* (London: Macmillan, 1880), 495.

23   Ibid., 496.

24   D. E. Hutchins, 'Extra-Tropical Forestry: Being Notes on Timber and Other Trees Cultivated in South Africa and in the Extra-Tropical Forests of Other Countries', *Agricultural Journal of the Cape of Good Hope* [*AJCGH*], 26, 1 (1905), 18–9.

25   For the history of the rise of botanic gardens in the British world, see D. P. McCracken, *Gardens of Empire: Botanical Institutions of the Victorian British Empire* (London: Leicester University Press, 1997); R. Drayton, *Nature's Government: Science, Imperial Britain, and the 'Improvement' of the World* (New Haven, CT: Yale University Press, 2000).

26   G. L. Shaughnessy, 'A Case Study of Some Woody Plant Introductions to the Cape Town Area', in I. A. W. Macdonald, F. J. Kruger and A. A. Ferrar (eds), *The Ecology and Management of Biological Invasions in Southern Africa*, 41, cited in van Sittert, '"The Seed Blows About in Every Breeze": Noxious Weed Eradication in the Cape Colony, 1860–1909', *Journal of Southern African Studies*, 26 (2000), 660.

very many of these seeds ever grew into mature, seed-producing trees, but even a small percentage of successful results would have led to substantial ecological changes.

From the 1850s onward, Australian botanists worked closely with Cape and Natal correspondents to encourage the acclimatisation of Australian trees in the two colonies. Ferdinand von Mueller, the government botanist for Victoria from 1853 to 1896, sent seeds and provided advice to botanists, farmers, and foresters there for over 40 years.[27] When Mueller's life ended, Joseph Maiden, director of the Sydney Botanic Gardens and Herbarium from 1896 to 1924, became the leading Australian botanist who corresponded with Cape and Natal botanists.[28] Maiden worked as the official seed collector for the Cape Colony from 1896, when the Agriculture Department, at the request of Hutchins, established a direct relationship with him.[29]

Seed introductions in the Free State and the South African Republic occurred primarily through private, not state, networks. Richard Wills Adlam was one of the few people in the Transvaal to pursue transnational botanical exchange with scientists.[30] The Fichardt family established an extensive plantation of various eucalypts near Bloemfontein in the 1860s, while Charles Newberry, a British immigrant to the Orange Free State, employed a forester and horticulturalist to plant a large number of trees on his Prynnsberg estate, in the Orange Free State near the border of then Basutoland, beginning in the summer of 1881–1882.[31] The lack of state action did not seem to slow the exchange of species between continents and colonies. Peer-to-peer networks and wealthy landowners who sometimes hired horticulturalists brought in the seeds available in the Cape.

---

27   P. MacOwan and E. Pillans, *Manual of Practical Orchardwork at the Cape* (Cape Town: W. A. Richards and Sons, 1896), 627–8.

28   See J. Frawley, 'Joseph Maiden and the National and Transnational Circulation of Wattle *Acacia* spp.', *Historical Records of Australian Science*, 21 (2010): 35–54.

29   See D. E. Hutchins to Undersecretary of Agriculture, 9 July 1896, F719, Department of Agriculture [AGR] 722, National Archives of South Africa Cape Town [NASA-CT]; Undersecretary for Agriculture, Cape Colony, to Undersecretary for Mines and Agriculture, New South Wales, 6 August 1897, F719, AGR 722, NASA-CT. Also see Frawley, 'Joseph Maiden'.

30   For a list of plants see the papers of R. W. Adlam, BC 815, UCT. Adlam moved to the Transvaal to work for Alois Hugo Nellmapius as a horticulturalist on his Irene Estate before serving as the first curator of Joubert Park in Johannesburg from 1893 to 1903. For Natal's wattle industry, see H. Witt, '"Clothing the Once Bare Brown Hills of Natal": The Origin and Development of Wattle Growing in Natal, 1860–1960', *South African Historical Journal*, 53 (2003): 99–112, 100–6. For eucalypts in the Transvaal, see R. J. Poynton, *Tree Planting in Southern Africa*, Vol. 2: *The Eucalypts* (Pretoria: Government Printer, 1979), 882, 14–5. By 1910, at least 34 species of *Eucalyptus* had been introduced to the Transvaal.

31   Arthur Emmanuel Fichardt, President of the Bloemfontein Chamber of Commerce, to the Dominions Royal Commission, *Royal Commission on the Natural Resources, Trade, and Legislation of Certain Portions of His Majesty's Dominions. Minutes of Evidence Taken in the Union of South Africa in 1914, Part I* (London: HMSO, 1914), 290; Charles Newberry, in a letter in 1887 to the Christian Express, reported having planted 350,000 trees (of which just 200,000 to 250,000 had survived, many species proving unsuited to this site), including 40 species of eucalypt, 'many' species of pine including *Pinus radiata* (found to be 'the best'), Italian poplars, and willows. See *Forestry News*, 3/78 (September 1978), 4–5.

People hoped trees that grew in the Cape would grow in other places, such as the Transvaal, which needed timber for the gold mines.[32] In 1893, Sammy Marks, a wealthy Jewish financier and industrialist, appointed a German horticulturist, Otto Brandmuller, to begin the afforestation of his estate Maccauvlei, south of Johannesburg on the Vaal River. Brandmuller started by planting 100,000 common oak trees, but the plantation eventually grew to nearly 2,000 hectares, with *Pinus radiata* and *P. pinaster* the main species.[33] Newberry, too, used species proven in the Cape. Hutchins noted that as late as 1890, 'we were positively assured that all the trees that grow at the Cape would succeed at Johannesburg'.[34]

Many exotic species introductions succeeded. This reshaped urban and rural landscapes. By the late nineteenth century, eucalypts, wattle, willow, poplar, and syringa punctuated Southern Africa's landscape as far north as Rhodesia. In 1897, James Bryce described the entire countryside of Southern Africa as being dotted by exotic trees:

> [One] finds them now everywhere, mostly in rows or groups round a house or a hamlet, but sometimes also in regular plantations. They have become a conspicuous feature in the landscape of the veldt plateau, especially in those places where there was no wood, or the little that existed has been destroyed. Kimberley, for instance, and Pretoria are beginning to be embowered in groves of eucalyptus; Bulawayo is following suit; and all over Matabililand and Mashonaland one discovers in the distance the site of a farm-steading or a store by the waving tops of the gum-trees.[35]

The trees Bryce and other travellers noted were used primarily for domestic purposes, such as shade, firewood, and other non-commercial uses.

Successful nineteenth-century acclimatisation happened as a matter of chance as well as experience with a small set of species easily propagated, rather than from a methodical, rigorous selection process that identified species and characteristics that would be best suited to different regions in Southern Africa. There was widespread global interest in planting exotic trees, especially those from Australia. Southern Africa just happened to be one of the few places where they seemed to grow freely, while elsewhere, with a few exceptions, people found *Eucalyptus* difficult to propagate.[36] The reason why some Australian

---

32   J. J. Kotzé and C. S. Hubbard, *The Growth of Eucalypts on the High Veld and South Eastern Mountain Veld of Transvaal*, Forest Department Bulletin No. 21 (Pretoria: Government Printing Office, 1928), 59, 6.

33   Vaal Triangle History—Maccauvlei: www.vaaltriangleinfo.co.za/history/maccauvlei/ (accessed 2 June 2013).

34   D. E. Hutchins, 'Forestry in South Africa', in W. Flint and J. D. F. Gilchrist (eds), *Science in South Africa: A Handbook Review* (Cape Town: T. Maskew Miller, 1905), 18.

35   J. Bryce, *Impressions of South Africa* (London: MacMillan and Co., Limited, 1899), 29–30.

36   Ibid.

species acclimatised in Southern Africa was because the two regions share similar climatic ranges. Settlers vaguely understood that both regions shared similar climatic qualities, but they had little knowledge about how to select species from one comparable climate and introduce it into another similar climate. This lack of knowledge led to many unsuccessful introductions and caused great frustration.

The need for timber in Southern Africa: The minerals revolution

The opening up of diamond fields near Kimberley in the early 1870s sparked a minerals revolution that transformed 'a slow-growing and impoverished section of the world' into a modern industrialised state.[37] It was the minerals revolution that thrust forestry into the foreground of the rural economy, driving and forming a forestry distinct in its history from forest histories elsewhere. The consequences of the mineral revolution to forestry was profound. The colonial forestry sector, which was characterised by a negative balance of trade and unsustainable harvesting of indigenous forests, was gradually replaced by a modern forestry sector that produced large volumes of plantation-grown timber for the domestic economy and, eventually, for export markets.

The early years of diamond mining gave advance warning of the long reach of minerals into South Africa's forests. Not only did developments in Kimberley rapidly deplete the timber in the surrounding arid savanna, but its demands for structural and energy timber, the knock-on demand for wagon woods, and the later demand for sleepers and fuel as the railways extended to serve the mines, stimulated the activities of the woodcutters in the Knysna forests and elsewhere.[38] The demand for wood from Kimberley 'caused these forests to be worked out to such an extent that in 1876 the forests west of the Kaaimans River, a few miles east of George, were declared closed to all workings and remained closed until 1925'.[39]

---

37   S. Marks, 'War and Union, 1899–1910', in R. Ross, A. K. Mager, and B. Nasson (eds), *The Cambridge History of South Africa Volume II* (New York: Cambridge University Press, 2011): 157–210, 128. B. Freund, 'South Africa: The Union Years, 1910–1948—Political and Economic Foundations', in *The Cambridge History of South Africa Volume II*: 211–54, 211.

38   J. D. M. Keet, 'Historical Review of the Development of Forestry in South Africa', (Pretoria, c. 1970), 52–3. MS available online: www2.dwaf.gov.za/webapp/resourcecentre/Documents/Publications_And_Media/Keet_Forestry_History_page_41-66.pdf.

39   F. S. Laughton, *The Sylviculture of the Indigenous Forests of the Union of South Africa with Special Reference to the Forests of the Knysna Region*, Forestry Series No. 7 (Pretoria: Department of Agriculture and Forestry, 1937), 22.

The growth of the goldfields of the Witwatersrand, following the discovery of the Main Reef in 1886, intensified the extraction of timber, transported by wagon 500 km from the Soutpansberg to the north, and nearly 1,200 km from Knysna. The Witwatersrand kindled demand for timber and railway development. Though railway construction created a demand for indigenous timbers for rail sleepers—Hutchins reports the sale of '100s of thousands' of yellowwood sleepers from the Knysna forest[40]—completion of the rail line between Kimberley and Johannesburg in 1892 allowed the delivery of imported timber at prices lower than road-delivered indigenous timber, and virtually eliminated the demand for wagon wood. Suddenly, the market for indigenous timber fell, and extraction from the forests of Woodbush and the Soutpansberg slowed to a halt (except for the last high-value sources in remoter enclaves).[41]

Knysna woodcutters, no longer able to sell the timber from the diverse species stipulated in their licences, switched to scarcer, high-value yellowwood and stinkwood for the smaller furniture and joinery market, undermining the Forestry Department's carefully constructed management plans by further degrading the forests.[42] The sudden surge in cheaper rail-borne timber imports created the opportunity for private enterprise, while at the same time aggravating the public concern about their effects on the country's balance of payments, so that economic and financial factors weighed in with conservation motives to accelerate a shift toward the domestic production of timber from planted forests. The newly integrated Union rail system opened the Witwatersrand market to the wattle growers in Natal and the eastern Transvaal, as well as new eucalypt growers, stimulating further investment in plantations.

The mines, on the Reef as well as on the goldfields of Lydenburg and Barberton, demanded 'spectacular' quantities of timber, satisfied initially from an assortment of sources, including imports, timber from the indigenous forests, savanna trees such as Knobthorn (*Acacia nigrescens*, now *Senegalia nigrescens*), Leadwood (*Combretum imberbe*), and Marula (*Sclerocarya birrea*), as well as early yields from plantings of eucalypts of different species.[43] Between 1893 and 1898, mining expenditure on timber on the Witwatersrand more than doubled.

---

40   Hutchins, 'Forestry in South Africa', 1.
41   Ibid., 25; J. C. Scheepers, 'An Ecological and Floristic Account of the Vegetation of Westfalia Estate on the North-Eastern Transvaal Escarpment' (MSc Thesis, University of Pretoria, 1966), 4; J. Tempelhoff, 'Die Ontginning van Noord-Transvaal Se Houtbronne in Die Negentiende Eeu and Vroeë Bewaringsmaatreëls', *South African Forestry Journal*, 158 (1991): 67–74.
42   Laughton, *The Sylviculture of the Indigenous Forests of the Union of South Africa with Special Reference to the Forests of the Knysna Region*, 25. Also Cape of Good Hope, *Report of the Superintendent of Woods and Forests for the Year 1889 (Part 1)* (Ministerial Department of Crown Lands and Public Works, 1889), 5.
43   This account of developments in the mining-timber market draws on H. A. Read, 'Mining Timber on the Witwatersrand', *Empire Forestry Journal*, 8 (1929): 248–62. See also E. T. E. Andrews et al., 'Minerals and Mining', in *A Survey of the Resources and Development of the Southern Region of the Eastern Transvaal Lowveld* (Barberton: Lowveld Regional Development Association, 1954): 58–74, 68.

By 1911, the next year of record, expenditure was double that of the late 1890s, and by 1918 it had redoubled. During World War I, prices of imported timber rose to double the pre-war prices; by 1928 around 400,000 tonnes of timber poured down the mines annually on the Reef (see Figure 1).

Figure 1. Timber supports in a mine on the Reef, 1920s.
Source: H. A. Read, 'Mining Timber on the Witwatersrand', *Empire Forestry Journal*, 8 (1929): 248–62.

Plantations for local production sprang up at private initiative. H. A. Read cites the report of J. Klimke, State Mining Engineer of the South African Republic, for the year ending 31 December 1894: 'Tree Planting for Mining Purposes', in which he promotes plantation production as a substitute for timber imports:

> As the mining industry is rapidly extending, the consumption of wood increases, sawn lumber for buildings and timbering of shafts is almost entirely imported from abroad … On this account the cultivation of trees is considered an important branch of agriculture, and in consequence plantations are laid out

in several places, as it is of importance for mining requirements the most useful and best trees should be planted ... [such] information ... best acquired from practical foresters.

Klimke's document appends a report by G. Genth, a forester from Saxony employed to develop the 'extensive plantations on the farm Braamfontein', in the present Johannesburg, in which Genth summarises experience with the 850-acre project (see Figure 2).

The earliest plantings of *Eucalyptus grandis* for commerce appear to be those by Conrad Plange and Heinrich Schulte Altenroxel soon after they bought land near Tzaneen in 1893,[44] followed after the South African War by trials by the Transvaal government. These were forerunners to major mining-timber enterprises such the investment by the Exchange Yard of Johannesburg, a warehousing subsidiary of Rand Mines, which processed substantial mining-timber volumes but soon included timber processing for the mines in its business.[45] Afforestation with eucalypts to supply mine-support timber expanded in the later centre of sawtimber forestry, Sabie, from around 1910[46] and the volumes of mine-support timber consumed grew steadily to its peak of over 1 million tonnes per year a few decades later.[47] Since the South African War, private plantations accounted for most of the rapid growth in area afforested for the following 90 years, and have always exceeded the state's forests.[48]

---

44   Supplement to the *Letaba Herald*, 20 October 1989. This land later became the Westfalia Estate in Hans Merensky's ownership. See Scheepers, 'An Ecological and Floristic Account of the Vegetation of Westfalia Estate on the North-Eastern Transvaal Escarpment'.

45   'Notes on the "History of Lotzaba Forests", with emphasis on Tzaneen Region', typescript, n.d., Merensky Timber Limited, Tzaneen.

46   Transvaal Gold Mining Estates, the largest gold miner in the area, with 20,000 acres of land in its concessions, entered the timber industry around Sabie from the early 1900s, experimenting with small wattle plantations; by 1927 the company had 1,457 hectares under plantation, making a profit of 12,000 pounds sterling: by the 1930s, its area under plantation in the Pilgrim's Rest district was 12,000 ha. See A. W. Greenstein, J. P. Kleynhans, and P. J. A. Loseby, 'Timber Resources', in *A Survey of the Resources and Development of the Southern Region of the Eastern Transvaal Lowveld* (Barberton: Lowveld Regional Development Association, 1954), 86–91; S. Schirmer, 'Enterprise and Exploitation in the 20th Century', in A. Delius (ed.), *Mpumalanga History and Heritage* (Pietermaritzburg: University of KwaZulu-Natal Press, 2007), 522, 291–347, 305–7.

47   Mine-support timber consumption amounted to 1.55 million tonnes in 1993, and still accounts for 700,000 tonnes of South Africa's annual wood consumption: Institute for Natural Resources, *Pilot State of the Forest Report: A Pilot Report to Test the National Criteria and Indicators, March 2005* (Pretoria: Department of Water Affairs and Forestry, 2005), 6.

48   The 1918 Agricultural Census showed 0.5 per cent of Mpumalanga under timber (c. 23,000 ha), and in 1993 11 per cent (515,000 ha), all other agricultural sectors having grown: Schirmer, 'Enterprise and Exploitation in the 20th Century', 292; D. Reekie, 'The Wood from the Trees: *ex libri ad historiam pertinentes cognoscere*', *South African Journal of Economic History*, 19 (2004): 67–99.

Figure 2. A plantation of *Eucalyptus globulus* in Braamfontein.
Source: Department of Agriculture, Forestry and Fisheries, Pretoria; photographer unknown.

The limits of acclimatisation

If Australian trees disproportionately shaped hopes regarding tree planting in the second half of the nineteenth century, they also helped to dispel the settlers' naive idea that they could so easily change nature. Like many speculations of the late nineteenth century—railways to nowhere, grandiose agricultural schemes, and mining busts—Australian trees proved as much a mirage as a reality. They were the El Dorado of forestry in the late Victorian era, a dream of convinced people trying to change the world. The failure and consequences of poorly selected trees led foresters in the Cape Colony to rethink the theoretical and practical basis of forestry in Southern Africa.

By the last decade of the nineteenth century, angry farmers were complaining bitterly about the advice given by botanists and state foresters. The trees that they had planted with great hope too frequently died young, did not produce usable timber, and often seemed to dry out natural springs, marshes, and streams. Of all species planted widely in the mid to late nineteenth century, none attracted as much initial enthusiasm followed by criticism as *Eucalyptus globulus*, the Blue Gum. Starting in the late 1820s, settlers planted the species across all of Southern Africa. The species was so widely planted in the nineteenth century that, despite not being planted for most of the twentieth century, many South Africans still call any species of *Eucalyptus* a 'blue gum' or 'bloekomboom'. The species grows best in a narrow climatic range similar to its cooler native habitats in New South Wales, Victoria, and Tasmania, but in South Africa settlers planted it in the semi-desert Karoo, on mountains, and in the subtropical interior.

By the early 1900s, farmers were berating those who recommended the Blue Gum. One farmer, P. H. Pringle, told readers of the *Agricultural Journal of the Cape of Good Hope* (*AJCGH*) how he had planted numerous genera and species of exotic trees, but found that the 'least satisfactory of the lot is the Bluegum'. People complained that their trunks twisted, rendering timber useless except for firewood.[49] Trials of *E. globulus* sleepers found them to split and crack, 'useless for railway purposes'.[50] For a time, De Beers did not buy them for use as supports in the diamond mines because of the structurally unsound nature of the timber.[51]

A similar sense of frustration existed throughout Southern Africa. On the Highveld, the newly established plantations did not succeed in the long run, although they did supply mines with timber some of the time. Read records how 'rigorous winters'[52] required the choice of better adapted species, and how the outbreak of the bark-borer larvae of the beetle *Phorocantha semipuncata* and the snout beetle *Gonipterus scutellatus* caused growers to fell Highveld plantations

---

49   E. E. Ogston, 'The Twisting of Blue Gums', *AJCGH*, 22 (1903), 216.

50   Sim, *Tree Planting in Natal*, 161, Sim summarises official and private concerns about *E. globulus* timber defects, pp. 156–61.

51   Farmer, 'De Beers and Blue Gum Wood', *AJCGH*, 22 (1903), 352.

52   For an account of the effects of severe winters around the end of the nineteenth century, see K. Showers, 'From Forestry to Soil Conservation: British Tree Management in Lesotho's Grassland Ecosystem', *Conservation and Society*, 4 (2006): 1–35, 10. Showers writes of Lesotho, but the conditions applied to the whole of South Africa's interior plateau; for a historical analysis of the incidence of severe winters, see S. W. Grab and D. J. Nash, 'Documentary Evidence of Climate Variability during Cold Seasons in Lesotho, Southern Africa, 1833–1900', *Climate Dynamics*, 34 (2010): 473–99.

prematurely, and this with the greater productivity from plantations in the humid subtropical regions of Natal and the eastern and northern Transvaal put paid to the Highveld schemes.

Figure 3. A plantation of *Eucalyptus globulus*, George, c. 1910.

While the early plantations at Tokai and Worcester are most often mentioned, initiatives between George and Knysna, Fort Cunningham, and elsewhere in favourable climates were equally successful.

Source: Department of Agriculture, Fisheries and Forestry; photographer unknown.

Failure forced botanists to study the native biogeography and natural variation of species in order to understand the biology and ecology of valuable species. But the unique evolutionary history of Australia's flora proved challenging even for the most seasoned botanists on the continent. *Eucalyptus* classifications proved particularly troublesome. Genetic variations and local environmental influences can cause two trees of the same species to produce different leaf shape and growth forms, leading to misidentification. Equally, the difficulties in distinguishing between many species similar in their morphologies made it difficult for botanists to trust the species determinations of their helpers. Australian collectors notoriously misclassified the species and provided poor geographical information on the regions from which they sourced seeds. Joseph Maiden discussed this problem candidly in a letter to the Agriculture Department of the Cape Colony: 'I cannot place your order in the hands of nurserymen, as their collectors are not at present sufficiently educated in regard to the difficult genus *Eucalyptus* to enable me to trust their naming'.[53] In another letter, Maiden told Hutchins to tell him '[i]f any unusual proportion of the seeds fails to germinate, or if the seeds appear to be wrongly named, or to be under names different to those under which you have previously received them'.[54]

Hutchins maintained a detailed correspondence with botanists in Australia, notably Maiden, who over a 10-year period offered him advice on what species to plant in the Cape Colony that matched the species' native climates. In their first exchange, Maiden decided to include *Eucalyptus saligna* (later properly identified as *Eucalyptus grandis*) in the shipment of seeds to Hutchins: 'It [*Eucalyptus saligna*] is not on your list, and you need not therefore pay for it unless you choose, but the expense is trifling'.[55] Hutchins asked to continue receiving this species, and Maiden's 'trifling' expense eventually became the most widely planted species of *Eucalyptus* in South Africa from the 1930s until today.[56]

The same species planted in Southern Africa often looked different than it did in Australia, making many published botanical guides useless. 'No Genus is so perplexing as *Eucalyptus* in the matter of discrimination of species', the Cape Town botanist Peter MacOwan wrote, 'especially when as here, they have grown in fresh woods and pastures new, different from their Australian home, and have taken on a new habit'.[57] Questionable classifications led many settlers to call different species by the same name. MacOwan wrote an exasperated response to one settler who inquired about a specific species in 1894:

> The so-called popular names are the cause of endless wrangling and misunderstanding. Thus there are about twenty-five different White Gums, a dozen Blue Gums, several Black Wattles, several Golden Wattles, and every non-botanic grower vows that his particular blue or white or golden is the real one and the rest are bogus pretenders.[58]

Another problem quickly became apparent. Settlers who planted seeds did so with little knowledge of whether the species they (supposedly) selected would actually grow in the regions where they planted them. This is because few settlers (and also few scientists) appreciated the fact that most plants can only grow in a finite range of climatic, geological, and ecological situations. This was compounded by the time it took to recognise when a species had been improperly selected for a site. Trees could grow for a decade or more before showing signs of disease or other deficiencies. Hutchins noted that, '[most] trees, unless they

---

53   Maiden to Undersecretary of Agriculture, 3 November 1896, F719, AGR 722, NASA-CT.
54   Maiden to Undersecretary of Agriculture, 13 August 1897, F719, AGR 722, NASA-CT.
55   Maiden to Undersecretary of Agriculture, 3 November 1896, F719, AGR 722, NASA-CT.
56   R. J. Poynton, *Tree Planting in Southern Africa*, Vol. 2: *The Eucalypts*, 350–81.
57   P. MacOwan, 'Gum of *Eucalyptus*', *AJCGH*, 6 (1893), 32.
58   P. MacOwan, 'Australian Hedge Plant', *AJCGH*, 7 (1894), 40.

are altogether unsuited to the climate and soil do well for a few years, perhaps the first 20 or 30 years'.[59] One example was *Eucalyptus robusta*, which settlers planted in the dry western districts of the Cape Colony. The seemingly healthy trees grew for a time, but '[t]hen came the inevitable failure. As a native of the damp semi-tropics of East Australia it was quite out of its place in the ... climate of the Cape Peninsula or the dry Karoo'.[60] Insect pests, whether native to South Africa or having followed the eucalypts from Australia, aggravated the problem. Fichardt, speaking of experience with his plantation near Bloemfontein, reported persistent damage by a wood-boring insect that had switched from *Acacia karroo*, its native host, damaging eucalypt plantations, and complained of the lack of response from government officials to his reporting the problem.[61] *Eucalyptus viminalis*, widely planted on the Highveld of the interior plateau, around Johannesburg, as well as *E. globulus*, became infested by the defoliating Australian weevil, *Gonipterus scutellatus*, which, despite the success of later biocontrol, effectively eliminated *E. viminalis* and *E. globulus* as a commercial species in South Africa—an interesting case of fortuitous biological control, a companion to the natural biocontrol reported by Fichardt.[62]

Although *Eucalyptus* species proved the most difficult to select, would-be planters faced the same problems with almost every species and genus. They had little advice that detailed the native climates of an exotic species, let alone matching them with corresponding climates in the Cape Colony.

Climate and experimentation in the Cape, 1881–1910

The creation of a forestry department[63] within the Cape Agriculture Ministry in 1881 significantly shaped the future of tree planting in the colony. A number of state-sponsored scientific programs began during the 1880s, including veterinary science, agricultural research, and irrigation. Historians have debated the effectiveness, extensiveness, and widespread degree of public interest in

---

59   Hutchins, 'Extra-Tropical Forestry', 523.
60   Ibid.
61   Fichardt to the Dominions Royal Commission, 290.
62   P. DeBach and D. Rosen, *Biological Control by Natural Enemies*. 2nd ed. (Cambridge University Press, 1991), 178–81; see 108–9 on natural and fortuitous biocontrol.
63   The government entity for public forestry took different forms in this period: in 1876 the Cape government formed a separate Department of Plantations and Forests as a Divison of Agriculture, from 1881 to 1891 the function formed part of the entity called the Commissioner for Crown Lands, Mines and Agriculture, and so on. We use the phrase 'department of forestry' or 'forestry department' to refer generically to any one of the forms the function took.

various scientific and technical programs during this period.[64] What is clear is that, though many people agreed with the views of scientific experts, many members of the public, especially in rural areas, did not.[65] Understandably, many people chafed at legislation and policies informed by scientists that mandated certain actions, such as to clear weeds, dip sheep, or plant trees. It was the wealthier farmers who tended to advocate 'progressive' scientific reforms, which often required extensive capital to implement successfully. However, in many instances, conservation laws did little to change people's practices because the Agriculture Department was chronically understaffed and underfunded, a point that van Sittert argues is too often overlooked.[66]

Forest policy encountered similar obstacles, and professional forestry faced a precarious existence in the Cape Colony during its first two decades. The forestry department's finances were paltry considering the extent of Crown forest land and the requirements for afforestation. The department's fate was put into the hands of the colony's first Superintendent of Woods and Forests, Comte de Vasselot de Régné. Bringing the French aristocrat and his large family down to Cape Town in 1880 was one of the more flamboyant scientific hires in the British Empire during the second half of the nineteenth century. De Vasselot did not write in either English or Dutch, and had to have his reports and ideas on forest

64   See W. Beinart, K. Brown, and D. Gilfoyle, 'Experts and Expertise in Africa Revisited', *African Affairs*, 108 (2009): 413–33; K. Brown, 'From Ubombo to Mkhuzi: Disease, Colonial Science and the Control of *Nagana* (Livestock Trypanosomosis) in Zululand, South Africa, c. 1894–1955', *Journal of the History of Medicine and Allied Sciences*, 63 (2008): 285–322; K. Brown, 'Frontiers of Disease: Human Desire and Environmental Realities in the Rearing of Horses in 19th and 20th Century South Africa', *African Historical Review*, 40 (2008): 30–57; K. Brown, 'Poisonous Plants, Pastoral Knowledge and Perceptions of Environmental Change in South Africa, c. 1880–1940', *Environment and History*, 13 (2007): 307–32; K. Brown, 'Tropical Medicine and Animal Diseases: Onderstepoort and the Development of Veterinary Science in South Africa 1908–1950', *Journal of Southern African Studies*, 31 (2005): 513–29; K. Brown, 'Agriculture in the Natural World: Progressivism, Conservation and the State. The Case of the Cape Colony in the late 19th and early 20th Centuries', *Kronos Special Edition on Environmental History*, 29 (2003): 109–38; K. Brown, 'Political Entomology: The Insectile Challenge to Agricultural Development in the Cape Colony 1895–1910', *Journal of Southern African Studies*, 29 (2003): 529–49; K. Brown, 'Cultural Constructions of the Wild: The Rhetoric and Practice of Wildlife Conservation in the Cape Colony at the Turn of the Twentieth Century', *South African Historical Journal*, 47 (2002): 75–9; K. Brown, 'The Conservation and Utilization of the Natural World'. For Africa more generally, see J. Hodge, *Triumph of the Expert: Agrarian Doctrines of Development and the Legacies of British Colonialism* (Athens, OH: Ohio University Press, 2007). On the origins of expertise in natural science, see F. Albritton-Jonsson, 'Rival Ecologies of Global Commerce: Adam Smith and the Natural Historians', *American Historical Review*, 115 (2010): 1342–63.

65   For discussion of conflicts between believers in water divining and their 'progressive' critics, see L. van Sittert, 'Nation Building Knowledge: Dutch Indigenous Knowledge and the Invention of White South Africanism, 1890–1909', in D. Gordon and S. Krech (eds), *Indigenous Knowledge and the Environment in Africa and North America* (Athens, OH: Ohio University Press, 2012): 94–109. On the conflict between 'progressive' sheep farmers and those reliant on 'traditional' knowledge for animal disease management, see N. Visser, 'A Space for Conflict: The Scab Acts of the Cape Colony, circa 1874–1911' (PhD Thesis, University of Cape Town, 2011), 196–9.

66   See van Sittert, 'National Building Knowledge'.

management translated into English.[67] De Vasselot brought with him the stiff French forestry tradition: he wanted foresters to wear blue military uniforms as they did in France and he mentioned in an official report how collecting wood illegally in France could be punished by the 'penalty of death'.[68] De Vasselot made profuse and grand claims about forestry that he could never live up to.[69] His position of Superintendent of Woods and Forests was not filled when he left in 1891 on the pretext of cost, a fate similar to that of the equally grandiose John Croumbie Brown.[70] After leaving, one prominent forester suggested that de Vasselot served merely as a 'figurehead' of forestry, a statement that is hard to deny when examining his limited role in discussions of forestry even during his tenure. Nevertheless, de Vasselot left an important legacy in the form of the conservation system for indigenous forests.

De Vasselot hired a number of people to work under his direction during the 1880s. He was fortunate that they needed little guidance from him. He hired foresters who had studied forestry in Europe (Hutchins and James Rawbone), others who had gained Scottish botanical training (Thomas Sim and Charles Legat), one who studied surveying at the South African College (Henry Fourcade), another who had qualified in land surveying in India and learnt plantation forestry there (Joseph Storr Lister), and even a former chamberlain of the Emperor of Austria (Johan Baron de Fin); Captain Christopher Harison (appointed after leaving military service), and de Fin had managed forests in the Eastern Cape since 1856 and 1865, respectively. These people did the real work of forestry, with de Vasselot touring the conservancies, offering advice and maintaining an extensive correspondence (in French) with foresters in the four conservancies. He retained the services of A. W. Heywood to translate his documents, and Heywood developed a knowledge of forestry that led him in time to a role as Conservator of Forests.[71]

67   Le Comte de Vasselot de Régné, *Introduction of Systematic Treatment to the Crown Forests of the Cape Colony: Summary of Rules and Instructions* (Cape Town: W.A. Richards and Sons, 1885); Comte de Vasselot de Régne, *Selection and Seasoning of Wood* (Cape Town: W.A. Richards and Sons, 1885). All of Vasselot's correspondence was in French. His manuscripts are maintained in Stellenbosch University's library.

68   Cape of Good Hope, *Report of the Superintendent of Woods and Forests* (Cape Town: W. A. Richards and Sons, 1882), 29.

69   Instead of bringing in the £235,000 which de Vasselot predicted forestry could eventually make, forestry brought in a mere £7,680 of revenue in 1881. De Vasselot's main accomplishments were the demarcation of Crown forests and the enhancement of the selection system for the management of indigenous forests, and he encouraged early plantation experiments, but had little understanding of climatic requirements of exotic trees. For instance, he wrote, 'Forests ought to be a mine of gold to the Colony; while the plantations and re-foresting of mountains will, in conjunction with hydraulic works, turn to the best account, the rainfall of the country. Irrigation would then be easier, and agriculture a veritable mine of diamonds': *Report of the Superintendent of Woods and Forests for the Year 1882* (Cape Town: W. A. Richards and Sons, 1882, 1883), 13.

70   Brown, 'The Conservation and Utilisation of the Natural World', 430.

71   Heywood's daughter Lillian married J. D. M. Keet, later a major figure in South Africa's afforestation program (see Chapter 4).

The first decade of state forestry in the Cape was one of slow progress and hard-fought battles. A variety of people—ranging from poor woodcutters in Knysna, groups of sawyers in the frontier regions, settlers on the Cape Flats, leaseholders, to Africans in the eastern Cape—contested the demarcation and policing of forest boundaries. Foresters lived in remote regions, were poorly paid and housed, lacked adequate funds to pursue large projects, had a hard time starting plantings and complained frequently about how difficult it was to stop people from trespassing and taking wood illegally. Yet the demarcation of property with beacons and fences, and their protection by an increasing number of guards, did lead to a slow acquisition of forest land that foresters deemed valuable for economic or conservation purposes, accompanied by the enforcement of forest laws, often through jailing for up to three months' hard labour.[72] The *Cape Forest Act* (no. 28) in 1888, based on the *Madras Forest Act* of 1882, gave foresters the legislation required to reserve forest land. Forest reservation was part of an attempt to 'wean' Africans off indigenous forests and to force them to purchase timber from plantations (usually in the form of wattle) or to grow it themselves. State forestry alienated prior forest rights of indigenous Africans.

Tree planting was often promoted positively through 'subsidies and competitions', but legislation and policing made it mandatory for some people to plant trees.[73] On the Cape Flats, lessees of Crown land were legally obligated to plant trees, usually provided free of charge, while many were fined for not doing so. Foresters dealt with a penurious agriculture minister and parliament by using convict labour.

Establishing plantations was the principal pillar of forest policy, along with the conservation of indigenous forests and catchment protection.[74] From the outset, Joseph Storr Lister played the leading role in establishing state plantations in the Cape. He was born in Cape Town in 1852 and as a teenager went to work on a tea estate in Darjeeling, India, in 1870 before qualifying in land surveying and Hindustani and becoming forest sub-assistant on the plantation at Changa Manga in the Punjab, then the frontier of professional forestry in British India. He also worked in the hill forests of the Himalayas, and undertook a six-month exploration of the Hazar region of present-day Pakistan. Henry Baden Powell and Berthold Ribbentrop mentored his forestry assignments, preparing him for his unexpected future role in South African forestry. Lister left India for England

---

72   See Tropp's account of the process of colonisation of forests. J. Tropp, 'Displaced People, Replaced Narratives: Forest Conflicts and Historical Perspectives in the Tsolo District, Transkei'. *Journal of Southern African Studies* 29 (2003): 207–33.

73   For the quotation see Showers, 'Prehistory of Southern African Forestry', 303.

74   De Vasselot's attention was given primarily to the management of indigenous forests, as seen in his publication on management and his forest reports. This is unsurprising given that plantations would take decades to grow, whereas indigenous forests already existed.

in 1874 for health reasons but was recruited by the Cape Colony government while there as the Superintendent for Plantations on the Cape Flats. After his return to the Cape he started by directing government drift-sand reclamation efforts at the Cape Flats in 1875.[75]

Lister founded many of the Cape's largest state timber plantations in the early to mid-1880s. Most significantly, Lister founded the Tokai arboretum and plantation to the south of Cape Town in 1883.[76] Tokai was the centre of experimental tree planting in Southern Africa from the mid-1880s to the 1900s—it had the largest number of tree species planted of any arboretum in Africa. Lister took care from an early period to disseminate knowledge to the farming and rural community. In 1884 the government published 3,000 copies of his *Practical Hints on Tree-planting in the Cape Colony* in Dutch and English for distribution throughout the colony.[77] The book listed species for tree planting, but offered little advice about the climatic suitability of species.

Lister was able to make new large timber plantations, such as Kluitjes Kraal, Tokai, and Worcester, from yearly provisions voted by parliament based on estimates of expenditure rendered acceptably low by including convict labour and thus reducing costs.[78] Convicts were required because of the Cape forestry department's paltry budget. By 1886 there were 49 prisoners on an average day at the Kluitjes Kraal plantation.[79] Foresters relied upon the use of convicts for some other plantations, such as at Fort Cunningham and Tokai.[80] At Fort Cunningham, many of these labourers were probably people imprisoned with hard labour (for up to three months) for trespassing and stealing from forests.[81] Foresters expressed little concern about using convict labour, because they saw the work as improving the colony's environment and economy, and they believed it helped to teach labourers, many of them African, skills and work

---

75   C. L. Wicht, 'Figures in South African Forestry: Joseph Storr Lister I. S. O.: Founder of Modern Forestry in South Africa', MS undated, Wicht Papers, South African Forestry Research Institute [SAFRI] Archives, Council on Science and Industrial Research [CSIR], Pretoria; M. H. Lister, 'Joseph Storr Lister, the First Chief Conservator on the South African Department of Forestry', *Journal of the South African Forestry Association*, 29 (1957): 10–8. That year Hutchins made one of the first climatic comparisons, arguing that Table Mountain was similar to climates in the Himalayas. See Cape of Good Hope, *Report of the Superintendent of Woods and Forests for the Year 1882* (Cape Town: W. A. Richards and Sons, 1883), 15, 17.

76   Cape of Good Hope, *Report of the Superintendent of Woods and Forests for the Year 1883* (Cape Town: W. A. Richards and Sons, 1884), 19.

77   See, for example, J. Storr Lister, *Practical Hints on Tree Planting in the Cape Colony* (Cape Town: W. A. Richards and Sons, 1884).

78   Cape of Good Hope, *Report of the Superintendent of Woods and Forests for the Year 1884* (Cape Town, 1885), 7.

79   Cape of Good Hope, *Report of the Superintendent of Woods and Forests for the Year 1886* (Cape Town, 1887), 13.

80   Ibid., 2–3; Sim, 8.

81   These convictions skewed towards Africans. Cape of Good Hope, *Report of the Superintendent of Woods and Forests for the Year 1885* (Cape Town, 1886), 36. See the next year 'The year has been marked by the great impetus given to forest work and tree-planting in employing convicts to afforest...', Cape of Good Hope, *Report of the Superintendent of Woods and Forests for the Year 1886*, 28.

ethic. Forest law was drawn up irrespective of race, but prosecutions, especially in the eastern Cape, were almost entirely among the African population. In Knysna, prosecutions were generally of poor whites. On the Cape Flats, white settlers received fines for not planting trees.

De Vasselot and Lister established the earliest plantations sometimes without reference to the suitability of species to climate or site.[82] One of their first plantations was at Beaufort West, a town located in the middle of the Great Karoo. Lister hoped that: 'If it can be proved practically, that Plantations in the Karroo [sic] can be grown without the aid of irrigation, there would then be a prospect of converting that dreary desert into a smiling land'.[83] This view was bolstered by de Vasselot's firm belief that forests positively influenced climate. Lister received de Vasselot's approval to make a plantation, and also tried encouraging local residents to plant trees. Residents at Beaufort West told a disappointed Lister that they had insufficient water for irrigation and could not be induced to plant trees.[84] Despite the disagreement of locals, the department put money into establishing a plantation near the town in the early 1880s, but it proved to be an example that did little to encourage suspicious farmers.[85] A drought in 1883 caused the town reservoir to dry up and the trees to die, and Lister decided reluctantly to abandon the experiment.[86] This failure imparted a key lesson for foresters that they followed from then on: climatic considerations had to be taken into account when establishing plantations.

Henry Fourcade, a Frenchman who had arrived with his mother in the Cape in 1880, soon demonstrated a particular interest in climate and tree planting. Fourcade grew to be a complex personality, described as 'a legendary character', but an 'enigma' with 'a forbidding manner', a 'man of mystery'.[87] He studied land surveying in Cape Town at the South African College, and although he made important innovations in land survey technique through his later career, he chose employment in forestry for his initial career, doing extensive field surveys for the demarcation of forests. In 1882, he joined the forestry department, working under the direction of de Vasselot. Fourcade spent much of his time at Knysna in the Midlands Conservancy, and was appointed the Conservator in 1885. In 1889, Fourcade was seconded by the Cape government to Natal to survey the forests of the colony there and to offer recommendations on the establishment of

---

82   De Vasselot seemed to know little about the importance of climate when planting trees. For instance, he recommended the Mediterranean stone pine (*Pinus pinea*) be planted in the all-year rainfall region of Knysna. Cape of Good Hope, *Report of the Superintendent of Woods and Forests* (Cape Town, 1882), 15. For his general recommendations on plantations see p. 41.

83   See Cape of Good Hope, *Report of the Superintendent of Woods and Forests for the Year 1882*, 18.

84   Ibid., 18.

85   Cape of Good Hope, *Report of the Superintendent of Woods and Forests for the Year 1883*, 6, 22.

86   Cape of Good Hope, *Report of the Superintendent of Woods and Forests for the Year 1884*, 7.

87   C. D. Storrar, 'The Phenomenal Dr Fourcade', *South African Forestry Journal*, 146 (1988): 73–83. doi:10 .1080/00382167.1988.9630362.

a forest conservancy.[88] After surveying the forests of the colony, he compiled his findings into a detailed report. In addition to recommending measures for the protection of indigenous forests and the development of plantations, he offered a remarkably sophisticated bioclimatic matching. He noted:

> In introducing exotic, or even indigenous trees, in new regions, it is essential to choose species suited to the climate. The influence of climate, on trees, is much greater than that of soil, and many species which will thrive in almost any soil and with varying supplies of moisture, can only grow in a zone with certain definite conditions of temperature.[89]

By analysing the climate of Natal and matching this analysis to the climates of other countries, he identified equivalent regions elsewhere. Careful, detailed accounting of the ecology and utility of the many tree species in matched climates of the southern and northern hemispheres[90] yielded a list of candidate species for local trial in Natal. He mapped the expected annual temperatures across Natal's hilly and mountainous terrain from the estimated changes with altitude, using the technique of mapping of isolines originating with Alexander von Humboldt.[91] Fourcade argued: 'When the climatic conditions are favourable, exotic trees may become acclimatised; that is to say if the species is not yet fully adapted to the climate and the soil, it produces a better fitted variety after a number of generations'.[92] Fourcade's report was the clearest early expression of the views that Cape foresters would espouse more frequently in the 1890s and 1900s.

---

88   H. G. Fourcade, *Report on the Natal Forests* (Pietermaritzburg: W. Watson, 1889), 197.
89   Ibid., 79–83.
90   Ibid., Appendix VII.
91   G. T. Cushman, 'Humboldtian Science, Creole Meteorology, and the Discovery of Human-Caused Climate Change in South America', *Osiris*, 26 (2011): 19–44, 26.
92   Fourcade, *Report on the Natal Forests*, 79–83.

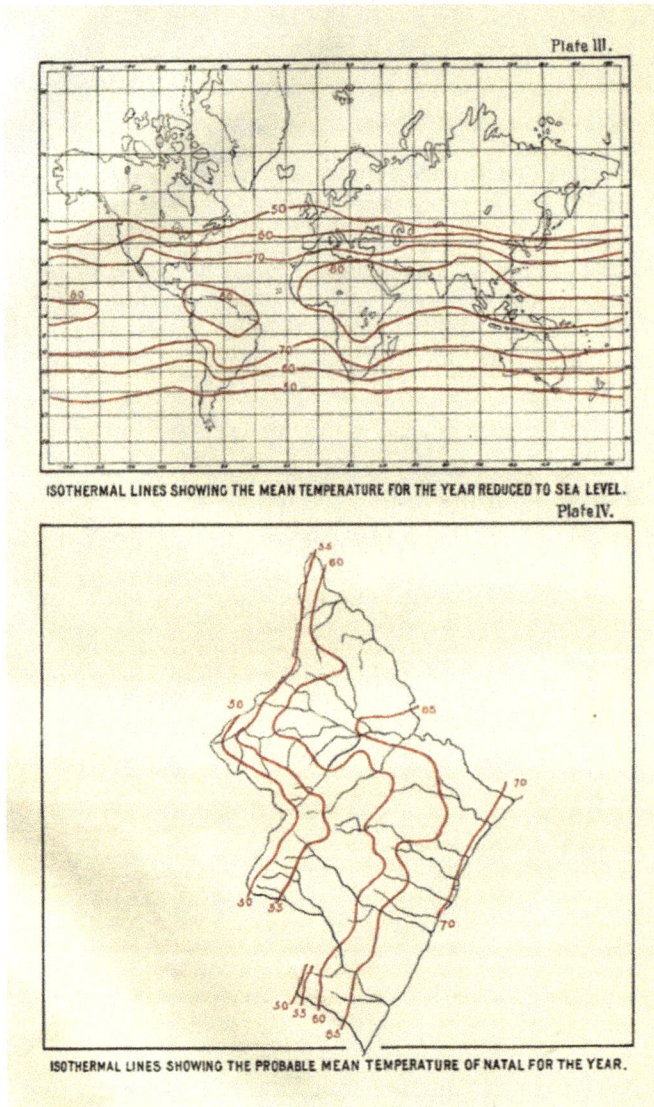

ISOTHERMAL LINES SHOWING THE MEAN TEMPERATURE FOR THE YEAR REDUCED TO SEA LEVEL.

ISOTHERMAL LINES SHOWING THE PROBABLE MEAN TEMPERATURE OF NATAL FOR THE YEAR.

Figure 4. H. G. Fourcade's climatic maps for Natal, with temperature isolines derived from topographic analysis.

Fourcade followed the climate mapping procedures developed by Humboldt, and had read William Ferrell's 1886 *Recent Advances in Meteorology* to inform his bioclimatic appraisal of source regions for trees in South Africa.

Source: H. G. Fourcade, *Report on the Natal Forests* (Pietermaritzburg: W. Watson, Printer to the Natal Government, 1889), Plates III and IV.

Fourcade's recommendations for tree planting in Natal came to naught, as he decided not to stay in Natal because of its climate and its small white population relative to indigenous African residents.[93] In 1891, the Natal Legislature brought out a German forester, Friedrich Schöpflin, who stayed for three years before leaving dissatisfied with his prospects in the colony.[94] Schöpflin demarcated a number of Natal's forests, which he entrusted to white district foresters and native guards. He did not focus on creating plantations of timber, although by that time the midlands of Natal already had thousands of hectares of wattle planted on private land.[95] At the end of his tenure, the Minister of Lands and Woods decided on the basis of an internal recommendation to transfer the district foresters to save costs, and to stop demarcating forests.[96] Natal's failed attempt to instil a conservancy indicates wider attitudes towards forestry in Southern Africa at the time.

Climate received greater attention with the arrival of Hutchins, an Englishman, in 1884. An individualistic forester, Hutchins blended European education with knowledge of Indian forestry while demonstrating considerable concern for the economic development of the Cape Colony during his tenure from 1884 to 1906. He had studied forestry at the L'Ecole Nationale des Eaux et Forêts, in Nancy, France, in the early 1870s before moving to India to work for the Indian Forest Service.[97] He worked with Australian trees in plantations in 1881 in the southern highland town of Ootacamund in the Madras Presidency.[98] He was mentored by Dietrich Brandis, the first Inspector General of Forests in India, and maintained a considerable interest in forest law, geography, climatology, and history. Hutchins took an active role in the Eastern Conservancy of the Cape when he arrived in King William's Town in 1883. He quickly took to reforming the forest management in the Conservancy, arresting trespassers and calling for the government to take over forests on lands owned and controlled by Africans.[99] In his second official report in 1886, Hutchins began recording meteorological observations, making observations on the climate of different regions of the Cape, and discussing a series of extensive trials of species.[100]

---

93   Hutchins to Fourcade, 10 July 1890, Fourcade Bequest BC 246, C5, UCT.
94   SGO, NASA-PMZ.
95   Hutchins, 'Forestry in South Africa', 14: Hutchins estimated 'not less than' 5,000 acres of Black Wattle.
96   Minute Paper 6002/1893, SGO, NASA-PMZ.
97   For a short biography see W. K. Darrow, *David Ernest Hutchins: A Pioneer in South African Forestry* (Pretoria: Department of Forestry, 1977).
98   See D. E. Hutchins, *Report on Measurements of the Growth of Australian Trees on the Nilgiris* (Madras: Government Press, 1883).
99   Brown, 'The Conservation and Utilisation of the Natural World', 428; these convictions skewed towards Africans. Cape of Good Hope, *Report of the Superintendent of Woods and Forests for the Year 1885*, 22–66. Hutchins was keen on forest law. He was instrumental in the drafting and lobbying the Cape Parliament to pass the Cape Colony's 1888 *Forest Act*, which was modelled on the 1882 *Madras Forest Act*, a piece of legislation that was passed when Hutchins worked in Madras.
100   Cape of Good Hope, *Report of the Superintendent of Woods and Forests for the Year 1886*, 35–40. Hutchins credited his relationship with MacOwan for much of his knowledge of botany.

Over time, Hutchins greatly strengthened the bioclimatic approach to afforestation, and adopted a straightforward mantra to direct plantation forestry: 'fit the tree to the climate'.[101] Hutchins read widely, drawing heavily from meteorologists and botanists (especially those in Australia).[102] He argued passionately that climatology and species' native climatic ranges were the most important subjects for foresters in the Cape to study, and ensured that the curriculum for higher education in forestry at Tokai included a course on climatology. Hutchins's ideas on climate appeared in a multi-part essay, 'Extra-Tropical Forestry', published from 1905 to 1906 in successive issues of the *AJCGH*. This is probably the clearest expression of the ideas embodying the Cape model. Here, he argued that the Cape Colony had an 'extra-tropical' climate, meaning a climate of regions near, but not in, the tropics, and characterised by seasonal dryness, abundant sunshine, and variable annual rainfall. Cape colonists, Hutchins wrote, should select exotic trees from other extra-tropical regions. He noted specifically that there was a southern hemispheric extra-tropical zone, 'the sea-level climate between about latitude 23° and latitude 43° which embraced southern Africa, Australia, Argentina, southern Brazil, Chile and northern New Zealand'.[103] To pinpoint extra-tropical regions directly comparable to the Cape, he analysed rainfall averages and patterns, altitude, average temperatures, light, and humidity.

Cape foresters tried to convince members of the public about the importance of climatic fitness and species selection to improve their chances of growing trees. Lister, Hutchins, and C. B. McNaughton wrote public pamphlets on tree planting.[104] Articles in the *AJCGH* by foresters and the government botanist, MacOwan, informed readers about how to select species, and what trials had succeeded and failed. Members of the public could write in to describe their own experiences in the *AJCGH*. Climatic considerations were paramount in these tracts. A 1904 booklet for farmers by McNaughton informed them: 'Forest species may be grown far from their natural habitat provided that the local climate is similar to that to which they are naturally accustomed'.[105] Hutchins also wrote a pamphlet and numerous articles in popular and official magazines to recommend what species should be planted in different regions throughout the Cape.[106]

---

101   Hutchins, 'Extra-Tropical Forestry', 521.
102   For the books Hutchins ordered which discussed Australian meteorology and botany, see documents in F791, AGR 723, NASA-CT.
103   Hutchins, 'Extra-Tropical Forestry', 19.
104   Lister, *Practical Hints*; D. E. Hutchins, *A Chat on Tree Planting with Farmers* (Cape Town: W. A. Richards and Sons, 1902); C. B. McNaughton, *Tree Planting for Timber and Fuel* (Cape Town: Townshend Taylor and Shashall, 1903).
105   McNaughton, *Tree Planting for Timber and Fuel*, 4.
106   Hutchins, *A Chat on Tree Planting with Farmers*.

Cape silviculture, contrary to being 'very much in its infancy at the turn of the century', began to emerge as a recognisable discipline in the 1900s.[107] By the early 1900s, foresters began to observe the results of experimental trials established in the 1880s and 1890s. The empirical information derived from successful species and sites provided foresters with a clear sense of how the climatic fitness and biology of specific species interacted with the climate and soil of particular sites in the Cape. By that time, Cape foresters had gathered detailed observations of rainfall patterns, temperature ranges, climatic cycles, and soil types of the colony. Hutchins was an active fellow of the Royal Meteorological Society. In 1888, he published a treatise that predicted climatic cycles based upon an analysis of sunspots and rainfall and temperature records.[108] He reported that the Cape experienced 12-year climatic cycles of relatively high and relatively low rainfall. His assessment is surprisingly similar to current meteorological models of the Cape's climatic cycle. Also useful was research by Frederic Juritz during the early to middle part of the 1890s that led to the publication of a complete soils analysis of the largest forest plantations in the Cape.[109]

Criticisms of forestry in the Cape Colony

From the early 1880s, foresters in the Cape suffered from public and professional criticism about the methods they used to create new plantations. Attempts to create plantations in arid regions—such as the failed Beaufort West plantation—did little to help bolster rural regard for the competence of foresters. MacOwan may well have had this example in mind when he told a parliamentary committee in 1882, in relation to the appointment of a Minister of Agriculture, that 'farmers are not very prone to take the advice of theoretical strangers. They ridicule the idea of experts from Europe being able to show them anything, and maintain that they [the experts] are practical failures'.[110]

Many members of the public questioned the value of forestry. Private nurseries complained about community and state nurseries that subsidised seeds.[111] People living near forests resented the imposition of the 1888 *Forest Act*, which limited access to forests. For the first 25 years of the existence of the colonial forest department, it faced questions about its contribution to the public good.

---

107   Brown, 'The Conservation and Utilisation of the Natural World', 430.

108   D. E. Hutchins, *Cycles of Drought and Good Seasons in South Africa* (Wynberg: Wynberg Times, 1888), 37–112.

109   C. F. Juritz, *A Study of the Agriculture Soils of Cape Colony* (Cape Town: T. Maskew Miller, 1910).

110   Cape of Good Hope, Report of the Select Committee Appointed by the Legislative Council and Report Upon the Appointment of a Minister of Agriculture, C.2-1882,6. As cited in Beinart, *The Rise of Conservation*, 126.

111   Botanic gardens competed with private nurseries, who protested that the state or city council gardens could charge below market rates. See Sim, *Tree Planting in Natal*, 18.

In 1906, Lister, then the Chief Conservator of the colony, noted in his report 'the department is more and more able to vindicate its existence as being for the general welfare. Unfortunately, self-seeking interests still continue to predominate in some quarters'.[112]

A penny-pinching parliament and Department of Agriculture made it difficult for foresters to complete their program of work. In 1906, the department commenced to raise public loans directly through parliament in order to afford plantation work because parliamentary appropriations varied so much from year to year that many plantations failed or suffered otherwise for lack of sufficient funds.[113] That decision came on the heels of a serious retrenchment in government in 1905. That year the Cape government decided to quit paying Joseph Maiden to send seeds from Australia in order to save money.[114] Seemingly trivial purchases, such as specialised books, received intense scrutiny from James Currie, the Undersecretary of Agriculture, who approved expenditure requests. Hutchins constantly demanded books and seeds, something that frustrated Currie. In late 1897, Currie denied Hutchins's request for three books on Australian meteorology. An impassioned Hutchins wrote back to Currie: 'In South Africa with its variety of trees and climates, meteorology and the climate requirements of each tree are the most important study for foresters'.[115] Currie begrudgingly ordered the books for Hutchins.

By the mid-1900s, foresters in Britain, Europe, and India became aware of the aggressive afforestation program then underway in the Cape. Many foresters knew about the Cape specifically because of the prolific writing of Hutchins. He famously wrote an article in *Nature* in 1902 arguing that *Eucalyptus* plantations could replace coal as a source for fuel.[116] The world's newspapers reported this work, no doubt reaching most professional foresters, who very likely scoffed at a plan that had no comparable example in the world at the time.

---

112   Cape of Good Hope, *Chief Conservator of Forests for the Year Ending 30th September, 1906, including Report on Railway Sleeper Plantations for the Calender Year* (G39/1907), 2.

113   Cape of Good Hope, *Report of the Superintendent of Woods and Forests for the Year 1884*, 7. This policy only changed in 1906. See Cape of Good Hope, *Chief Conservator of Forests for the Year Ending 30th September, 1906, including Report on Railway Sleeper Plantations for the Calender Year*, 1. As of 1906 the Forestry Department raised money by loans through government.

114   Joseph Maiden to Undersecretary of Agriculture, 3 November 1896, F719, AGR 722, NASA-CT; Undersecretary for Agriculture to Agent General for the Cape of Good Hope, 7 June 1905, F719, AGR 722, NASA-CT.

115   Hutchins to Undersecretary of Agriculture, 13 December 1897, B559/6, AGR 723, NASA-CT; Undersecretary of Agriculture to Hutchins, 8 January 1898, A27, AGR 723, NASA-CT; Hutchins to Undersecretary of Agriculture, 20 January 1898 (quotation in latter) B19/6/98, AGR 723, NASA-CT. He also promoted this view publicly; see Hutchins, 'Extra-Tropical Forestry', 521. Currie also quibbled about granting the leave that was outstanding to K. A. Carlson, who had requested leave to attend the forestry course at Cooper's Hill; Carlson, having persisted in the hope of official approval, could eventually proceed to Cooper's Hill when Schlich insisted on giving him a personal loan: K. A. Carlson, *Transplanted: Being the Adventures of a Pioneer Forester in South Africa* (Pretoria: Minerva Drukpers, 1947), 101–4.

116   D. E. Hutchins, 'Misuse of Coal', *Nature*, 66 (1902): 246–7.

Hutchins's eccentric personality and prolific writings (which were sometimes seen as the work of a dilettante) did little to help the Cape's wider reputation among empire foresters. Stories of Hutchins jumping into a stream in full clothes to cool off in the midday heat and of stopping passing trains to use their hot water to brew a pot of tea likely filtered their way around imperial forestry networks.

The idea of plantation forestry was often intensely contested, at least in respect of apparently unwarranted claims of its potential in the Southern African colonies. Certain foresters at Oxford mocked Hutchins. C. B. McNaughton, a former Cape forester who studied at Cooper's Hill, reported that foresters at Oxford 'ridiculed' Hutchins and the Cape forestry department. 'One man said the really only good thing we did that he knew was to free India of Hutchins'.[117] McNaughton was highly critical of Hutchins and Lister, two foresters he felt peddled lies to people of the Cape about the prospects for growing trees in the colony. He told Fourcade in a letter: 'I really regret I did not sever my continuation with the Forest Dept years ago when I saw that ... the public could be kept duped and blinded with ... impossible promises and statements'.[118] He said that Natal would 'regret her enormous speculation in wattle and the Cape her many ... plantations'.[119]

McNaughton's comments echoed similar concerns expressed by William Thiselton-Dyer, director of Kew Gardens, when he received a 10-tonne seed order from Hutchins. He wrote to the Secretary of Agriculture for the Cape that, 'I am obliged to remark that the instructions have been drawn up with want of practical knowledge ... some [species on the list] ... are actually unknown'.[120] This called into question, officially, Hutchins's competence. Hutchins was not put off by these allegations. Hutchins wrote back to the Secretary of Agriculture: 'the remarks of the Director need not cause surprise. The Kew establishment can have had but a limited experience of the supply of forest seeds. Of forestry proper they have no knowledge either theoretical or practical'.[121] Hutchins explained the distrust of Cape forestry methods in his essay in the *AJCGH* by noting that:

> Forest Meteorology in Northern Europe is without the practical importance that it possesses in the extra-tropical parts of the world, and its study has been neglected in Europe, with the result that after the failure of many unsuitable trees, all introduced trees have been decried.[122]

117 See C. B. McNaughton to Henry Fourcade, 9 December 1909, Fourcade Bequest BC 246, C7, UCT.
118 C. B. McNaughton to Fourcade, 9 December 1909, Fourcade Bequest BC 246, C7, UCT.
119 Ibid.
120 Agent General of the Cape of Good Hope to Secretary for Agriculture, 14 September 1896, enclosing W. C. Thiselton-Dyer to Sir David Tennant, 12 September 1896, No. 117, AGR 725, NASA-CT.
121 Hutchins to Undersecretary for Agriculture, 21 October 1896, B566, AGR 725, NASA-CT.
122 Hutchins, 'Extra-Tropical Forestry', 517.

Establishment and spread of the Cape model

By the 1900s, Cape foresters had developed a program of afforestation that reflected local environmental considerations and originated out of widespread failures to introduce and grow exotic trees in Southern Africa. Unmoved by foreign criticisms, and increasingly able to dictate the terms on which they created plantations, foresters in the Cape continued to consolidate their research and experiments. The Cape model further expanded in 1902 and after, when foresters from the Cape Colony took positions as the heads of state forestry programs in Natal, the Transvaal, and the Orange Free State. In 1909, Colin C. Robertson, a Yale-educated forester then working in the Orange Free State, wrote in an article, 'In the other branches of the science of Forestry, we [South Africans] can look to some other countries, and particularly to Germany … but the scientific naturalisation of exotic trees has so far received comparatively little attention in these countries'.[123] Robertson, an expert on species selection and climatic comparisons, noted that, 'probably more experimental planting of exotics has been carried out here than in any other part of the world'.[124] How that model travelled to the Transvaal, Orange Free State, and Natal is the subject of the next chapter.

123   C. C. Robertson, 'Some Suggestions as to the Principles of the Scientific Naturalisation of Exotic Forest Trees', *South African Journal of Science*, 6 (1910), 219.
124   Ibid.

Chapter 2

Forestry in Reconstruction South Africa: Imperial Schemes, Colonial Realities, c. 1901–1905

The British military and political annexation of the former South African Republic (ZAR) and Orange Free State at the end of the South African War (1899–1902) integrated the former Boer republics within British South Africa and the wider British Empire.[1] Among a broader suite of reforms, reconstruction officials established government forestry programs in the Transvaal and Orange Free State (renamed the Orange River Colony (ORC) from 1900 to 1910).[2] Though a few historians have briefly discussed forestry, studies of the reconstruction period have not yet documented and analysed the establishment of professional forestry in the Transvaal and ORC.[3] Forestry received considerable attention from reconstruction leaders, especially Alfred Milner, High Commissioner

1   Historiographical debates focus on the strategic and economic imperatives causing the war's origins, development and conclusion. For the strategic argument see, R. Robinson and J. Gallagher, *Africa and the Victorians: The Official Mind of Imperialism* (London: Macmillan, 1961); T. Pakenham, *The Boer War* (New York: Random House, 1979); A. Porter, *The Origins of the South African War: Joseph Chamberlain and the Diplomacy of Imperialism, 1895–99* (London: St Martins Press 1980); I.R. Smith, *The Origins of the South African War, 1899–1902* (London: Longman, 1996).
2   We use the name Orange River Colony in this chapter to describe official titles and the official government during reconstruction. The name Orange Free State and Free State are also used.
3   Forestry is usually mentioned in passing when discussing agricultural policy and reconstruction. See D. E. Torrance, *Strange Death of the Liberal Empire: Lord Selborne in South Africa* (Montréal: McGill-Queens University Press, 1996), 18; P. Cranefield, *Science and Empire: East Coast Fever in Rhodesia and the Transvaal* (Cambridge: Cambridge University Press, 1991), 55; M. Meredith, *Diamonds, Gold, and War: The British, the Boers, and the Making of South Africa* (New York: Public Affairs, 2007), 484; H. Giliomee, *The Afrikaners: Biography of a People* (Charlottesville: University of Virginia Press, 2003), 265.

of South Africa and Governor of the Transvaal (1901–1905). Milner and his 'Kindergarten' of elite British advisors saw forestry as a key pillar of rural reconstruction. [4]

Milner had a free hand from Colonial Secretary Joseph Chamberlain to construct a new imperial South Africa. Milner sought first and foremost to make a modern, capitalist, *British* South Africa where all residents—Briton, Afrikaner, or African—showed loyalty to the Crown and empire; where Briton worked with but dominated the Boer; and where whites dominated politically over indigenous Africans.[5] Milner sought to 'irrevocably transform' South Africa, modernising it according to liberal principles that would create the secure conditions within which the business class could make a profit.[6] The two former Boer republics he regarded as backward peasant states, best transformed by swamping the rural Afrikaner through an agrarian program based on mass immigration of a progressive British yeomanry. Cheap labour for mines and industry was to be secured through a common native policy and migrant labour system, designed to underpin racial capitalism for the following 100 years.[7]

The Milner administration faced the immediate task of repatriating 200,000 Boer farmers and their families, half of whom were still in the concentration camps, and the 60,000 Africans in camps.[8] On this basis, his administration set about restarting the mining industry, expanding agricultural productivity by facilitating resettlement of land, and creating modern scientific and technical departments to aid farmers. The restored mineral revolution drove renewed and accelerated demand for timber. Agricultural settlement laid the conditions for conflict about water resources. The sudden emergence of plantations made them an immediate scapegoat.

Many historians have argued that Milner's grand reconstruction plans—such as the attempt to populate the country with Britons, to make English the dominant language, and to break the power of Afrikaners—were political and social failures, what Darwin describes as an 'imperial fantasy' and Denoon a

---

4   S. Marks and S. Trapido, 'Lord Milner and the South African State Reconsidered', in M. Twaddle (ed.), *Imperialism, the State and the Third World* (London: British Academic Press, 1992): 80–94.

5   A. Thompson, 'The Language of Loyalism in Southern Africa, c. 1870–1939', *English Historical Review*, 118 (2003): 617–50, 635.

6   We acknowledge that wider strategic concerns drove British imperial policymaking but suggest that the economic interests played a driving role in reconstruction efforts, especially in areas relating to mining. See R. Hyam and P. Henshaw, *The Lion and the Springbok: Britain and South Africa Since the Boer War* (Cambridge: Cambridge University Press, 2003), 4–5.

7   P. E. Louw, *The Rise, Fall, and Legacy of Apartheid* (Westport, CT: Praeger Publishers, 2004), 1–26.

8   Ibid., 12.

'grand illusion'.[9] Yet other historians have challenged the idea that Milner's plans failed by pointing out that reconstruction, though politically damaging for Milner and unsuccessful in terms of the overall anglicisation of South Africa, did effectively lay the foundations for capitalist expansion, imperial integration, and white supremacy; all outcomes that Milner sought.[10] Scholars tend to agree that agricultural policy fulfilled many of the aims desired by Milnerites, namely increased production, land resettlement, and the expansion of white farmers' agricultural and pastoral reach.[11]

Efforts to create a modern forestry program during reconstruction also had its successes, if success is defined as Milner achieving some of his goals, namely encouraging afforestation in the Transvaal and ORC, reserving what remained of indigenous forests, and maintaining the supply of timber to mines. By 1904, professional foresters directed newly formed state forestry programs in the ORC and Transvaal. They immediately began reserving forests, establishing trial plantations, and working closely with farmers to encourage small-scale tree planting. Early limited efforts laid the foundation for a gradual shift of power from Cape foresters to those in the Transvaal. Milner predicted such developments when he noted in early 1902 that in many areas, 'the Cape will be overshadowed by the enormous development of the Transvaal'.[12] Knowledge gained in the Cape helped lay the foundation for the northern and eastern parts of the Transvaal to become the centre of the country's plantation forest and processing industry after 1910. At Union, the government established a single Forestry Department, headquartered in Pretoria, the new executive capital of the Union of South Africa. Transvaal concerns and officials played a disproportionate role in shaping national policies from 1910.

But forestry was also a great disappointment, at least to Milner, one of the great forestry enthusiasts. From his vantage point when he left the country, forestry seemed to be yet another failing reconstruction effort.[13] An early start at tree planting had been aborted in the ORC. His attempt to find an elite British

9   For the quote see J. Darwin, *The Empire Project: The Rise and Fall of the British World-System, 1830–1970* (Cambridge: Cambridge University Press, 2009), 249; D. Denoon, *A Grand Illusion: The Failure of Imperial Policy in the Transvaal Colony during the Period of Reconstruction, 1900–1905* (London: Longman, 1973). Milner himself recognised that the British could never be supreme throughout the entire countryside. See I. Waag, 'Rural Struggles and the Politics of a Colonial Command: The Southern Mounted Rifles of the Transvaal Volunteers, 1905–1912', in S. Miller (ed.), *Soldiers and Settlers in Africa: 1850–1918* (Leiden: Brill Press, 2009): 251–86, 253.

10   S. Marks and S. Trapido, 'Lord Milner and the South African State', *History Workshop Journal*, 8 (1979): 50–80.

11   J. L. Thompson, *A Wider Patriotism: Alfred Milner and the British Empire* (London: Pickering and Chatto, 2007), 94; P. Rich, '"Milnerism and a Ripping Yarn": Transvaal Land Settlement and John Buchan's Novel *Prester John*', in B. Bozzoli (ed.), *Town and Countryside in the Transvaal* (Johannesburg: Ravan Press, 1983).

12   Milner to Lewis Michell, 29 January 1902, in C. Headlam (ed.), *The Milner Papers (South Africa) 1899–1905* (London: Cassell Press, 1933), 403.

13   D. E. Hutchins, 'Forestry in South Africa', 1.

forester to direct the Transvaal's new forestry program had failed. Forestry was chronically underfunded by governments in both colonies. Milner left South Africa worried about the institutional reforms he had helped set in motion. In his farewell speech, in Johannesburg on 31 March 1905, Milner stated that he believed, 'Nature intended wide tracts of South Africa to be forest country', but he admitted that he worried that forestry funding might be cut after he left: 'unless people can be awakened to their vital permanent interests [i.e. the planting of trees], the first responsible Ministry which has a difficulty in squaring the Budget will starve the whole thing [state forestry] to death'.[14] That same year, Natal abolished the position of conservator, a vacancy that remained until the re-establishment of a conservancy after Union in 1910.

Putting aside questions about the 'balance sheet' of reconstruction for a moment, one notable aspect of forestry during the reconstruction period was that Cape foresters came to direct state forestry throughout South Africa. By 1904, foresters who had at one time worked in the Cape headed all four colonial forestry departments. The spread of Cape foresters across Southern Africa was by no means an inevitable development. The deployments occurred because of a series of contingent events that were not part of Milner's grand plans. Quite to the contrary, Milner had initially focused on bringing in experts from outside of South Africa. The ways by which Cape foresters came to direct all of South Africa's forestry departments illuminates the constraints Milner faced in his reconstruction efforts, and provides the groundwork to understand post-1910 developments in the Transvaal and the rest of the country centred on afforestation.

What is clear from actions of imperial officials in the Transvaal and ORC is that colonial progressives from the Cape and Natal figured little in early discussions about determining the formation of major scientific institutions and the hiring of professionals to run them. Reconstruction officials did not set out to purposefully mirror 'the scientific and administrative approach to rural development which had already started in the Cape' because they did not look to the Cape as a unique model.[15] The approach to rural development in the Cape—including the creation of agriculture, forestry, and veterinary departments, and the building of railways—was based on earlier precedents in India, Egypt, Australia, New Zealand, and Canada. This expression of state science in South Africa, comparatively, was not novel in the empire at the time. For instance, forest laws and institutional structures drew heavily on British–Indian precedents. But many intellectual movements (such as the strong

---

14   Lord Milner, *The Nation and the Empire: Being a Collection of Speeches and Addresses* (London: Constable, 1913), 87.
15   Brown, 'Tropical Medicine and Animal Diseases: Onderstepoort and the Development of Veterinary Science in South Africa 1908–1950', 516.

emphasis on bioclimatic modelling) *were* unique, and in the actual practice of forestry, Cape foresters departed greatly from their professional counterparts in Europe or India. Scholars seeking to find the 'origins' of forestry in South Africa, and elsewhere in the empire, should be careful to discern between institutions and ideas, theory and practice.

Reconstruction officials sought to establish a modern government forestry department in the Transvaal and ORC as part of the wider tool kit of colonial state-making. Milner and his acolytes did not look to the Cape for a model forestry department—they looked first and foremost to India. As Barton demonstrates, the prestige of forestry in India allied with a general shift towards bureaucratic, utilitarian naturalist expertise led colonial elites throughout the British Empire to create forestry departments and demarcate forest reserves.[16] India had the largest forestry department in the empire and the most prestigious forestry educational program in Britain, at the Royal Indian Engineering College at Cooper's Hill (succeeded by the forestry school at Oxford after 1905). Though the influence of Indian foresters varied, the reconstruction government in the Transvaal and ORC clearly looked first and foremost to India.

Milner was not interested in South Africa as such, but rather in empire. He and his Kindergarten did not view Southern Africa's population or its intellectual elite to be particularly innovative or inspiring.[17] To a large extent, these attitudes reflected Milner's views of Cape society. Not long after his arrival in 1897, Milner's 'distaste for Cape Town soon became well known' among the city's British and Afrikaner elites.[18] He worked with Cape progressives, but as much as a way to oppose the Afrikaner Bond (the anti-imperialist political party) as to champion local progressive causes.[19] Milner showed little care for the Cape's political elite after the war when he proposed to suspend the Cape's constitution in order to harmonise constitutions among Britain's South African colonies and to keep control of the Bond.[20] Even with close friends, Milner always focused on the empire. Torrance points out that '[Milner] did not subordinate his interest to those of Johannesburg', even though he was personally friends with the mining leaders of the Transvaal Progressive Party.[21]

---

16   Barton, *Empire Forestry and the Origins of Environmentalism*.
17   Many Cape leaders were less than enthusiastic about Milner's imperial pretensions. See S. Dubow, 'Colonial Nationalism, The Milner Kindergarten and the Rise of "South Africanism" 1902–10', *History Workshop Journal*, 43 (1997): 53–86, 56–7.
18   J. L. Thompson, *Forgotten Patriot: A Life of Alfred, Viscount Milner of St. James's and Cape Town, 1854–1925* (Madison, NJ: Fairleigh Dickinson Press, 2007), 113.
19   Thompson, *A Wider Patriotism: Alfred Milner and the British Empire*.
20   Giliomee, *The Afrikaners*, 267.
21   Torrance, *Strange Death of the Liberal Empire*, 19.

Throughout reconstruction, Milner and his Kindergarten looked consistently *outside* of southern Africa for expertise and advice; they only hired people from the Cape for higher-level appointments when they failed to get a 'first-rate' man from abroad. As an informal rule, British officials tried to bring in high-level administrators and judges from Britain and technical staff from India and Egypt. This is because appointments in those colonies were seen as being more prestigious than in the settler colonies; thus the men were considered to be of a higher calibre and better social standing. From the perspective of London or a would-be imperial scientist, the Cape Colony was a far less prestigious appointment than India, governed by the empire-like Indian Office, or Egypt, governed by the elite Foreign Office. Milner brought in a colleague from Egypt, William Wilcocks, to survey the country's irrigation needs with a view to creating a productive agricultural economy. The head of the new Transvaal Department of Agriculture was a Colonial Office appointee from Britain, Frank S. Smith.[22] And, unlike judges previously appointed in the Cape Colony, the judges for the Transvaal and ORC were 'distinguishable from their colleagues because they were born in England'.[23] Milner specifically asked Chamberlain for 'a good man from Home' to fill these judicial roles.[24] 'Home' was frequently on the minds of Milner and his inner circle.

Forestry in the Transvaal and ORC was seen as no different than agriculture, irrigation, or law: attempts to establish new forestry programs focused on recruiting top-rate men from outside of Southern Africa. But, as it turned out, Cape foresters were eventually selected to direct the forestry department in the Transvaal, ORC, and Natal. From 1902 to 1905, Cape-based foresters founded new forestry departments in Natal, Transvaal, and the ORC, and drafted reports for Southern Rhodesia, Basutoland, Kenya, and Sierra Leone. The botanist and forester Thomas Sim went to Natal in 1902, K. A. Carlson moved to the ORC in 1903, and Charles Legat moved to the Transvaal in 1904. That Cape foresters came to direct the whole of forestry in Southern Africa was not in the Milnerite plan. Contingency and competence, rather than imperial design, explains how Cape foresters came to dominate tree-planting efforts throughout the entire subcontinent by 1904.

---

22   For Smith's biography and background see, Cranefield, *Science and Empire*, 54–5.
23   S. D. Girvin, 'The Influence of an English Background on Four Judges Appointed to the Supreme Courts of the Transvaal and Orange River Colony, 1902–10', *Tijdschrift voor Rechtsgeschiedenis/Legal History Review*, 62 (1994): 145–63, 146.
24   Ibid., 153.

Forestry in reconstruction Southern Africa

The South African War acted as a great shock to the entire Southern African region. War mobilised over 100,000 foreign troops, caused the destruction of the agricultural backbone of the Transvaal and ORC, and in its aftermath integrated the mining wealth of the Rand with the pastoral economies in the Cape and Natal. The war took a high toll on the country's environment, as troops pillaged and burned farms, cut down trees, and depopulated the countryside by putting people into concentration camps. A series of pre-war events exacerbated the economic and environmental change. Successive waves of drought, coupled with the rapid spread of the epizootic disease, the Rinderpest (1896–1899), led to a collapse in the subcontinent's wider cattle economy, killing up to 95 per cent of all African cattle in parts of what is now South Africa.[25] The entire rural economy of the Transvaal and Free State required rebuilding.

Forestry was one of the key pillars of Milner's reconstruction plans.[26] Discussions of forestry focused not only on establishing a proper system of forest conservation, but it also included proposals for the settlement of British farmers, who would themselves help to embower with trees the countryside of both former Boer republics.[27] A recurring question was how to get the 'right sort' of person to run state forestry programs, and how to find 'good' British stock farmers to help develop the countryside. These farmers would help provide a political and economic counterweight to disgruntled rural Afrikaners. Anglicisation drove imperial politics during the reconstruction period, and beyond.[28]

The development of mining, and the subsequent building of railways, caused a precipitate change in tree planting. The Cape administration established railway plantations of eucalypts at Epping, Beaufort West, and Worcester for sleepers and fuelwood. Displacement of wagon transport by the railways brought down the price of imported timber, rendering extraction from indigenous forests unprofitable. Farmers and mining houses began with speculative plantations for mine-support timbers. An example of the latter is the Maccauvlei plantation, established on land where major coal deposits were found in the Orange Free State, south of the Vaal River, in 1878. A German horticulturalist appointed

---

25   P. Phoofolo, 'Face to Face with Famine: The BaSotho and the Rinderpest, 1897–1899', *Journal of Southern African Studies*, 29 (2003): 503–27, 503–8. Also see E. Kreike, 'De-Globalisation and Deforestation in Colonial Africa: Closed Markets, the Cattle Complex, and Environmental Change in North-Central Namibia, 1890–1990', *Journal of Southern African Studies*, 35 (2009): 81–98.

26   Milner to Selborne, 30 November 1899, in Boyce (ed.), *Crisis of British Power*, 97–8. Cited in Thompson, *A Wider Patriotism*, 88.

27   See Chapter 8, on the Tzaneen Government Estate.

28   See K. Fedorowich, 'Anglicisation and the Politicisation of British Immigration to South Africa, 1899–1929', *Journal of Imperial and Commonwealth History*, 19 (1991): 22–46.

in 1893 started by planting 100,000 common oak trees, but the plantation eventually grew to nearly 2,000 hectares, with *Pinus radiata* and *P. pinaster* the main species; the estate later had its own forest railway and electrically powered sawmill.[29] More important was the early 1902 investment in afforestation by the then Transvaal Gold Mining Estate, which marked the beginning of serious corporate investment in plantation forestry. In Natal and the Transvaal, Black Wattle plantations began to expand, and many eucalypt plantations grew on the interior plateau, near the mines.[30]

Official policy discussions regarding the establishment of a program for afforestation began in Johannesburg just before reconstruction began. Richard Adlam, former curator of Johannesburg's Joubert Park, got in touch with Milner through Herbert Maxwell, a British MP with a strong interest in forestry.[31] Adlam proposed that 'in my judgment no tract of country is so well suited for Forestry on a large scale as the High Veld', an area he said encompassed all around Johannesburg.[32] Adlam's letter reached the reconstruction government officials in Johannesburg in early 1902, months before peace was concluded. Milner discussed the letter with his close officials, telling them that, 'I have got this matter very much at heart, and am distressed that I have not been able to do anything yet'.[33] He recommended to Lionel Curtis, Town Clerk, that Curtis consider increasing funding to the gardens of Joubert Park. Milner also agreed to meet with Adlam to discuss the issue personally.[34]

Milner's interest in forestry reflected wider concerns of colonial Southern Africans who wanted to put each colony on a firm economic footing after the war ended. Forestry had received a good deal of attention in Natal during the war. In 1900, Maurice Evans, MP for Durban, called in parliament for the creation of a forestry department for the colony.[35] As the war was winding down in early 1902, the Natal government sent its entomologist, Claude Fuller, to inspect the Cape Colony's forestry program.[36] Fuller recommended that Natal bring Joseph

---

29   R. L. Leigh, *Vaal Triangle History*, 1968, www.vaaltriangleinfo.co.za/history/maccauvlei/ (accessed 2 June 2013).

30   Appearance of the eucalyptus snout beetle on the Highveld in the early 1920s rendered these plantations marginal, so that mine-timber plantations subsequently concentrated in more favourable climates, remote from the mines. See DeBach and Rosen, *Biological Control by Natural Enemies*, 178–81.

31   For a brief biography of Adlam, see M. Fraser (ed.), *Johannesburg Pioneer Journals, 1888–1909* (Cape Town: Van Rieebek Society, 1986), 33.

32   R. Adlam to H. Maxwell, 15 November 1901, CO 201, National Archives of South Africa Pretoria [NASA-P].

33   Milner, 'Forestry on the High Veld', 10 February 1902, Governor's Office [GOV] 613, PS 360/01, NASA-P.

34   Ibid.

35   T. Sim, *Report of the Conservator of Forests* (Pietermaritzburg, 1903), 3–4. Evans was born in Manchester, UK, before emigrating to Natal in 1875 to establish a farm. For a biography of Maurice Evans see G. M. Frederickson's introduction in, M. S. Evans, *Black and White in the Southern United States: A Study of the Race Problem in the United States from a South African Point of View* (New York: Longmans, 1915), x–xi.

36   Sim, *Report of the Conservator of Forests*.

Storr Lister, then Conservator of the Eastern Conservancy, to report on how to best establish a conservancy. Lister visited Natal from May to July 1902 before writing an influential report that laid out the outlines for the colony's future department. He advised the government in Natal to hire Thomas Sim, a Scottish immigrant who had trained as a botanist in Scotland, England, and America before emigrating to the Cape in 1889 to take up a role as the botanist at the King Williams' Town botanic garden. He joined the Forestry Department in 1894, and by 1898 had worked his way up to the position of District Forest Officer for King William's Town.[37]

Once appointed as Conservator in Natal, Sim sought to reserve indigenous forests and help establish government and private timber plantations. Drawing on his experience working with foresters such as Hutchins and Fourcade, Sim advocated a bioclimatic comparative method for selecting possible exotic trees to plant. He noted the general 'un-reliability of European text-books under South African conditions'.[38] He warned that 'even the behaviour of the numerous species can hardly be predicted from their well-known behaviour elsewhere. Success under similar climatic and thermal conditions is the safest guide when dealing with unknown exotics'.[39] But the Natal government abolished the Forestry Department and his position as Conservator three years later, in 1905; this was the second time in Natal's history when the attempt to start a forest conservancy was abandoned. Sim stayed in Natal, establishing a nursery in Pietermaritzburg, although he later travelled to England to represent South African timber growers and went to Portuguese East Africa (Mozambique) to write a report for the government on the forests there. Like Hutchins, Sim was a prolific author who worked hard to popularise ideas of forest conservation amongst educated English-reading audiences.[40] The closure of Sim's conservator post in Natal was a clear indicator of the fragile security of conservation reforms in Southern Africa at the time.

Around the same time Sim came to Natal, Hamilton Gould Adams, Lieutenant Governor for the ORC, brought over a 'young man'[41] from the Cape Forestry Department to plant trees, which were seen as needed because of the devastating toll of the war on the countryside of the ORC. But Adams grew dissatisfied with the person, who lacked sufficient 'energy', and he was summarily dismissed

---

37   See T. R. Sim, 'Memoir Vol. 1', Stellenbosch University Engineering and Forestry Library.

38   Sim, *Tree Planting in Natal*, 8.

39   Ibid.

40   After leaving his post he authored numerous articles and books. See T. R. Sim, 'Recent Information Concerning South African Ferns and their Distribution', *South African Philosophical Society*, 16, 1 (1905): 267–300; *The Forests and Forest Flora of the Cape of Good Hope* (Aberdeen: Taylor and Henderson, 1907); *Forest Flora and Forest Resources of Portuguese East Africa* (Aberdeen: Taylor and Henderson, 1909); and *Tree-planting in South Africa* (Pietermaritzburg: Natal Witness, 1927).

41   We have been unable to identify this subordinate official in the existing historical record.

after planting approximately 100,000 trees.[42] Milner derided the experiment and the person.[43] British officials in the Transvaal and ORC took a lesson from this initial mistake: they tried later to only hire experienced professional foresters, preferably those from outside of the Cape.

British administrators in Johannesburg and Bloemfontein worked closely together to design a forestry program focusing on afforestation and the protection of remaining indigenous forests.[44] Milner and his Kindergarten all agreed that the Transvaal, a region more ecologically and climatically diverse than the ORC, required its own conservator and forestry department. But they remained undecided about the future of the ORC, situated on a high plateau with extreme temperature fluctuations and dominated by windswept, treeless grasslands. They went back and forth on whether the seemingly inhospitable ORC should have its own forestry department and conservator or whether the reconstruction governments for the two former Boer republics should pool their resources.

Sir Harry F. Wilson, Colonial Secretary for the ORC, initially discussed the issue with the Director of Kew Gardens, William Thiselton-Dyer, and communicated his discussion to Adams. Thiselton-Dyer encouraged Wilson to pursue a larger survey, similar to Willcock's 1901 *Report on Irrigation in South Africa*.[45] Thiselton-Dyer told Wilson that he knew a forester in India who would be suitable for such a role.[46] Frank Smith, Director of Agriculture in the Transvaal, cast doubt on Thiselton-Dyer's idea of undertaking an extensive survey and questioned the quality of his proposed candidate. Smith noted that Thiselton-Dyer 'would like to rule the British Empire' by offering advice and promoting his candidate. Smith commented wryly that 'his selections have many of them been anything but happy'.[47] Instead, Smith contacted Dr William Somerville, a former lecturer in forestry at Edinburgh University, to inquire about the right person.[48] In his views, Smith concurred with other British officials about the need to import a first-rate man: 'I should like a man versed in Continental and Indian methods'.[49] But failing this, he was 'inclined to obtain from the Cape

---

42   Curzon to Hely-Hutchinson, 22 June 1903, No. 711-F, CO 201, NASA-P.

43   Milner, 'Forestry on the High Veld', 10 February 1902, GOV 613, PS 360/01, NASA-P.

44   The High Commissioner for the Transvaal and Orange River Colony and the Governor of the Orange River Colony sat in the Transvaal until 1907. Milner presided over all three roles until 1905.

45   Adams to Milner, 29 October 1902, CO 201, NASA-P; Smith to Adam, 7 October 1902, CO 201, NASA-P. See W. Willcocks, *Report on Irrigation in South Africa* (Johannesburg, 1901). Also see Beinart, *The Rise of Conservation*, 176–82.

46   Adams to Milner, 29 October 1902, CO 201, NASA-P.

47   Smith to Adams, 7 October 1902, CO 201, NASA-P.

48   Somerville was part of a broader Scottish 'lobby' that shaped forestry in India, the British colonies, and back in Britain. See J. Oosthoek, *Conquering the Highlands: A History of the Afforestation of the Scottish Uplands* (Canberra: ANU E Press, 2012), 38–9.

49   Smith to Adams, 7 October 1902, CO 201, NASA-P.

because they have already had considerable experience there of forestry upon a large scale, and under conditions somewhat similar to those obtaining in this country'.[50]

Milner himself was undecided on whether 'the afforestation of the Orange River Colony is a question of sufficient importance to justify the engagement of a good man independently of the Transvaal'.[51] Yet he told Adams that he 'should certainly not raise any objection' if Adams saw it as a sufficiently important issue to appoint a conservator specifically for the colony.[52] Milner recommended that Adams invite Hutchins to write an interim report on the colony.[53] Milner's recommendation was based on his own decision to invite Hutchins to the Transvaal. This was part of a plan he had devised earlier that year: if he could not hire a forester straight away, he noted, 'I should engage a man at once—provisionally—if I could find him'.[54] Hutchins was seen as the 'right sort' of forester to tour the ORC; Milner told Adams that '[Hutchins] has European training, Indian experience and recently great experience and success in South Africa'.[55] Turning down the offer of Hutchins, Adams informed Milner that he believed the best plan would be to get a 'thoroughly good man for the Transvaal', who could be loaned to the ORC.[56]

As a result of this decision, the government sought to engage a leading Indian forester for the Transvaal. Milner sent a formal request to Lord Curzon, Viceroy of India, to loan the Transvaal one of India's leading foresters, Robert S. Troup, Deputy Conservator of Forests in Burma.[57] Curzon declined, noting that his own staffing was 'insufficient' at the time.[58] At the same time, Hutchins had approached Troup privately. He laid it on thick, describing the job as 'one of the best, in my opinion, since modern Forestry has been known amongst Englishmen'.[59] Hutchins's letter caused a stir at the highest level of the Indian government. Curzon discovered the exchange and wrote an inflamed letter to Walter Hely-Hutchinson, Governor of the Cape Colony, complaining about Hutchins's 'very serious breach of official etiquette'.[60] Curzon protested that Hutchins's letter would lower Troup's morale because he could not accept the Transvaal's unofficial offer. Hutchins possibly did act officially (though not through the 'official' channel). The rule-bound and haughty Curzon may have

---

50   Smith to Adams, 7 October 1902, CO 201, NASA-P.
51   Milner to Adams, 20 October 1902, CO 201, NASA-P.
52   Milner to Adams, 20 October 1902, CO 201, NASA-P.
53   Milner to Adams, 20 October 1902, CO 201, NASA-P.
54   Milner, 'Forestry on the High Veld', 10 February 1902.
55   Milner to Adam, 20 October 1902, CO 201, NASA-P.
56   Adam to Milner, 23 October 1902, CO 201, NASA-P.
57   Governor of Transvaal to Governor of India, 2 June 1903, CO 201, NASA-P.
58   Governor of India to Governor of Transvaal, 11 June 1903, CO 201, NASA-P.
59   Hutchins to R.S. Troup, 25 February 1903, CO 201, NASA-P.
60   Curzon to Hely-Hutchinson, 22 June 1903, No. 711-F, CO 201, NASA-P.

used the pretence to keep one of India's most respected foresters. Indian officials had become frustrated at losing their best officers as a result of other colonies constantly poaching them.

In the wake of the Troup fiasco, Milner and his entourage had to go back to the drawing board. Meanwhile, Hutchins had toured the Transvaal on instruction of the Milner administration, which he detailed in his 1903 *Transvaal Forest Report*. His recommendations effectively laid down the scientific, legal, and administrative blueprint for the future Transvaal forestry department. He recommended copying the *Cape Forest Act* of 1888 and establishing an extensive network of plantations and forest reserves. His *Transvaal Forest Report* strengthened Hutchins's reputation among Colonial Office administrators, who later drew on his services in Kenya, Australia, and New Zealand.

On his Transvaal tour, Hutchins was unimpressed with what he saw. He criticised the overcutting of indigenous forests during the 'Kruger régime',[61] by both indigenous Africans and Boers. Africans came under scrutiny for practising slash-and-burn agriculture and destroying forests.[62] Hutchins was prejudiced against the Boers. He lamented the destruction of indigenous forests at Woodbush by the 'old Boers'.[63] To control both groups, Hutchins recommended demarcating forest reserves, policing them strictly, and passing forest legislation. Hutchins, who lived in India during the debates leading up to the passing of the 1882 *Madras Forest Act*, believed that over time—and with judicious policing—both the Afrikaner and African would adjust to state forest conservancy.

Hutchins noted that most attempts to plant trees as commercial ventures had failed. However, the promise of plantations that would pay their way was evident. A number had been created around Johannesburg after the discovery of gold. Miners recognised that the high cost of imported mining-timber could be overcome by production from plantations. Buoyed by the fast growth of some species, they raised money and invested in plantations. In his report of 1905 Hutchins noted:

> Unfortunately, altogether fallacious estimates were based on the profits to be realised from these plantations. The rapid growth of isolated and avenue trees was taken as a basis for the growth of trees in masses. Sufficient allowance was not made for the reduced growth consequent on the increased drain on subsoil moisture when trees were planted in dense forest. It was often assumed that so many trees planted per acre would leave a nearly equal number of trees to fell at the final cutting; and, worst of all, there was little climatic selection.[64]

---

61   D. E. Hutchins to R. S. Troup, 25 February 1903, CO 201, NASA-P.
62   Hutchins, Transvaal Forest Report, 21.
63   Ibid., 17.
64   Hutchins, 'Forestry in South Africa', 18.

He found that early attempts at plantations had proved to be disappointments.

Figure 5. The landscape in the vicinity of Woodbush, c. 1910.
Isolated patches of indigenous forest are seen growing scattered amongst grassland. The site was later converted to a pine plantation.
Source: Department of Agriculture, Fisheries and Forestry, Pretoria; photographer unknown.

Hutchins argued that planters in the Transvaal needed to select the right trees for each region. Prior to his arrival, there had been no systematic program of tree planting. He was familiar with the *ad hoc* nature of tree planting having witnessed a similar situation in the Cape: 'the Transvaal, like others of the South African Colonies, has planted its trees entirely neglecting this most important consideration of climatic fitness'.[65]

His report of 136 pages included 83 pages devoted to species recommended as suited to the different parts of the colony. While most of these were species intended for other purposes than timber, he advocated planting eucalypts and Mexican pines in the Woodbush Range east of Pietersburg (Polokwane). Hutchins wrote that 'the countries to which one would naturally look to furnish trees for the Transvaal are not winter rainfall areas such as the Mediterranean and California, but summer rainfall areas such as Mexico'.[66] Not only did Hutchins want timber plantations, he sought to sow self-propagating trees throughout the

65   Hutchins, *Transvaal Forest Report*, 124.
66   Ibid., 18.

colony, where possible. With this in mind, Hutchins recommended planting self-propagating pines and wattles, along the south-eastern border of Swaziland.[67] He pointed to expanding populations of cluster pine (*Pinus pinaster*) at the Cape as an example of how this could proceed.[68]

The Transvaal government appointed the Cape forester Charles Legat in 1904 to work as the first Assistant Chief Forester for the Transvaal, then under the Department of Agriculture.[69] Legat specifically followed Hutchins's advice to establish government nurseries to supply seeds to government plantations and private individuals. Legat emphasised the uniqueness of the Transvaal, contrasting its environment to that of the Cape, where he first worked. Echoing Hutchins's ideas, he distinguished South African conditions from those in Europe:

> In Great Britain and other European countries there is no necessity for the Government to undertake work of this sort, because transplants of forest trees can be bought from private firms as cheaply as they could be produced by the Government, and because afforestation there, from a climatic point of view, is not such a pressing need as here ...[70]

By 1905, the Department of Agriculture had received a land grant from the city of Pretoria to found a nursery at Irene and had purchased land to establish nurseries and small timber plantations at Ermelo, Lichtenburg, Belfast, Potchefstroom, and Gemsbokfontein.[71]

The timber found by trial and experience to suit Reef mining conditions best had small dimensions, six inches being ideal. Timber of this size could be grown quickly and profitably in plantations of Saligna Gum (*Eucalyptus grandis*) and Black Wattle (*Acacia mearnsii*) with little silvicultural treatment or cost; private growers in Natal and especially mining corporates in the Transvaal quickly produced enough to supplant imported timber for the mines.

While early work on establishing plantations proceeded, protecting the indigenous forests presented a different set of challenges. Work began on demarcating forests and regulating forest access. Protective measures caused friction between foresters and a variety of groups who utilised the indigenous forests. Race and class exacerbated, but did not cause, tensions relating to the onset of state forestry in South Africa. Historical interpretations of forestry in South Africa have tended to emphasise how *white* foresters 'technically colonised' forests traditionally used

---

67   Hutchins, *Transvaal Forest Report*, 6.
68   Ibid.
69   Hutchins, 'Forestry in South Africa', 6.
70   Conservator of Forestry, *Transvaal Department Agricultural Report 1903–4* (Pretoria: Government Stationary Office, 1905), 322.
71   Transvaal Department of Agriculture, *Annual Report of the Director of Agriculture 1st July 1905 to 30th June 1906* (Pretoria: Government Stationary Office, 1907), 14–5.

by *black* Africans without recognising that state forestry also faced considerable resistance from rich and poor whites.[72] It must be remembered that the expansion of state forestry was controversial throughout the world, including in Europe, the Americas, and Australasia. While it is undoubtedly true that the colonial state fundamentally transformed indigenous forest usage and rights, the principles of sustainable forest usage, which guided forestry, reflected the belief that Southern Africa lacked enough indigenous timber to sustain the population and economy as a result of settlement and the minerals revolution. Foresters often supported imperialism and settlement, but they saw themselves foremost as conservators who protected forests from the rich and poor.

From the perspective of foresters, the actions of extractive capitalists and poor Africans alike had to be regulated. This put them at odds with the mining lobby, land commissions, and agents employed to look after the interests of Africans. In the aftermath of the war, the Lands Commission and the Agricultural Department wanted to maintain a free flow of timber to the gold mines because they were 'anxious not to cripple the mining industry', then South Africa's biggest export and source of taxes.[73] Adam Jamison, Commissioner for Lands, allowed the Transvaal Gold Mining Company Ltd to cut timber near Pilgrims Rest, Lydenburg, freely from July 1903 until such time as a legal licensing regime was in place.[74] Yet foresters sought to curtail this free-for-all, which they believed had led to the loss of most indigenous forest cover. A letter from the District Forester to the company in September 1903 told them, 'in future no person will be allowed to cut timber on Crown Lands without being duly provided with the necessary authority'.[75] The company, in response, complained to the Land Commissioner about its high taxes and called on foresters to focus instead on stopping Africans from 'burning down the bush between Pilgrims Rest and Sabie'.[76] In response, Legat pointed out that the company had miscalculated its taxes. He lamented that at the time foresters had 'no power' to legally regulate whether Africans burnt land on their own property, though he deplored it as an 'evil'.[77] Despite appeals by business owners for foresters to regulate Africans but not white businesses, Legat worked to bring all forest users, black and white, under state control.[78]

---

72   See Tropp, *Natures of Colonial Change*, 1–7; Showers, *Imperial Gullies*. For East Africa see T. Sunseri, *Wielding the Axe: State Forestry and Social Conflict in Tanzania, 1820–2000* (Athens, OH: Ohio University Press, 2009).

73   A. Jamison to E. F. Bourke, 3 July 1903, G1938/4, Transvaal Agriculture Department [TAD] 435, SANA-P.

74   A. Jamison to E. F. Bourke, 3 July 1903, G1938/4, TAD 435, SANA-P.

75   Acting Forestry Assistant to E. F. Bourke, 22 September 1903, G1938/4, TAD 435, SANA-P.

76   E. F. Bourke to Commissioner of Crown Lands, 30 November 1903, TAD 435, SANA-P.

77   C. Legat to F. Smith, 20 January 1904, TAD 435, SANA-P.

78   See the analysis of Charles Edward Lane Poole's time at Woodbush during the reconstruction period in J. Dargavel, *The Zealous Conservator: A Life of Charles Lane Poole* (Perth: University of Western Australia Press, 2008), 16–8. Lane Poole was a British forester who disliked his time in Southern Africa.

The ORC presented a different set of challenges, largely environmental.[79] The colony had little tree cover, neither indigenous nor exotic, a reflection of the extreme fluctuations in rainfall and temperature and the evolutionary history of the fire-dominated grassland biome. Many of its rural inhabitants did not see tree planting as a profitable enterprise, although many farmers had planted trees for local wood, shade, and shelter. According to the first conservator, K. A. Carlson, many farmers living in rural areas objected to tree planting because they believed that forests encouraged ticks or they argued that trees could not grow in many areas without irrigation.[80] There had been some larger-scale attempts by wealthier farmers to plant trees in well-watered sites. The largest of them was located on Fichardt's farm outside Bloemfontein, established around 1865; on Charles Newberry's Prynnsberg estate near the then Basutoland, which he established from 1881 onward;[81] and on the Vereeniging Estates near the Vaal River (see earlier).[82] These were isolated initiatives, and if rural residents had overly negative attitudes, it was equally the case that Carlson overestimated the region's ability to produce commercially valuable timber, that is, beyond timber necessary for local usage.

Figure 6. A farmstead in the southern Orange River Colony, c. 1900, after destruction by the British military forces.

Settlement in such a treeless landscape motivated foresters to promote tree planting as a means to mitigate the harsh climate.

Source: Boer War Museum, Bloemfontein; photographer unknown.

---

79   T. Smith to 25 July 1902.

80   K. A. Carlson, *Forestry in the Free State, Bulletin 6, Union of South Africa* (Pretoria: Government Printer, 1912), 1.

81   *Dominions Royal Commission*, 290; Adam to Milner, 23 August 1902, CO 201, NASA-P.

82   Carlson, *Forestry in the Free State*, 4.

Figure 7. An example of an eroded gully at Ladybrand in the eastern Orange River Colony, c. 1902.

It was the general belief at the time among officials that erosion such as this was attributable to European land use (see, for example, Fourcade, *Report on the Natal Forests*, 16, and the 1923 Report of the Drought Commission). Scenes such as this supported the 'forestry enthusiasts' in their campaign to employ afforestation for environmental improvement.

Source: Department of Agriculture, Forestry and Fisheries, Pretoria; photographer unknown.

An early appointment of a forester had led to the establishment of 100,000 young trees, most of them from species that were 'known to thrive in the country, but experiments are already being tried with other valuable trees'. After the war ended, the government encouraged a number of 'good British settlers' who migrated into the colony, to plant trees.[83] Frank Smith, the Transvaal Secretary for Agriculture, hoped to give saplings to farmers to eventually make 'a considerable portion of the country wooded'.[84] But reconstruction officials remained disappointed with their achieving little for their efforts to encourage the immigration of skilled farmers who would populate rural areas.

ORC officials decided finally to bring in the Cape's Conservator of Forests, Lister, to survey the countryside and recommend on how to establish a department. The decision to bring in Lister was an about-turn on the initial policy of joint management of the Transvaal and ORC, and followed the failure to recruit

---

83   Milner to Adam, 20 October 1902, CO 201, NASA-P.
84   Smith to Adams, 7 October 1902, CO 201, NASA-P.

Troup, while recognising that forestry in the ORC would necessarily focus on experimenting with exotic trees in Southern African conditions. In 1904, Lister made recommendations to Wilson about how to create a forestry program.[85] Wilson then recruited Carlson, a Norwegian who worked as a forester in the Cape, who arrived in the ORC to direct forestry efforts there.

Carlson worked alongside Colin C. Robertson, who was appointed in late 1903.[86] Robertson was the well-connected nephew of H. F. Wilson, the Colonial Secretary (1901–1907), former private secretary for Chamberlain, and member of the secret Cambridge 'Apostle' society.[87] Wilson's influence no doubt helped to build on Adam's earlier actions, but by no means did he 'introduce' forestry in the colony (indeed, forestry had begun as part of the earliest stage of reconstruction in mid-1902), as suggested by Carlson, who was probably seeking to flatter either Wilson or Robertson.[88] Robertson always showed great talent as a forester, was described as 'a monument of integrity', and like so many of his peers, had a reputation as 'a prodigious worker', but his family connections may help explain his later rapid rise up the ranks of the Union Forestry Department's Research Branch. Robertson went to Yale University for his Master's degree in 1905, returning in 1907.

In 1906, during his summer vacation while at the Yale forestry school, Robertson conducted a reconnaissance of Mexico over several months, primarily with a view to finding tree species potentially suited to the ORC, but also with the requirements of the other provinces in mind. This responded to Hutchins's recommendations (and perhaps Fourcade's too) and followed on a 'hurried' visit to the same country that G. A. Wilmot made in 1905. Robertson's approach was thorough, and he consulted leading taxonomists of the conifers, including G. A. Shaw at the Arnold Arboretum at Harvard, and the 'veteran collector of the Mexican flora', C. G. Pringle, visited forests in several states of Mexico, and brought back seed of candidate species. His unpublished report 'Notes on the Trees of Extra-tropical Mexico' served as his dissertation for his Master's degree at Yale. Robertson planted trials of the Mexican pines on his return, and some of the trees are still to be seen in the arboretum in Bloemfontein.[89]

---

85   H. F. Wilson to High Commissioner, 29 December 1904, enclosing J. Storr Lister, 'Forest Officers: Scientific Training Of', 29 November 1904, TAD 540, 1181/06, NASA-P; K. A. Carlson, *Transplanted: Being the Adventures of a Pioneer Forester in South Africa* (Pretoria: Minerva Drukpers, 1947), Chapter 7.

86   H. Cox to C. Robertson, 13 January 1903, CO 201, NASA-P.

87   On Wilson, see W. C. Lubenow, *The Cambridge Apostles, 1820–1914: Liberalism, Imagination, and Friendship* (Cambridge: Cambridge University Press 1998), 158.

88   Carlson, *Forestry in the Free State*, 1.

89   C. C. Robertson, 'The Cultivation of Mexican Pines in the Union of South Africa, with Notes on the Species and their Original Habitat', reprinted from *Empire Forestry Journal* (1933), 1–83, 2–3; C. E. Legat, 'The Late Mr. C. C. Robertson', *Journal of the South African Forestry Association*, 15 (1947): 6–7; letter Mrs D. M. Rowswell to H. A. Lückhoff, early 1982, Sabie Forestry Museum.

Forestry in 1905

Work in forestry had only just begun when it was threatened by the storm raised by Milner's controversial policies. Milner had arrived in the Cape Colony in 1897 with the support of Conservatives and Liberals alike, but his close relationship with Chamberlain, and the decisions he made as Commissioner engendered anger among Afrikaners and Africans, and disloyalty among politicians in Britain. Back in Britain, the political winds drove against Chamberlain, who went into the 1905 election campaign proposing to abandon Britain's free-trade status to move to a system of imperial preferences and Federation. Within South Africa, Milner's reputation suffered greatly from a number of political blunders; the most disastrous being his decision to import Chinese workers into the mining industry beginning in 1904, the last year of his appointment. This inflamed the former British *uitlanders*, an Afrikaans term for 'foreigners' who worked in the Transvaal, as well as Afrikaner nationalists, who were represented increasingly by the Het Volk Party, an organisation attempting to provide political and economic power to Afrikaners, especially poorer ones. Boers in rural areas bristled at overt and secret attempts to anglicise rural districts.[90] Hyam and Henshaw argue that 'Milner's long-term legacy was the poisoning of Anglo-South African relations for fifty years'.[91]

With his popularity flagging, Milner worried that forestry was perceived by many Transvaal residents to be just another of 'Milner's fads', something ultimately to be discarded when he left.[92] Natal's decision to disestablish forestry that year did little to assuage this concern. On top of this, forestry had received very little funding from the Transvaal government for that year. When one considers the vast sum of money spent on reconstruction—around £19 million—the fact that Milner complained resentfully that Transvaal's Parliament could not appropriate Milner's desired £100,000 per annum on afforestation shows how unimportant forestry was to Transvaal's political and economic elite.[93] Milner would at that time have agreed heartily with J. M. Powell's description of Australian colonial forestry: 'recognisable forestry—I mean as a scientific and technical field—was nonetheless a minor enclave, where lonely and frustrated inhabitants basked from time to time in the imperial vision'.[94]

---

90   Darwin, *The Empire Project*, 248. Plans included having Rhodes and other Randlords purchase land quietly in rural districts to provide farms for Britons.
91   Hyam and Henshaw, *The Lion and the Springbok*, 54. Also see J. Benyon, *Proconsul and Paramountcy in South Africa: The High Commission, British Supremacy and the Sub-continent, 1806–1910* (Pietermaritzburg: University of Natal Press, 1980).
92   Milner, *The Nation and the Empire: Being a Collection of Speeches and Addresses*, 86.
93   Ibid., 86.
94   J. M. Powell, *An Historical Geography of Australia* (Cambridge: Cambridge University Press, 1988), 37.

Forestry in the period of reconstruction following the South African War had no well-shaped policy or institutional structure. However, the perception of the need for a reliable domestic supply of timber—to supply the mines, to build the railways, to substitute for imports, and to alleviate the destructive extractive pressures on the small and diminishing indigenous forests—was growing in strength and clarity, and the option of afforestation was becoming obvious to many. Foresters with experience in the successes and failures of afforestation trials in the Cape had strengthened their leadership role across what was to become the Union of South Africa, and had begun energetically to expand trials and forest nurseries at suitable sites throughout the territory. Fortuitously or not, attempts to import heads of the new forest departments from the imperial network had failed, leaving the field open to local expertise. From the 1890s, the Cape government had strengthened its capacities by sending candidates to Cooper's Hill for graduate training, while others had joined the service after receiving an education overseas at their own initiatives.[95] These formed the core of the professional forestry capacity that served the department of forestry in the Union government, assuring the South Africanisation of forestry, a process reinforced by a coherent policy of forest education, the subject of the following chapter.

---

95   Cape foresters sent to Cooper's Hill included C. B. McNaughton, J. S. Henkel. R. Burton, and K. A. Carlson. Those educated at their own inititaive included C. C. Roberston and G. A. Wilmot (Yale) and E. J. Neethling. See also Chapter 4.

Chapter 3

Educating a Nascent 'South African' Forestry Corps, 1880–1932

Poorly funded though the colonial forest services may have been, their first constraint during and after reconstruction of the Southern African economies following the South African War was the shortage of properly educated forest scientists and managers. The responses to this shortage led to several lines of development. First, the short-lived South African College School of Forestry provided education in forest science. It was based at Tokai in southern Cape Town, a facility that after a hiatus of 20 years was succeeded by a new Department of Forestry at the University of Stellenbosch in 1932. Second, the School for Foresters—from 1911 housed at Tokai in the facilities vacated by closure of the South African College course, and then from 1932 at Saasveld—provided training in the management of the forest estate (a diploma program). Third, training in field forest management in the native territories (another diploma program, but for African candidates) was initiated in 1904 at what became the Swartkops College and, later, at Fort Cox in the then Transkei.

Between 1911 and 1932, a carefully directed program of government sponsorship created a small but crucial stream of new professionals with advanced degrees from overseas forestry schools, all of whom soon joined the leadership in South African forestry. Whether through design or exigency and luck, the combination of a brief injection of 'Cape forestry', and the drive to create a new balanced South African economy, produced a vigorous and innovative forest sector. A strong emphasis on the climatic and other environmental determinants of forest productivity, and a community of forceful, heterodox, clear-thinking, and scholarly figures, committed to forest protection and development, marked this dispensation. Key systems of forest management imported from Nancy and India (via Cooper's Hill) were adapted to South African requirements and then transmitted in the training of managers of the forest estate at the School for Foresters.[1] The diaspora of aspirant South African foresters enabled the subsequent education of foresters in South Africa. These initiatives created a diverse corps of scientists and managers in a country with little forestry expertise. This group later took on the mission of forest protection and forest resource development with unusual innovation, and they guided the forest hydrology program needed to shape the pattern of afforestation that followed.

Though the Tokai degree course was short-lived, it served to crystallise thinking about policy for forest education in South Africa, thinking that influenced the direction of later investments in students sponsored by government, as well as creating the beginning of a corps of modern South African forest scientists who were closely attached to South Africa and who played leading parts in forestry policy and development. Tokai graduates, schooled through a curriculum designed for a creative engagement with South Africa's complex climates and landscapes, formed minds sympathetic to the country which would support the comprehensive program of forest science that developed over the succeeding decades. But an institution which contained the potential to become one of the world's leading centres of forest science soon foundered on the twists and turns of South Africa's politics, and closed in 1911. After the short life of the Tokai school, there followed a coherent program of sending promising South Africans to forestry schools overseas—chosen in the light of the education needed for South Africa's conditions—who together with Tokai graduates created modern forestry in South Africa.

A 'South African' School of Forestry

The opening of South Africa's first school of forestry in Cape Town and Tokai on 27 February 1906 was a significant day for local politicians and foresters.[2] Five students from the Cape Colony, one student from the Transvaal, and another from the ORC enrolled for the first intake. During their first year, students studied at the South African College and in the second year they focused on practical and theoretical forestry at Tokai, a Crown forest just to the south of Cape Town, with an extensive plantation and arboretum. The South African College in Cape Town and the Cape Colony's Forestry Department jointly managed the school.

---

1   These elements included the hierarchical system of forest policy development, working plans, and annual plans of operations, and figures such as J. S. Henkel were instrumental in their transfer to South Africa.
2   This work uses 'South Africa' to refer both to the Union of South Africa from 1910–1961 and also to the general region of Southern Africa, including then Southern Rhodesia (Zimbabwe) and Basutoland (Lesotho). Foresters and officials often used these terms interchangeably to describe both geographies and political entities. Note that the government opened the Tokai School for Foresters in 1911, for the purpose of technical training in the management of forests, a school that did not offer a university-level qualification. This school transferred to Saasveld, near George, in 1932.

David Ernest Hutchins presided as the school's Professor of Forestry. No one was happier about the school's founding than him. Just less than a year earlier he argued passionately for the founding of the school because there was 'no Forest School in the Southern Hemisphere, nor, in fact, any purely extra-tropical school of forestry, imparting its instruction in English'.[3] The newly founded school of forestry was the first ever opened in South Africa and the southern hemisphere.

Foresters in South Africa such as Hutchins hoped that the school's opening portended a healthy future for forestry education in South Africa. The school focused specifically on South African environmental conditions and had the support of governments in the Cape Colony, ORC, and Transvaal. Students from three of the four South African colonies attended, with the hope of more. Southern Rhodesia promised to send students once it founded a forestry department. But despite the initial burst of enthusiasm surrounding the school's opening, it closed down only five years later in 1911, never to be reopened. The Union government started a school for the technical education of field forest managers at Tokai that year, but South Africa remained without a university school of forestry to train its officers until Stellenbosch University founded a Department of Forestry in 1932.

This chapter resuscitates the school's history and explains its importance for the larger environmental histories of South Africa and the British Empire. The motive among leading South African politicians and foresters to create a school of forestry in the country was that they wanted a school where South Africa's future forestry officers could gain practical and theoretical experience of forestry in local conditions rather than going to Britain, Europe, or India, where climates and environments differed. They wanted students to focus on local conditions because they believed that the subtropical indigenous forests of the country dictated an approach to silviculture that differed from the temperate forests of Europe. They recognised that the future of forestry in South Africa depended on the creation of plantations of exotic trees rather than on timber from existing indigenous forests, and that this required an 'extra-tropical' thrust to forestry education. The success of the plantations in the Cape Colony begun in the 1880s and 1890s was hardly assured as many of the early attempts failed. Instead of being able to look to Europe or even India for examples of how to grow plantations of exotics in Africa, foresters had to learn from trial and error. With the rapid growth of forestry departments in the other colonies, where climates and environmental conditions differed greatly from the Cape, officials wanted foresters with practical and theoretical experience of the country's diverse conditions.

---

3   See D. E. Hutchins's report, in L. S. Jameson, Minute, 13 March 1905, Lieutenant Governor (LTG) 94, No. 96/79, NASA-P.

This chapter shows how the school's opening happened during the last years of the Cape Colony's dominance over forestry in South Africa, whereas its closing exemplifies the migration of power from the Cape north to the Transvaal that happened just after Union in 1910. However, the school's closure was not only caused by a larger national shift of power northward. The school faced serious problems beginning in its first year that handicapped its future. In 1907, Hutchins moved to Kenya to work for the Colonial Office, leaving the school without a professor. The academic leadership at the South African College became critical of the school because none of them were experts of forestry and they wanted the Cape government to provide money to hire a new professor. The lack of financial help from governments in Natal and the ORC made the school's continuation an almost impossible proposition without increased support from the Cape Colony or the Transvaal. Eventually, the foresters in the Transvaal started to criticise the school. Without national, financial, or university support, the school closed its doors forever on a unique moment in South Africa's environmental and scientific history.

The history of the Tokai school of forestry fits a broader pattern of conflict that surrounded the founding of new forestry schools in the British Empire.[4] Starting and maintaining a new school of forestry proved difficult. It allowed for the expression of strong political, scientific, and ideological divisions. In the colonies, funding often proved divisive because schools usually received funds through complex interstate arrangements that depended upon consensus and goodwill, something often in short supply during tight fiscal periods. The first imperial arrangement to educate British forest trainees bound for India began in the late 1860s and ended in 1885 with the appointment of William Schlich as Professor of Forestry at the Royal Indian Engineering College at Cooper's Hill. Many leading British foresters and Indian officials prompted this development because they wanted a national school to inculcate common British cultural values.[5] This school closed in 1905 and the institution transferred to Oxford University.[6] The same story is true in Australia. A school of forestry opened initially at the University of Adelaide in 1910 only to be shut down with the opening of the Australian Forestry School in 1927.[7] The Australian Forestry School ran into continual conflict because its principal from 1927 to 1945, C. E. Lane Poole, battled with state foresters and officials who disliked his national and

---

4   This resembles a broader pattern: see B. M. Bennett, 'The Consolidation and Reconfiguration of "British" Networks of Science, 1800–1970', in B. M. Bennett and J. M. Hodge (eds), *Science and Empire: Knowledge and Networks of Science Across the British Empire 1800–1970* (Basingstoke: Palgrave Macmillan, 2011).

5   B. M. Bennett, 'A Networked Approach to the Origins of Forestry Education in India, 1855–1885', in *Science and Empire*, 81–4.

6   See Parliament, *East India (Forest Service) Correspondence Relating to the Training of Forestry Students: Presented to both Houses of Parliament by Command of His Majesty* (London: HMSO, 1905).

7   See M. Roche and J. Dargavel, 'Imperial Ethos, Dominions Reality: Forestry Education in New Zealand and Australia, 1910–1965', *Environment and History*, 14 (2008), 529–31.

imperial bias, and strong personality.[8] Despite following this general pattern, the South African school's history is distinct from British world forestry schools because it opened with a unique local mission.

The climate of forestry education in South Africa, 1880–1906

South Africa's four colonies had no school of forestry for the training of officers before 1906. Forestry officers studied abroad or worked as assistants in nurseries, plantations, and gardens before becoming officers in the Cape, the only colony to have a continuous state forestry program from the early 1880s to Union in 1910. The Comte de Régné De Vasselot was the first professional forester to work in South Africa.[9] De Vasselot, instead of advocating that the Cape Colony create a school of forestry to train foresters, an infeasible proposition given the paltry size and finances of the fledgling Cape forestry department at the time, asked the Parliament of the Cape Colony to import trained foresters into the colony.

During the 1880s and 1890s, the Cape Colony's Department of Agriculture showed little interest in paying local students to study forestry abroad as a means of building up its forestry program. Records indicate that in 1892, the department sent C. B. McNaughton to Cooper's Hill in England. This was an exception to the department's rule of not giving financial assistance for students studying abroad. Charles Currie, the Undersecretary for the Cape's Agricultural Department, refused an applicant in 1898, noting that 'the aiding of Forest Officers going through a course of instruction at Cooper's Hill is not a recognised practice'.[10]

Though no formal system of education existed, the Cape model of forestry, marked by a coherent culture and unique practice of forestry, determined the curriculum for the future school of forestry. Of all Cape foresters, Hutchins, the future professor of the school, directed many of these climatic comparisons and experimental trials. Hutchins's enthusiasm and boldness gained him both the admiration and criticism of foresters around the world. He saw himself as the *de facto* 'conservator' and leader of forestry in the Cape.[11] Hutchins was widely

---

8   B. Bennett, 'An Imperial, National, and State Debate: The Rise and Near Fall of the Australian Forestry School, 1927–1945', *Environment and History*, 15 (2009): 217–44.

9   Brown, 'The Conservation and Utilisation of the Natural World', 420–1.

10   For Henry Fourcade see file 935, Colonial Secretaries Office (CSO) 1181, National Archives South Africa Pietermaritzburg [NASA-PMB].

11   Darrow, *David Ernest Hutchins*, 12–3.

respected for his education and experience and became a natural choice to head the school when it opened in 1906. His erratic behaviour, though, endangered the school not long after it opened.

Debating a South African school of forestry

The expansion of state environmental scientific programs in South Africa during the early twentieth century was part of state reconstruction in the aftermath of the South African War. British reconstruction politicians who took over the governments of the former South African Republic and Orange Free State started to discuss the possibility of creating a school of forestry for the whole of South Africa with the governments of Natal and the Cape. Afforestation was one of the important political topics during the reconstruction period (see Chapter 2).[12] This was because sparsely wooded grassland and savanna covered most of the former Transvaal and ORC. Alfred Milner and his British 'Kindergarten' of advisors argued for the extensive afforestation of the grasslands of South Africa. From his High Commissioner's office in the city, Milner viewed the grasslands surrounding the city as future forests.[13]

British elites in South Africa agreed widely about the need to create a centralised political union, something that a national school would help to achieve. But the relations between the colonies before the Union of South Africa in 1910 made coordination among colonies difficult. Each colony had its own government and Governor (the ORC and Transvaal had the same Governor from 1902 until 1907). Finances remained separate. Each colony had its own departmental structure and culture, with distinct local social, economic, linguistic, and ecological conditions.

The perceived future demand for trained forest officers prompted the first serious discussions about a school. Two reports by Joseph Storr Lister in 1903 brought the question of a national school to the attention of the ORC government and the British High Commission in Johannesburg.[14] The first, a report on forestry in the ORC in 1903, contained his recommendations on how to develop a forestry department for the ORC. Lister discussed the scientific competence desirable for forestry officers required to staff the ORC forestry department in a letter of 29 November 1904. In this letter he re-emphasised the importance of scientific

---

12   See Brown, 'The Conservation and Utilisation of the Natural World'; for Lesotho see K. Showers, 'From Forestry to Soil Conservation: British Tree Management in Lesotho's Grassland Ecosystem', *Conservation and Society*, 4 (2006): 1–35; for the importance of wattle in Natal see Witt, '"Clothing the Once Bare Brown Hills of Natal"'.

13   For Milner's discussion of Basutoland see Showers, 'From Forestry to Soil Conservation', 12–5.

14   For Lister's request to visit in 1902 and the correspondence related therein see F3659, AGR 701, NASA-CT.

training for the future of South African forestry.[15] He believed that '[to] meet future requirements I submit the time has arrived for the establishment of a South African forest School or Training Depot'.[16] A variety of suitable locations existed throughout the Cape Colony, such as the large plantations at Fort Cunynghame, Tokai, or Kluitjes Kraal. He also remained open to the idea of locating the school within an agricultural college in another colony of South Africa. He wanted the school located in South Africa because it offered 'obvious' advantages. It would provide access to 'promising students' who 'could not afford to go abroad'.[17] Locating the school in South Africa would also allow students to learn methods 'applicable to the peculiar conditions of South Africa, which differ from those which prevail in Europe and America'.[18]

The question of where to locate the school had a strong environmental dimension. Lister suggested that it should be located in the Cape Colony because it had the oldest and most advanced state plantations. Despite having existing managed forests, the environmental conditions of the Cape were actually less conducive to forests than elsewhere: much of its land received less rain than the subtropical climates of Natal and Transvaal, which were better suited to the commercial production of trees. Despite these advantages, foresters were still unsure about what species would grow best,[19] and neither Natal nor the Transvaal had large plantations or extensive natural forests on which to base a school of forestry. The ORC never factored as an important location because few believed the environments there could support large-scale afforestation.[20] Nevertheless, Lister preferred the Cape as the location for a training centre.

The government of the ORC contacted Milner in Johannesburg late in December 1904 regarding Lister's suggestion.[21] Milner forwarded the dispatch from the ORC to Sir Richard Solomon, the Lieutenant-Governor of the Transvaal, on 11 January 1905.[22] Milner wrote that he 'attached the greatest importance to the proposal to establish a School of Forestry in South Africa'.[23] Milner noted that the Cape Colony, Natal, and Southern Rhodesia had also contacted him about the foundation of a school. The letter raised the possibility of founding the school in the Transvaal at an agricultural extension farm near Johannesburg called Frankenwald, which the wealthy mining magnate Alfred Beit had donated to the

15   J. Storr Lister, Forest Officers: Scientific Training of, 29 November 1904, in H. F. Wilson to A. Milner, 29 December 1904, Transvaal No. 96/79, LTG 94, NASA-P.
16   Ibid.
17   Ibid.
18   Ibid.
19   See. Hutchins, *Transvaal Forest Report*.
20   See Carlson, *Transplanted*, Chapter 7, 'Orange River Colony'.
21   H. F. Wilson to A. Milner, 29 December 1904, Transvaal No. 96/79, LTG 94, NASA-P.
22   Milner to R. Solomon, 11 January 1905, Transvaal No. 96/79, LTG 94, NASA-P.
23   Ibid.

colonial government. Frankenwald had few trees and was primarily grassland. (These grasslands later became a site where University of Witwatersrand researchers such as John Phillips and Eddie Roux studied grassland ecology.)

Charles Legat, the Conservator of Forests in the Transvaal, wrote to Frank Smith, the Transvaal's Secretary of Agriculture, laying out his view that the foundation of a school in South Africa 'is of the utmost importance'.[24] However, he did not support the idea of making a 'self-contained' school of forestry.[25] He recommended making forestry an adjunct to an existing university science program: 'The Forest School site would therefore have to be wherever the teaching University is to be situated, probably Johannesburg'.[26] The practical training would then take place at Frankenwald. But his letter left open the possibility of also locating the school in the Cape, subject to a meeting of all the country's conservators who could decide the question together.

Smith agreed with Legat's view that the school should be associated with an existing university science curriculum and indicated that the Frankenwald location would be eminently suitable.[27] Smith wrote:

> Schools of Forestry and Veterinary Science are Institutions which are urgently required in South Africa as there are so many problems connected with these subjects which are more or less peculiar to this Sub-Continent, and which can only be satisfactorily studied and investigated on the spot.[28]

Smith suggested that all students study the same broad 'scientific principles' before studying their specialty, such as forestry.[29]

The jockeying for the school's location continued into early 1905. L. S. Jameson, the Progressive Party Prime Minister of the Cape Colony, transmitted a report written on 26 February by Hutchins, then the Conservator of the Cape's Western Conservancy, on the possibility of founding a school of forestry in South Africa, and laying the outlines for a future school.[30] Although Hutchins himself studied at Nancy, he strongly criticised the largely 'impractical character of their instruction'.[31] The high costs of travelling to and attending foreign schools also made any arrangement other than a South African school prohibitive. Only five students from the Cape Colony's Forestry Department had attended foreign schools, and the one student then abroad at Yale was paying his own way.

---

24   C. E. Legat to F. B. Smith, 25 January 1905, Transvaal No. 96/79, LTG 94, NASA-P.
25   Ibid.
26   Ibid.
27   Smith to Jameson, 26 January 1905, Transvaal No. 96/79, LTG 94, NASA-P.
28   Ibid.
29   Ibid.
30   Jameson, Minute, 13 March 1905, Transvaal No. 96/79, LTG 94, NASA-P.
31   Ibid.

Hutchins believed that with the current and future size of forestry departments in South Africa, a school training upwards of 19 students would not produce too many foresters.

Hutchins again emphasised the unique climatic and environmental conditions that foresters in South Africa faced. Much of the country was marked by aridity, divergent patterns of rainfall seasonality, and ecological conditions vastly different than Britain, Europe, or India, the other leading centres of forestry. Hutchins envisaged the South African school as the beacon of forestry knowledge in the southern hemisphere and the English 'extra-tropical' world, i.e. regions with Mediterranean to subtropical climates. He noted, '[t]here is at present no Forest School in the Southern Hemisphere, nor, in fact, any purely extra-tropical forest school, imparting its instruction in English'.[32] He argued that no existing school could supply the blend of theory with the practical experience of South Africa's unique conditions required for foresters in the region. A South African school would have to focus on the various 'climatic conditions' of the country because the 'trees suited to each area differ widely'.[33] Hutchins concluded his report with an appeal for the school: 'Such a Forest School would be the only English institution of its kind dealing with extra-tropical forestry, and as such could probably count on considerable private support not only in South Africa, but also from the Australian Colonies'.[34]

Hutchins requested that the school be located in Cape Town. He discounted the Frankenwald location because it lacked the native forests and plantations needed to accommodate practical study. Johannesburg, he admitted, offered the best 'endowments'.[35] As an investor in gold, Hutchins recognised that South Africa's wealth and power was gravitating slowly to the north.[36] But the Frankenwald site left much to be desired. After visiting the location, Hutchins surmised, 'that with every care and a liberal expenditure it could not, within half a century, offer the practical instruction and demonstration attainable at Tokai and Ceres Road'.[37] By contrast, forestry in the Cape 'is a quarter of a century ahead of forestry in the Transvaal'. The South African College with the corresponding plantations and arboretum at Tokai 'offer the best facilities'.[38]

32   Ibid.
33   Ibid.
34   Ibid.
35   Ibid.
36   D. E. Hutchins to Fourcade, 2 February 1887, Fourcade Bequest, UCT.
37   Jameson, Minute, 13 March 1905, Transvaal No. 96/79, LTG 94, NASA-P.
38   Ibid.

Making a school at the Cape

The Cape Colony determined the fate of a future school of forestry by announcing in December 1905 that the Cape Colony's forestry department and the South African College in Cape Town would support the opening of a school of forestry in Cape Town and Tokai. The scientific teaching facilities would be based at the South African College and the practical work in the plantations and arboretum at Tokai just a few miles to the south.[39] The school would educate forest officers through a two-year specialised course in forestry. For those without the requisite scientific background, the South African College offered an extra year of preliminary study. At Tokai, a newly created reading room and library catered to forestry students. The bulletin emphasised: 'The Tokai arboretum which now comprises the largest collection of timber trees in South Africa, affords unique opportunities for practical instruction in silviculture'.[40] Students would gain practical experience around Cape Town by exposure not only to the forestry practices at Tokai but also the driftsand reclamation project in the Cape Flats and at the plantation and arboretum at Ceres Road. Students would also work in the indigenous forests near Knysna–George. At the end of this course, graduates would receive a certificate or diploma signed by the College and the Chief Conservator of Forests for the Cape Colony. In 1906, the school sought initially to enrol 10 resident students: five from the Cape and five from elsewhere in South Africa.

The curriculum reflected the long-standing interests and experience of Hutchins and other Cape foresters rather than the traditional continental education for foresters (see Box 1). It focused on subjects seen to have local applications, such as climatology, local geology, and silviculture for South African conditions. The courses emphasised climate to a greater extent than at Cooper's Hill or elsewhere. This reflected the strong belief among Cape foresters in the primacy of climate and exotic silviculture. The course also featured a section in forest geography and history, a keen interest of Hutchins who often lectured publicly on the subject.

39   Cape Colony, *Department of Agriculture Bulletin*, No. 1383 (Cape Town, 1905).
40   Ibid.

COURSE OF STUDY AT THE SOUTH AFRICAN COLLEGE AND TOKAI

First Year Forestry Courses

Botany (2 terms physiology, 2 terms mycology, 2 terms Forest Botany), 6 hours a week
Chemistry (soils and plants, 2 terms), 6 hours a week
Climatology and Meteorology (1 term), 3 hours a week
South African Geology (1 term), 3 hours a week
Survey and Elementary Engineering, 6 hours a week
Forestry (Lectures and Field Work), 6 hours a week

Second Year Forestry Courses

Forestry (Lectures and Field Work), 6 hours a week
South African Arboriculture and Silviculture, 4 hours a week
Climatic influence on forestry, 2 hours a week
Forest Entomology, 3 hours a week
Forest Law, 1 hour a week
Forest Geography and History, 1 hour a week

Box 1. Course of study at the South African College and Tokai.

The staff of the newly formed school of forestry included Hutchins as professor of forestry and lecturer in forestry geography and history; G. A. Wilmot, freshly returned from his study at Yale, as assistant lecturer and demonstrator in forestry and lecturer in forest management and forest law; L. Peringuey as lecturer in forest entomology; J. C. Beattie as professor of physics and lecturer in climatology and meteorology; and H. H. W. Pearson, as professor of botany. Other South African College professors served as teachers: P. D. Hahn (chemistry), H. Payne (engineering), Andrew Young (mineralogy and geology), W. S. Logeman (modern languages), and Lawrence Crawford (mathematics).

The Cape Colony government promoted the school's case to the High Commissioner and the other colonies. In a letter dated 24 January 1906, Jameson, the Cape's Prime Minister, requested that the High Commissioner forward information regarding the South African College Forestry Course at Tokai to the governments of the ORC, the Transvaal, Natal, and Southern Rhodesia.[41] The Cape government supported the new program, Jameson wrote, in the hope that it 'will serve the needs of all the States of South Africa'. He hoped for 'co-operation' from the other governments.[42]

Cooperation, however, could not be won so easily. Natal ministers did not support the idea of creating an independent school of forestry, although they supported the idea of creating a larger technical college, which taught forestry as one of

---

41   Jameson to R. Selborne, 24 January 1906, Transvaal Agriculture Department (TAD), 540, No. G.1181/06, NASA-P.
42   Ibid.

its disciplines.[43] The ORC did not plan on contributing to the scheme, although it later did send one student.[44] It remained up to the Transvaal government to determine the viability of the Cape's proposed program.

The High Commissioner queried the Transvaal government about whether it favoured the proposal by Hutchins or Milner. Whereas Milner supported the idea of a school at Frankenwald, Lord Selborne, the High Commissioner of South Africa (1905–1910) who replaced Milner, instead suggested that the Cape Colony offered a location where 'better practical instruction and demonstration is obtainable'.[45] After discussion, the Transvaal government leaned towards the Cape Town location. An internal government minute that recommended sending students to the college in Cape Town noted that the facilities at Tokai were 'superior to any in the Southern Hemisphere available'.[46] The Transvaal government wrote to the Imperial Secretary and High Commissioner in January, informing them that 'there is only room for one Forest School in South Africa'.[47] The Private Secretary informed the High Commission 'that it is not proposed to consider further the proposal to establish a Forest School at Frankenwald' as previously raised by Milner in January of 1905.[48] On 27 March 1906, the Transvaal Department of Agriculture committed to sending one officer to undertake the two years of study.[49]

The school commenced teaching on 27 February 1906 with six students. Only two students from outside of the Cape Colony attended: one each from the Transvaal and the ORC. The Director of Agriculture supported the school in a letter to the Prime Minister, noting 'that it is deserving of all the support that can be accorded it by this Colony'.[50] Jan Smuts, the Education and Colonial Secretary of the Transvaal government, offered his support to the Cape government for the school, even urging the Transvaal government to give more publicity to the school as requested by the Cape government.[51]

---

43   H. W. Hamilton Fowle to Solomon, 20 December 1905, No. G.II81/05, LTG 94, NASA-P.
44   Ibid.
45   Hamilton Fowle to H. E. Clark, 20 December 1905, No. G.II81/05, LTG 94, NASA-P.
46   Transvaal Department of Agriculture Minute Paper, 2 January 1906, No. G.II81/05, LTG 94, NASA-P.
47   Clark to Imperial Secretary, 31 January 1906, No. G.II81/05, LTG 94, NASA-P.
48   Ibid.
49   Lawley to Selborne, 27 March 1906, No. G.II81/05, LTG 94, NASA-P.
50   Smith to D. C. Malcom, 21 March 1907, No. G.II81/06, LTG 94, NASA-P.
51   Minute 45, 11 April 1907, No. G.II81/06, LTG 94, NASA-P.

Troubles with the school

The school ran into trouble in 1906 when Hutchins decided to take a Colonial Office appointment in East Africa to write a report on the forests of Kenya.[52] This decision, prompted by his not being appointed by the Cape government as the Cape Colony's Chief Conservator of Forests in 1905 (the Cape government chose Lister for the position), left the school without a professor of forestry.[53] Foresters and government officials outside the Cape Colony, especially in the Transvaal, soon began to question the viability of the school. A lecturer, Wilmot, became the new head of the school in 1907.

In June – July 1908, the ministers of the Cape Colony tried to put the school 'on a more permanent basis' by integrating it with the forestry departments of the other South African colonies and attempting to hire a new professor of forestry.[54] But, for this plan to work, all the colonies would need to offer more money and send more students. The Cape recommended appointing a professor of forestry who would work under the Board of Management, the governing body of the school comprised of representatives of the academe of the College and the colonial administration. Further, the professor of forestry was to be committed full time to teaching. The Cape would provide £499, Transvaal £299, ORC £199, Natal £199, Southern Rhodesia (Zimbabwe) £59, and Basutoland (Lesotho) £25—a total of £875. This proposal circulated to the colonial governments. Natal rejected the offer.[55] The ORC also turned down the Cape's plan and refused to send money.[56] Southern Rhodesia agreed to fund the school but supplied no students.[57] Basutoland approved the measures.[58]

It was apparent that the school lacked the full support of Natal and the ORC, and it still had no professor. The Board of Management and faculty at the South African College started to question the viability of the school because of lack of adequate support. In a memorandum sent to the Governor in 1908, the board made

---

52   For Hutchins's discussions regarding his retirement see Correspondence Relating to Hutchins Retirement, FOR 1, NASA-P.

53   See Darrow, *David Ernest Hutchins*, 12–3. The selection of Lister by the Minister of Agriculture strongly affected Hutchins, who otherwise supported the Tokai school and saw it as a continuation of his work on extra-tropical forestry. But Hutchins was eccentric, outspoken against the Department of Agriculture, and was not considered as suitable for the position as Lister, a more diplomatic forester. Hutchins noted tactfully that with Lister 'we shall obtain that share of the Ministerial ear which is so important to a minor Department in the struggle for existence'. See Hutchins to Fourcade, 10 October 1905, Fourcade Bequest, UCT.

54   Minute 1/330, 24 June 1908, 32/2/1908, GOV 1146, NASA-P; W. Hely-Hutchinson to Selborne, 2 July 1908.

55   Minute 7, Prime Minster Natal, 22 September 1908, 32/2/1908, GOV 1146, NASA-P; F.R. Moor – Minute No. 7 1908, 26 September 1908.

56   Minute 4144/08, Prime Minister's Office Orange River Colony, 25 September 1908, 32/2/1908, GOV 1146, NASA-P.

57   Hamilton to Selborne, 10 August 1908, PM 809/08, CO 73, NASA-PMB.

58   Sloley to Selborne, 13 August 1908, PM 809/08, CO 73, NASA-PMB.

a number of suggestions for improvements.[59] In it they 'emphasised the South African Character' of the college and asked the other colonial governments to share in the costs of running the school.[60] This scheme envisioned contributions from all of the South African colonies and even £50 from British East Africa. They asked that extra monies be raised to hire a professor of forestry.

Officials in the Transvaal also began to openly question the school in late 1908 and early 1909. In a letter to Lister, Legat complimented the school on the high quality examination results of the 1908 cohort of students. But he asked about the staffing of the school: 'I suppose you will get a very highly trained and experienced man for the post, as the prestige of the School will in the first instance rest mainly with the Professor of forestry'.[61] Legat's suggestion that the 'prestige' of the school rested on the professor of forestry, not the curriculum or the performance of the students, contradicted his praise of the high performance of the students at the beginning of his letter. As if to hint at the Transvaal's lack of support, Legat ended the letter by asking the question: 'Are you expecting Forest Students from Australia and New Zealand? Mr. Hutchins thought it quite likely those Colonies would send men to be trained'.[62] Legat quietly hinted that all was not well with the school's future.

Lister's reply to Legat began by expressing his frustrations at the failure to find an ideal professor of forestry. He supported Wilmot, describing him as 'a clever, good, all around fellow. He has gentlemanly manners, is popular with the students and is an excellent lecturer'.[63] Instead of worrying about his lack of training, he privately worried that 'we cannot expect him to retain the post permanently'.[64] He 'was at a loss' to know what forester to select as the professor.[65] He remained 'a little nervous' about bringing in a forester from India or Europe without adequate knowledge of South African conditions. Knowledge of South Africa remained of the highest priority owing to the unique conditions of the country and South Africa's reliance on plantations of exotic trees, a science still in its infancy.[66] Lister envisioned the future of forestry in South Africa as focusing on the 'formation and management of plantations of exotic trees and, notwithstanding past experience, Forest Officers for many years will have to continue to more or less feel their way by constant and systematic experiments'.[67]

---

59   See discussions in the memorandum that was recirculated in 23 May 1910, No. G.II81/06, LTG 94, NASA-P.
60   Ibid.
61   Legat to Storr Lister, 23 December 1908, No. G.II81/06, LTG 94, NASA-P.
62   Ibid.
63   Storr Lister to Legat, 6 January 1909, No. G.II81/06, LTG 94, NASA-P.
64   Ibid.
65   Ibid.
66   Ibid.
67   Ibid.

Despite forebodings, the Cape government did not officially recognise any serious flaws in its School of Forestry as late as April 1909, three years after the school's founding. In a minute sent to all governments, Walter Hely-Hutchinson noted happily that the Transvaal, Rhodesia, and Basutoland all supported the school, although Natal and the ORC did not offer financial assistance.[68] In the letter, he supported Wilmot as the principal lecturer of the school, suggesting that it would be better to keep his services than to import a foreign forester. First, the Cape Colony could not afford to pay a large salary with the current funds provided by the other colonial governments. Second, the government considered that the 'services of a local officer with experience of South African conditions will compensate for those of a Forest Officer from another country'.[69] Locality trumped universal scientific study. Not only would Wilmot stay on as a lecturer, he would be given the 'entire control and management' of the School of Forestry.[70] He would live at Tokai and work as a District Forest Officer, tending the plantations on top of his teaching duties. Thus he would have an 'adequate salary' and room to pursue his research.[71] In closing, Hely-Hutchinson noted that '[m]inisters cannot too strongly reiterate the importance of maintaining the School of Forestry on its present satisfactory basis and trust that the general scheme proposed will be approved'.[72]

A letter from Legat on 17 April 1909 still officially supported the school, asking the Director of Agriculture to provide money to send a student to the School of Forestry.[73] But only a little over a week later on 25 April, Lionel Taylor, then acting Conservator of Forests for the Transvaal, wrote a letter of protest to the Director of Agriculture with the consent of Legat. Taylor offered a 'strong protest against the appointment of Mr. Wilmot' by offering his opinion 'that it is most desirable to appoint a highly trained and experienced expert from Europe who can come to this country with unbiased views'.[74] This turned the argument about the need for local experience on its head. Taylor blasted the Cape's forest department, suggesting that they were 30 years old but 'no nearer to the solution of many of their problems than they were when they started'. He inferred from Lister's comments to Legat that South African forestry remained dogged by problems with exotic plantations, especially of *Eucalyptus*. Taylor recognised the 'prejudice against getting a man with European experience only', but he instead suggested 'this is rather an advantage' because they could 'work on scientific principles without following the groove into which officers

68   Minute No. 1/116, 3 April 1909, No. G.II81/06, LTG 94, NASA-P.
69   Ibid.
70   Ibid.
71   Ibid.
72   Ibid.
73   Legat to Smith, 17 April 1909, No. G.II81/06, LTG 94, NASA-P.
74   Taylor to Smith, 15 April 1909, No. G.II81/06, LTG 94, NASA-P.

in the Cape Forestry Department have run for 30 years and which had led to no practical solution of vital problems'.[75] The South African College needed to appoint a leading forester 'if the Cape Forest School is to become the training ground for Forest Officers from Australia, New Zealand and other Colonies as Mr. Hutchins intended it to be'.[76] He ended the letter by suggesting that the £299 contribution of the Transvaal gave it the right to discuss the management of the school. Taylor called the Transvaal's support of the school into question.

The Transvaal government supported the school with £300 for the period 1 January 1909 to 30 June 1910.[77] Despite this outward show, the Department of Agriculture for the Transvaal drew upon Taylor's letter to craft a detailed internal criticism of the new terms of agreement. Louis Botha and his ministry urged officially that the Cape government appoint a professor of forestry 'who should have no duties beyond those relating to his office'.[78] The Transvaal administration sent its minute to the governments in Natal, the ORC, Rhodesia, and Basutoland.[79] At the same time, the Transvaal government asked the professor of botany at the South African College, H. H. W. Pearson, to serve as the Transvaal's local representative on the Board of Management.[80] This opened up a direct channel of communication between the faculty at the South African College and the government of the Transvaal.

In a letter to Smith, Pearson noted that the faculty members who taught the program worried privately about the quality of the forestry lectures and fieldwork: 'The Members of the Board actively engaged in teaching accessory subjects felt very strongly that the present arrangements are far from satisfactory and believe that Mr. Wilmot, the forestry lecturer, agrees with us'.[81] He noted that the South African College forestry staff, except for Wilmot, 'know nothing about forestry as such'.[82] The college's academic members felt that the school would not succeed unless an 'expert' forester, who taught full time and undertook no other duties, led the forestry program. In his letter, Pearson asked for the Transvaal's advice on what type of expert forester they wanted to head the program, the salary required to hire such a person, and how much more money the Transvaal would provide to hire them.

---

75   Ibid.

76   Ibid.

77   Director of Agriculture to the Colonial Treasurer, 23 September 1909, No. G.II81/06, LTG 94, NASA-P.

78   Minute No. 538, 31/1/1909, GOV 1200, NASA-P; Selborne to Hely-Hutchinson, 27 November 1909, 31/1/1909, GOV 1200, NASA-P.

79   Selborne to Governor of the Orange River Colony, 27 November 1909, 31/1/1909, GOV 1200, NASA-P. Selborne to Governor of Natal, 27 November 1909, 31/1/1909, GOV 1200, NASA-P.

80   Legat to Smith, 10 November 1909, No. G.II81/06, LTG 94, NASA-P; Smith to H. H. Pearson, 23 November 1909, No. G.II81/06, LTG 94, NASA-P.

81   Pearson to Smith, 8 March 1909, No. G.II81/06, LTG 94, NASA-P.

82   Ibid.

Smith's reply included copies of two reports by Legat, which he asked Pearson 'to treat as confidential'.[83] Legat's reports questioned the viability of the school in its current form, as well as Wilmot's ability to lead the entire school with such a high workload and a relative lack of experience. He offered an analysis of the failings of the school and a possible solution to its problems from his point of view. Highlighting Wilmot's lack of experience (he had had only two years of study at Yale), Legat wrote, 'I venture to say that in no other Forest School in the world is the teaching of the principal subject left to the care of an inexperienced junior'.[84] To remedy the current problem, he suggested hiring someone to work independently of the Cape's Forestry Department with a salary of around £750 per year. Legat suggested the services of H. Meyer of Munich University, since Meyer had a broad background, including experience in Europe, Japan, and North America, in addition to being an expert on the cultivation of exotic trees. In conclusion, Legat suggested that Pearson raise these issues with the school's board and the Cape government, if required. Smith agreed with Pearson's desire to establish the school on a 'more satisfactory basis'.[85]

The South African College continued to worry about the quality of the school. On 3 May, the Registrar of the College sent a resolution from the Council and Senate of the College.[86] Its first clause stated: 'The provision for the teaching of Forestry in the South African School of Forestry is inadequate'.[87] This set the stage for making changes to the program or for closing it down. The second clause stated: 'The minimum of staff necessary for the proper teaching of the various branches of this subject is one Professor and one Lecturer'.[88] Both of the teachers had to teach full time; they could not hold additional duties as forest officers.

The Registrar discussed in his letter the memorandum sent in 1908 from the academic staff to the Governor of the Cape Colony. This memorandum reminded the government that at least two teachers—one professor and one lecturer—were required to make the school of world standing, using examples from Europe, where schools had two to three professors, and Yale, with three professors and seven assistants. But the South African College employed only one lecturer who also worked as a District Forest Officer in charge of a large plantation. The Registrar suggested that the reputation of the school suffered badly from its mismanagement and the lack of a professor of forestry. Only four students graduated in 1909 and only four students had enrolled for 1910. The Transvaal

---

83   Smith to Pearson, 5 April 1909, No. G.II81/06, LTG 94, NASA-P.
84   Legat to Smith, 21 March 1910, No. G.II81/06, LTG 94, NASA-P.
85   Ibid.
86   Registrar of the South African College to the Council and Senate of the South African College,
3 May 1910, No. G.II81/06, LTG 94, NASA-P.
87   Ibid.
88   Ibid.

quit sending students in 1907, although it continued to contribute financially to the school. The Board noted open criticism of the school 'by more than one authority in the Transvaal'.[89] The memorandum painted a negative picture:

> Being the only school of Forestry south of the Equator it was hoped that the S.A. [sic] School would attract students from Australia where the Forestal Problems are more akin to those presented in South Africa. But while the school has not sufficient standing to command confidence in South Africa itself it cannot hope to be recognised by other states. It is a serious reflection upon Cape Colony, the Forest Department and the South African College that we should pretend to run a School of Forestry in which such utterly inadequate provision is made for the training of students in the most important subjects of a Forestry curriculum.[90]

Pearson continued to work as a liaison for the Transvaal on the board of the School of Forestry. He acted as the leader on the South African College's resolution, previously discussed and passed unanimously by the Council and Senate.[91] Pearson asked Smith whether he could raise this issue on behalf of the Transvaal at the next meeting of the School of Forestry Board of Management. The Board of Management of the South African School of Forestry decided to meet in the Chairman's office in Cape Town on 30 May to discuss the resolution of the Senate.[92] The Transvaal wired a letter to Pearson to encourage him to raise the Senate proceedings at the Board of Management meeting, calling for changes to the school or its closing.[93] Pearson confirmed that he would speak for the Transvaal at the meeting.[94]

Before the faculty could meet, the death knell of the school sounded when in October 1910 the Cape's *Government Gazette* announced that the school would no longer accept applicants for the upper grade, effectively closing the school's doors.[95] This resulted from internal decisions by the Cape government in coordination with the other provinces. Forestry conservators met with the new government ministers shortly after the Union in 1910.[96] There the conservators and government ministers decided to close down the school. The enthusiasm for the school that characterised the period from 1902 to 1908 had slowly dissipated. Political geography also helped to shape the decision to close the school. The centralisation of departments in Pretoria finally brought financial

---

89   Ibid.
90   Ibid.
91   Pearson to Smith, 17 May 1910, No. G.II81/06, LTG 94, NASA-P.
92   G. A. Wilmot to Legat and Smith, 23 May 1910, No. G.II81/06, LTG 94, NASA-P.
93   Smith to Pearson, 23 May 2010, No. G.II81/06, LTG 94, NASA-P; Smith to Pearson, 26 May 1910, No. G. II81/06, LTG 94, NASA-P.
94   Pearson to Smith, 24 May 1910, No. G.II81/06, LTG 94, NASA-P.
95   W. Ritchi, *The History of the South African College* (Cape Town: T. Maskew Miller, 1918), 567.
96   Union of South Africa, *Report of the Chief Conservator of Forests for the Year Ending 31st December, 1910, with which is Incorporated Report on Railway Sleeper Plantations for the Same Period* (Cape Town: Government Printer, 1911), 1.

and political power into the north and away from the Cape.[97] The four forestry departments amalgamated, organised after the Cape's structure, but with its new head office in Pretoria. Tokai, with its suitable plantations and arboretum, became the station for the School for Foresters until this facility moved to Saasveld, outside George, in 1932. The new South African government effectively ended its relationship with the South African College at the same time that the South African College and the school's Board of Management turned against the school.[98] So, ultimately, the school did not live up to Hutchins's hope that it would flourish and provide a centre for forestry in the southern hemisphere.

Figure 8. The Tokai forestry graduates, 1911.

Standing, left to right: A. E. Gower (caretaker, Tokai), H. J. Sankey, Vernon. Sitting, second row: A. J. O'Connor, Vernon, L. E. Taylor, A. Clarke. Sitting, front: G. A. Wilmot (District Forest Officer and Lecturer), John Spurgeon Henkel, E. J. O'Connor, and J. D. M Keet.

Source: George Museum; photographer unknown.

---

97   The same is true of botany, which saw the creation of a national herbarium in Pretoria. See the argument of L. van Sittert, 'Making the Cape Floral Kingdom: The Discovery and Defence of Indigenous Flora at the Cape ca. 1890–1939', *Landscape Research*, 29 (2003): 113–29, 120–1. White South Africans in the twentieth century increasingly identified with 'indigenous' plants. For key works see Pooley, 'Pressed Flowers'; S. Dubow, *A Commonwealth of Knowledge: Science, Sensibility, and White South Africa 1820–2000* (Oxford: Oxford University Press, 2006), 181–4.

98   Union of South Africa, *Report of the Chief Conservator of Forests for the Year ended 31st December, 1911, including Report on Railway Sleeper Plantations for the Same Period* (Cape Town: Government Printer, 1912), 1.

The Internationalisation of South African Forestry Education: 1911–1932

The new Union-wide Forestry Department determined to recruit its upper-level forest officers from South African students who studied abroad, initially from Rhodes Scholars who studied for Oxford University's diploma in forestry.[99] Two Rhodes Scholars were already studying at Oxford in 1910, whereas no students from anywhere other than the Cape enrolled for the South African College's 1911 session, a fact that helped to bolster the argument to close down the Cape Town school.[100] The small size of the classes at the South African College made its closing easier. The buildings and arboretum at Tokai continued to provide non-officer foresters the experience of local conditions required to pursue the creation and tending of plantations of exotic species.

At the closing of Tokai, the government decided to send its future forestry researchers overseas to study at leading forestry centres before returning to work in South Africa. Would-be trainees applied to the Office of the Public Service in order to be accepted for sponsorship as trainees for the Forestry Department. The government supported both English and Afrikaans-speaking applicants. Many applicants already had received a Bachelor degree in science from South African universities, such as Stellenbosch or the University of Cape Town, but others had not received university-level education; in the latter case, the candidate needed to undertake government-sponsored study at a South African university before proceeding overseas on graduation. On the basis of an employment contract with the Forestry Department, trainees went overseas for advanced degrees in forestry. Students were required to study key subjects, such as silviculture and forest management, and were encouraged to read widely in foreign literature while pursuing novel theoretical and empirical research. They would spend at least a month per year in the field learning 'practical' skills. Postgraduate students had some freedom to choose their specific area of study, although it had to be approved by officials in Pretoria, who closely monitored students' academic progress.

Importantly, the Forestry Department did not make a formal relationship with any single overseas institution. Rather, it created a highly flexible system that fostered intellectual innovation and offered a diversity of educational and environmental experiences. South African students studied at a variety of institutes, the three most important being Yale University, Oxford University, and Edinburgh University, but with some students at others, such as Tharandt and the University of Wageningen. The Forestry Department under the direction

---

99   Ibid., 2.
100   Union of South Africa, *Report of the Chief Conservator of Forests for the Year Ending 31st Dec, 1910*, 2.

of its research leader, Colin C. Robertson, consulted with overseas faculties and domestic foresters and students in order to select the best institution for each candidate. In this way the department was able to send students overseas to match training with the diverse problems of South Africa's forest sector. On their return to South Africa, foresters were directed to locations where their education could be put to best use. This flexibility ensured that South Africa's research officers could bring cutting-edge knowledge to bear on local problems, had personal contacts at leading institutions, and would stay abreast of new innovations internationally. Students studied in a variety of environmental, social, and intellectual contexts, which proved to be a significant factor shaping the research trajectories of research officers.

The department first looked to Oxford University, the host of Britain's most distinguished forestry program. The Oxford forestry program, founded in 1905, was directed by esteemed Indian foresters: first William Schlich, and then from 1920, Robert Troup. In 1924, an Imperial Forestry Institute was opened at Oxford to train forestry recruits from different parts of the British Empire. Oxford teachers sought to act as the centre of a vast web of empire forestry. Oxford's teaching staff trained foresters to manage a diverse variety of forests throughout the British Empire. A great focus was placed on learning European principles of silviculture and management as they had been developed in British India, where most of its staff had worked previously.[101] Other influential Oxford teachers included R. A. Fisher and Sir Harry Champion, who had served in the Indian Forest Service from 1915 to 1939, primarily at the Forest Research Institute at Dehra Dun. The most influential South African trainee to study at Oxford was Christiaan Wicht, although other leading foresters such as W. E. Watt and C. E. Duff also qualified there.

The other British school of choice was Edinburgh University, where BSc forestry degrees were granted from 1907. Edward Percy Stebbings as Professor of Forestry in 1920 invigorated the school. Stebbings was the former Inspector General of India, author of the classic three-volume *History of Forestry in India*, and an advocate of massive tree-planting campaigns. Stebbings had a chequered career as a forester. He distinguished himself in India by coming to grips with the history of the subcontinent, but he marred his reputation by calling for an Anglo-French tree-planting effort in the Sahel to halt the Sahara desert, which he perceived to be encroaching into savanna. Stebbings was known for training students 'of a more practically minded type'.[102] At Edinburgh, students also

101   P. Anker, 'Ecological Communication at the Oxford Imperial Forestry Institute', in C. Folke Ax, N. Brimnes, and N. Thode Jensen (eds), *Cultivating the Colonies: Colonial States and Their Environmental Legacies* (Athens, OH: Ohio University Press, 2011), 278.

102   M. Anderson, 'An Outline of Possible Developments in Higher Forestry Education', *Commonwealth Forestry Review*, 31 (1952), 196.

studied under Isaac Bayley Balfour and Frederick Orpen Brower, two of Britain's most influential botanists. Their perspectives, based on the Danish founder of the field of ecology, Eugenius Warming, proved inspirational for the forest ecologist John Phillips.

Yale was an equally important institution. Yale University's program focused primarily on North American conditions, but its professors had strong linkages to Europe, Southern Africa, and Asia. Yale's forestry program started in 1900 as a result of a generous endowment by the wealthy Pinchot family, becoming the flagship program for North American forestry. Its professors, deans, and graduates directed US Forest Service policies for much of the first half of the twentieth century. The faculty at Yale included James Toumey, the country's leading expert in forest botany and ecology of forest regeneration. Toumey acted as an important mentor for South Africans studying silviculture at Yale. Other leading silviculturists included Henry H. Chapman, an expert on the role of fire in the regeneration of southern pines. Foresters who studied under Toumey, along with other leading US foresters, included C. C. Robertson, Eardley Wilmot, Ian J. Craib, and J. J. Kotzé.[103] As a result of their education, Yale-trained foresters contributed to new, often controversial silvicultural methods from the 1900s to the 1950s.

South African students who studied abroad played a guiding role in the evolution of research and forest policy from the 1930s to 1960s. Although Tokai graduate A. J. O'Connor made a fundamental contribution to silviculture through his design of the so-called CCT trials (see pp. 102–103), foreign-trained students had the most prominent influence on South African research agendas from the 1930s to the 1960s.[104] Three foresters, in particular, shaped key understandings of ecology, silviculture, and hydrology that formed the basis of South Africa's broader forestry policy from the 1930s onward: John F. V. Phillips (BA in botany and forestry and PhD botany, Edinburgh), Ian J. Craib (BA and MSc, University of Cape Town; MSc and PhD forestry, Yale) and Wicht (BA geology and botany, Stellenbosch; MA forestry, Oxford; PhD forestry, Tharandt).

Their diverse experiences abroad shaped how they each responded to the local problems and national context which they faced on their return to South Africa in the 1920s and 1930s. Phillips was the first trained ecologist to be employed by the Forestry Department. His work in the Knysna forest led him to become a spirited critic of many Forestry Department policies, especially the formation of exotic timber plantations. Craib, working on the contrasting situation of

---

103   See Roberton and Wilmot in, *Yale Forest School, Biographical Record of the Graduates and Former Students of the Yale Forest School; with Introductory Papers on Yale in the Forestry Movement and the History of the Yale Forest School* (New Haven: Yale University Press, 1913), 136, 169.

104   Others who had major influence were Nils B. Eckbo and J. M. Turnbull, recruited directly from overseas, who worked on wood properties.

plantation silviculture, helped to improve the growth-rate and yield of private wattle and government pine plantations during the 1930s. Rather than working in indigenous forests, as did Phillips, Craib worked to develop exotic forests. Wicht pioneered South Africa's forest hydrology program at Jonkershoek from the mid-1930s to the 1960s. Wicht's post-1935 research agenda grew directly out of the conflict that arose between critics of plantation forestry, such as Phillips, and advocates, such as Craib.

Conclusion

South African foresters opened the first school of forestry in Cape Town and Tokai in 1906 because they believed deeply in the importance of training officers in local environmental conditions. But the realities of inter-colonial funding and the migration of power away from the Cape towards the Transvaal after 1910 meant that these ideals could not sustain the day-to-day operations of the school. The Cape and its Anglo leadership no longer dominated South African forest policy and education after 1910. When Hutchins left the school, the South African College faculty started to complain about the leadership of the school. The criticism by foresters in the Transvaal and the withholding of students and monies by both the ORC and Natal relegated the school to a slow, steady decline. The creation of a single Forestry Department in Pretoria after 1910 provided the final blow to the Cape Town and Tokai School of Forestry. For the next 20 years, the Forestry Department drew many of its higher officer classes from the former graduates of Tokai, from students who worked their way up the ranks from being managers in the field (after studying at the practical course in Tokai), and from the small group of South Africans who studied abroad.

The closing of the school in 1911 highlighted South Africa's unique political position as a nascent nation and a colony in the British Empire. Government ministers and leading foresters decided to rely upon a British world network, the Rhodes Scholarship, to educate many of its officer class. The Union-wide forestry department was still largely 'British' in the 1910s. But power slowly began to shift to the north. The Afrikaner political elite in the Transvaal had to support any forestry education program that hoped to survive. When a new school of forestry opened in South Africa in 1932 it was under the leadership of E. J. Neethling at a staunchly Afrikaner university, Stellenbosch University, and not at the Anglophone University of Cape Town.[105] With the joining of the Department of Forestry into the joint Department of Agriculture and Forestry in 1934, South African forestry became even more integrated with Afrikaner

---

105   In 1918 the South African College moved and became the University of Cape Town (UCT). The original campus in Gardens still is a functional part of UCT's campus.

agricultural politics. The closing of the school in Cape Town and Tokai signalled key shifts in power and geography that characterised the rest of the twentieth century—foresters in the Cape influenced, but did not direct, South Africa's future forestry policies.

The School for Foresters that found its home in the Tokai facilities from 1911 to 1932, and then transferred to Saasveld (again, after lengthy wrangles among officials about a preferred location), provided the training for the field forest managers—foresters, as opposed to forest officers—who saw to the secure management of each state (or private) forest, i.e. the planning and execution of forest operations, law enforcement, and the management of employee welfare. Without this corps, little of the forest policy for South Africa would have been feasible. With the establishment of the Department of Forestry at the University of Stellenbosch in 1932,[106] a relationship between the two institutions developed that saw many holders of forestry diplomas continue at Stellenbosch, to graduate and postgraduate levels, and assume key roles in South African forestry.

---

106   It was established, after several hesitant attempts by the university, through the efforts of F. E. Geldenhuys and on the direction of General J. C. Kemp, Minister of Agriculture and Forestry, following the recommendation of a consultative conference of government forest officers in Pretoria in 1931 and much debate in the University Senate: R. C. Bigalke, 'Fakulteit van Bosbou 50 Jaar Oud', *Forestry News*, 1/82 (1982), 1–3; Keet, 159.

Chapter 4
Afforestation: Politics, Labour, and Science, c. 1910–1935

After political Union in 1910, the newly formed Union government decided to create a single Forestry Department located in Pretoria, the new administrative capital of South Africa. The department drew on the 'system in force in the Cape Colony', which was then 'applied throughout' the country.[1] English-speakers from the Cape filled the upper echelons of forestry, with less-educated as well as educated younger Afrikaners often taking up junior or technical positions.[2] The creation of a forestry department paralleled the establishment of a number of other new departments and institutes created after 1910, including the Department of Agriculture (1911), Geological Survey of South Africa (1910), South African Institute for Medical Research (1912), and the Meteorological Office.[3]

Afrikaner politicians in the ruling South African Party (1911–1924) and then the Nationalist–Labour Pact government (1924–1934) both embraced forestry as a means to solving national economic, social, and environmental problems. This chapter focuses on how the policy of afforestation was used to aid in solving these problems. Ministerial decisions from 1916 onwards amplified the social dimension of afforestation, decisions that challenged the economic orientation of existing afforestation policies. A major shift occurred when in

---

1   Union of South Africa, *Report of the Chief Conservator of Forests for the Year Ended 31st December, 1910* (U.E. 30-1911), 1, 27.
2   Joseph Storr Lister became the Chief Conservator, while Charles Legat and K. A. Carlson, respectively, remained the heads of the Transvaal and Orange Free State. A. W. Heywood became the Conservator for Natal; J. D. M. Keet and E. J. Neethling were among early Afrikaners with qualifications in forest science, but many came through the Tokai school for foresters; see Chapter 3.
3   S. Dubow, 'A Commonwealth of Science: the British Association in South Africa, 1905 and 1929',
in S. Dubow, (ed.), *Science and Society in Southern Africa* (Manchester: Manchester University Press, 2000), 77.

1922 the South African Party, led by Jan Smuts, decided to undertake a policy of rural white labour settlements in order to solve the 'poor white problem', and to alleviate political pressure relating to the 1922 Rand Strike by white miners. On 25 May 1922, Smuts announced the settlement policy to parliament, emphasising the 'afforestation colonies' as well as irrigation projects and railways work.[4] The election of Hertzog's Nationalist government in 1924 saw forestry become an even more important part of social engineering for nationalist Afrikaners. Foresters were required to accommodate poor white workers at a time of fiscal austerity and increased afforestation. This new focus in departmental employment preferences for poor whites as forest workers—as opposed to African and coloured workers—led to considerable disquiet among foresters about the scheme.

Forestry as social engineering reached its peak in 1931 when General J. B. M. Hertzog appointed F. C. Geldenhuys, an economist without training in forestry, to direct and reorganise the Forestry Department to implement nationalist policies more fully. This move destroyed morale and threatened the unique intellectual trajectory that South African foresters had embarked on since the 1880s. His reorganisation of the department was deeply resented by forestry officers, including those who spoke Afrikaans and came from what may be described as an Afrikaner background.[5] Foresters were furious at the interference. Their titles were changed from Conservator to Inspector, a term that they interpreted to undermine their proper roles and be a demotion in rank. Geldenhuys's attitudes towards forestry differed greatly from those of his officers. He argued that in overworked indigenous forests, 'productivity could be very greatly increased by cutting out all the worthless species and planting up with exotics'.[6] This situation progressively undermined the proven institutions in state forestry, until Geldenhuys was removed following the report by a Commission of Enquiry appointed by Cabinet to investigate the state of affairs in the department. In 1934, the department merged into a new Department of Agriculture and Forestry with J. D. M. Keet as Chief of the Division of Forest Management.[7]

4   Grundlingh, '"God Het Ons Arm Mense Die Houtjies Gegee"', 43; T. R. Roach, 'The White Labour Forest Settlement Program in South Africa 1917–1938' (MA Thesis, University of the Witwatersrand, 1989), 284, 106.
5   Conservators expressed a 'unanimity' of criticism for the scheme, which was hastily implemented without full consultation. See Legat to G. D. Mentz, 21 November 1931, PSC 4/6, SDK 37, NASA-P.
6   Union of South Africa, *Annual Report of the Forest Department For the Year Ending 31 March 1931* (Pretoria: Government Printer, 1932), 14.
7   G. D. Mentz to P. Viljoen, 9 October 1934, PSC 4/6, SDK 37, NASA-P. Keet, 'Historical Review of the Development of Forestry in South Africa', 78, 82–3; H. B. Stephens in Olivier (ed.), *There is Honey in the Forest*, 148–9.

The rapid expansion of state plantations as a result of white resettlement policies created controversy outside of forestry. Farmers downstream from upland plantations complained bitterly about the progressive desiccation of streams and rivers as a result of tree planting. The well-trodden argument that trees encouraged rain and conserved water faced renewed criticism, especially in north-eastern Transvaal and the south-western portions of Cape Province where tree planting occurred near farms. These criticisms, as discussed in this chapter and the next, formed the basis for the eventual creation of the Jonkershoek Forest Influences Research Station in 1935, a program that can be traced back to an initial (failed) attempt in 1911 to measure the influence of tree planting in the Transvaal.

Afforestation: Overcoming the obstacles

Foresters understood South Africa's most pressing forestry problem after 1910 to be its severe shortage of softwood. Whereas foresters prior to the 1900s experimented widely with many species of trees, for diverse purposes, they began to focus on sawlog forests during the first decade of the twentieth century as government determined that its priority should be to create a secure domestic supply of sawtimber.[8] In 1904, Hutchins had written that the 'country must have pine plantations. The present importation of pinewood to South Africa must considerably exceed in value a million pounds sterling', a sum that was about half the export revenue from wool.[9] Again, in 1923, the Drought Commission noted that imports of timber and paper products accounted for more than half the value of South Africa's imports in the period 1918–1922, an annual average for the four years of just over £3 million,[10] the same value as in 1934, when timber imports amounted to 22 million cubic feet (about 622,000 cubic metres, of which 90 per cent was softwood lumber) at a cost of £3 million.[11] Given the decline of exports from South Africa in the 1930s, the country's considerable shipping costs, and its trade imbalance, this was a heavy burden from a product that could be grown domestically.[12] The imbalance further reinforced the perception of a South African economy with an excessively weak agrarian sector, little employment potential, and dangerous dependence on mines and minerals.

---

8   C. E. Legat, 'The Cultivation of Exotic Conifers in South Africa', *Empire Forestry Journal*, 9 (1930): 32–63, 6.

9   D. E. Hutchins, *The Cluster-Pine at Genadendal: Spreads Self-Sown*, Union Bulletin No. 4 of 1904 (Cape Town: Cape Times, 1904), 7; Beinart, *The Rise of Conservation in South Africa*, 14.

10   Union of South Africa, *Final Report of the Drought Investigation Commission*, 67.

11   Union of South Africa, *Proceedings 4th British Empire Forestry Conference* (Pretoria: Department of Agriculture and Forestry, 1935), 99. Domestic production of sawlogs was one million cubic feet, equivalent to about 400,000 cubic foot sawn.

12   K. Burley, *British Shipping and Australia 1920–1939* (Cambridge: Cambridge University Press, 1968), 40, 73–5.

The new Forestry Department inherited a patchwork of land with diverse veld and forest types amounting to nearly 650,000 hectares in area.[13] It controlled most of the country's indigenous closed-canopy forests and had a considerable range of grassland, shrubland (Cape fynbos), and woodland scattered with indigenous trees. Much of the country's plantations were in private hands, located in Natal (mostly wattle) and the Transvaal (eucalyptus and wattle), and grown for wattle bark and mine supports, but not for sawtimber.[14] The aggregate of government plantations amounted to 12,167 hectares, almost all in the Cape, with a few newly established ones in the Transvaal, Natal, and a smattering in the eastern Free State. The timber harvested from the state forests (both from indigenous forests and plantations) that year was only around 28,000 cubic metres, while imports were nearly 300,000, or 10 times domestic production.[15]

Soon after Union, Lister, the first Chief Conservator for the Union, called for a long-term, large-scale tree-planting program to augment the country's limited timber supply and to substitute for imports. He proposed this to the Commission to Inquire into the Conditions of Trade and Industry in 1912. The commission agreed and recommended 'the necessity of putting afforestation equal with agriculture ... if this country is to develop as it should'.[16] The need sharpened with the onset of World War I in 1914: timber imports 'collapsed' by 50 per cent or more from immediately before the war, and domestic prices of timber 'rocketed'.[17] This had the dual effect of spurring planting on and allowing the profitable sales of the department's first harvests of plantation-grown timber, derived from the railway plantation at Worcester and the 30-year-old *Pinus radiata* (then called *P. insignis*) from its Tokai plantation in 1918 and 1919.[18] The Conservator of Western Australia, Charles Edward Lane Poole, congratulated Legat on the success, describing the profit as 'magnificent', a sincere statement on his part given the difficulty in establishing plantations elsewhere in the British Empire.[19] The sale proved that, in certain circumstances, domestic plantations could justify the investment in the long-rotation sawlog regime.

---

13   For a review see, *Report of the Chief Conservator of Forests for the Year Ended 31st December, 1910* (U.G. 30-1911), 2–18.

14   W. D. Reekie, 'The Wood From the Trees: Ex Libri ad Historiam Pertinentes Cognoscere', *South African Journal of Economic History*, 19 (2004): 67–99, 73–5.

15   C. E. Legat, testimony to the Commission. See *Royal Commission*, 286–7. Quote from the Commission in Sim, *Treeplanting in South Africa, Including the Union of South Africa, Southern Rhodesia, and Portuguese East Africa*, 21.

16   Roach, 'The White Labour Forest Settlement Program in South Africa 1917–1938', 90.

17   Reekie, 'The Wood from the Trees', 76–7; M. Grut, *Forestry and Forest Industry in South Africa* (Cape Town: A. A. Balkema, 1965), 6–7.

18   Keet, 'Historical Review of the Development of Forestry in South Africa', 67–8; Grut, *Forestry and Forest Industry in South Africa*, 7.

19   Legat to Lane Poole, 24 January 1919, A988, FOR 312, NASA-P; Lane Poole to Legat, 10 April 1919, A988, FOR 312, NASA-P. The only other British colony with as advanced a plantation program was South Australia, which started establishing *P. radiata* plantations in the 1870s.

Figure 9. Pit sawyers, Amatole region, Eastern Cape Conservancy, 1906.

Timber from this source supplied, among others, Ballantyne's Wagon Works in nearby Keiskammahoek, for a time the largest wagon builder in South Africa. The logs are possibly of Hard Pear (*Olinia ventosa*).

Source: George Museum; photographer unknown.

Very soon after the establishment of the new department, in 1912, Lister created a central Research Section within it, based in Pretoria, to focus especially on finding and growing suitable exotic trees.[20] Robertson had pointed out in 1910 that despite the extensive research on exotics that had been pursued in South Africa, 'much of it has been of a comparatively small value, owing to lack of central organisation and continuity of direction, insufficient scientific method, incomplete records and unsuitable silvicultural treatments'.[21] Compounding this lack of silvicultural knowledge, scientists had not yet described or analysed the climate or geology of the Transvaal, Orange Free State, and Natal.[22] There was

20   Union of South Africa, *Report of the Conservator of Forests for the Fifteenth Months' Period ending 31 March 1914* (U.G. 5-13), 34; see also C. L. Wicht, 'Forestry Research – Elixir of the Industry', *South African Forestry Journal*, 53 (1965): 7–13.
21   C. C. Robertson, 'Some Suggestions as to the Principles of the Scientific Naturalisation of Exotic Forest Trees', *South African Journal of Science*, 6 (1910): 219–30, 219.
22   See Hutchins, *Transvaal Forest Report*, 136, 9; The first major study of rainfall was T. E. W. Schumann and J. L. Thompson, *A Study of South African Rainfall: Secular Variations and Agricultural Aspects*, University of Pretoria Series No. 1, 28 (Pretoria: Van Schaik, 1934).

no accurate Union-wide vegetation map.[23] Hydrological research had only just begun to assess the volume and flow of rivers and catchments. The hydrology of many rivers was poorly known.[24]

Despite the encouraging sales of the Worcester eucalypts and the Tokai pines, many scientific and technical problems stood in the way of creating successful future pine plantations. Early trials, both successes and failures, by Hutchins and other Cape foresters had provided a useful body of experience, but the lists of candidate tree species for new afforestation was long, the identity of species confused and uncertain, the performance of each in the country's widely diverse habitats unknown. A more rigorous system of climatic matching, species and genetic selection, and silviculture was required in order to ensure that a large-scale afforestation drive could succeed.[25] Forest researchers found it extremely difficult to predict growth of pines that produced suitable sawlogs. Uncertain choice of species and poor matching of species to site conditions resulted in unsatisfactory, unprofitable yields, despite sometimes spectacular growth rates on well-matched sites. But even where species were well-matched to a site, growth though rapid varied erratically among individual trees, with many trees unusable because of crookedness or other quality defects relating both to silviculture and to their genetics.[26] Foresters in South Africa (as well as globally) lacked knowledge of the genetic provenance of species—the problem of seed source—and how provenance affected performance in new environments. They also lacked the knowledge of the ideal spacing and thinning regimes for optimum growth.

Department-led research focused primarily on improving the success of plantations at every stage—from establishment to growth, harvest, and finally, processing. Robertson took charge of the Research Section in 1913 (later with Charles S. Hubbard and John M. Turnbull as assistants).[27] He developed an innovative program of silvicultural experiments and climate research, working later with A. J. O'Connor, who was educated at the Tokai school and author of the uniquely innovative Correlated Curve Trend (CCT) trials for the determination

23   The Botanical Survey of South Africa began only in 1919.

24   Twenty-five stream gauging stations had just become operational by 1905 in the Hydrographic Survey of the Transvaal.

25   Robertson, 'Some Suggestions as to the Principles of the Scientific Naturalisation of Exotic Forest Trees', 220. Sim, *Treeplanting in South Africa, Including the Union of South Africa, Southern Rhodesia, and Portuguese East Africa*, 8–10.

26   For example, Keet emphasised the problem during the 1935 British Empire Forestry Conference, noting that certain plantations of 200–300 acres of a poor strain of a species had produced 'only firewood' and that 'we have now very strict rules regarding this matter of seed collection, and as to the certificates that must be sent in by the foresters as to the origins, the form, the habits, and so on of the trees'; *Proceedings of the Fourth British Empire Forestry Conference*, 157.

27   Anon, 'Charles Hubbard's Distinguished Forestry Career', *Forestry News*, 2/79 (1979): 2–3; John Turnbull was to conduct revolutionary work on the determinants of wood strength in pine species.

of species- and site-specific espacement and thinning regimes (see below).[28] The Research Section actively pursued studies in botany, silviculture, climatology, ecology, genetics, and breeding, and the technological development of timber utilisation from novel forests. Experiments became increasingly professional, and were designed and coordinated centrally as a countrywide program.[29] This program was explicitly farsighted. For example, the CCT experiments were to be maintained for up to 50 years.

The department initiated research in wood technology in 1919 by hiring Nils B. Eckbo, a Norwegian graduate in forestry from Yale and wood technology from the US Forest Service Products Laboratory in Madison, to serve as its Timber Research Officer.[30] When Eckbo arrived in South Africa, domestic timber had a bad reputation on the South African market, despite the positive 1919 sale. He noted that South Africa's domestic buyers did not accept 'colonial' timber— they rejected lumber from early plantation harvests in favour of more expensive imports because, '[local wood] often warps, splits, or shrinks to such an extent as to cause serious losses … which means waste of material, waste of labour, and probably loss of customers'.[31] Eckbo embarked on research in Pretoria to discover how to season and improve plantation-grown timber so that it could be successfully processed and sold in South Africa, working in concert with colleagues such as Robertson and O'Connor, who dealt with genetics and silviculture, and Turnbull, who worked on the effects of silviculture on wood properties and timber quality.

In 1930, J. J. Kotzé, another Yale graduate, was appointed as Chief Forest Research Officer, and Nils Eckbo promoted to Chief Timber Investigations Officer.[32] Kotzé was later to express their philosophy as being that 'silviculture begins … in the Forest Products laboratory', and 'the closest liaison between silvicultural and forest products research' was needed so that '[t]he silviculturalist once assured that the timber produced will be satisfactory, can so regulate his treatment of the trees that the product will be turned out'. The wood technologist, in turn, should not be satisfied to work in the laboratory alone, but he must go to the stands occasionally so as not 'to lose touch with the factors of treatment and environment which have expressed themselves in the material he is called on to examine'.[33]

---

28   A. J. O'Connor, *Forest Research with Special Reference to Planting Distances and Thinning*, British Empire Forestry Conference, 1935 (Pretoria: Department of Agriculture and Forestry, 1935).

29   Robertson, 'Some Suggestions as to the Principles of the Scientific Naturalisation of Exotic Forest Trees', 219; Poynton, *Tree Planting in Southern Africa*, Vol. 2: *The Eucalypts*, 19.

30   Keet, 'Historical Review of the Development of Forestry in South Africa', 69.

31   N. B. Eckbo, *The Seasoning of South African Woods* (Pretoria: Government Printing and Stationery Office, 1922), 1.

32   Keet, 'Historical Review of the Development of Forestry in South Africa', 74–5.

33   *Proceedings of the Fourth British Empire Forestry Conference*, 168.

Working together, they started to coordinate and harmonise an overarching program that soon began to deliver highly novel solutions to the problems of the new plantation forestry.[34] Researchers quickly learnt that timber quality problems could only be solved by improvements throughout the system, from the genotype, through the silvicultural systems employed, and into the sawmilling and seasoning technology. This program was tightly interwoven with the afforestation program through regular joint conferences between researchers and professional managers, so generating the knowledge base for the development of plantations and the products they yielded.[35] The coherent approach to research and development in plantation forests endured for 80 years or more afterward.

Through this approach, officers working for the Forestry Branch from 1912 to 1935 solved many fundamental problems relating to climate mapping, silviculture, species identification, and wood quality. Foresters revived their interest in the genus *Eucalyptus* in the 1920s, growers in the private sector having rapidly expanded plantings of eucalypts. Problems with identification lingered because of their growth forms and because some species hybridised.[36] Exploration of the home-range potential of species suitable for South African conditions, following Hutchins's recommendations in his *Transvaal Forests Report*, began with Robertson's 1907 report to the department, 'Notes on the Trees of Extra-Tropical Mexico', which became his dissertation for his Master's degree at Yale.[37] This led to vigorous exchange in seed collections, of Mexican pines and other species, but provenance—the question of the influence of home-range genetic variation within the introduced species on their performance in South Africa's many sites—remained a problem.[38] To address this, the Minister of Agriculture, Sir Thomas Smartt, chose Robertson to visit Australia in 1924. Robertson toured Australia's forests for six months studying its flora and climate. His report offered an analysis of similarities and differences between Australia and South Africa with recommendations on the habitats and true classifications of species and genera.[39] These two initiatives heralded numerous overseas visits

---

34   This arrangement led to the amalgamation of the two entities to form the national Forest Research Institute in 1956.

35   The first countrywide conference on silviculture was held in 1907, leaving the question of thinning regimes in plantations 'undecided'; a key conference in 1935 resolved the major shift in silviculture to wide espacement and frequent thinning: Keet, 'Historical Review of the Development of Forestry in South Africa', 66, 86.

36   See District Forest Officer Butterworth, FBT 1/3, NASA-CT; Chief Regional Forest Officer Transkei, FCT 3/1/57, NASA-CT; FCT 3/1/60, NASA-CT; FCT 3/1/61, NASA-CT. Foresters worked to create comprehensive guides of *Eucalyptus* species in South Africa. See E. K. Marsh, 'A Key to the Species of Eucalyptus Grown in South Africa', *Journal of the Southern African Forestry Association*, 3 (1939): 16–64.

37   Robertson, 'The Cultivation of Mexican Pines in the Union of South Africa', 2–3.

38   Poynton, *Tree Planting in Southern Africa*, Vol. 2: *The Eucalypts*, 15.

39   C. C. Robertson, *Trees of Extra-Tropical Australia: A Reconnaissance of the Forest Trees of Australia from the Point of View of Their Cultivation in South Africa: A Report of a Tour in 1924* (Cape Town: Cape Times Limited; Department of Forestry, 1926), 1.

by South African foresters that progressively clarified the biogeography of foreign species—pines and other conifers from Europe, Asia and the Americas, and other conifers and eucalypts from Australasia, as well as other genera—and fed an experimental program that steadily improved predictions about which species and genetic varieties would grow best in South Africa, while developing the genetic stocks for further domestication in South Africa.[40]

Building on A. J. O'Connor's novel research design, Ian J. Craib began the research that was to make radical innovations in silviculture that improved the growth rates and yield of wattle. Craib had studied under James W. Toumey, one of America's leading silvicultural experts. The research for his PhD thesis focused on the effects of soil moisture on the growth of trees.[41] Under the influence of O'Connor, Craib experimented in Natal to test regimes that included planting trees with wider spacing and then thinning intensely at a younger age. Traditional plantation management spaced trees closer together and did not thin so aggressively at an early age. In 1934 he published a pioneering paper, 'The Place of Thinning in Wattle Silviculture and its Bearing on the Management of Exotic Conifers'.[42] When it was published, South African foresters saw the paper as an expression of 'some revolutionary views on certain aspects of silviculture'.[43] Craib argued for a fundamentally different management regime—a risky decision when the results of his methods could take 30 years to become apparent, but his regime was adopted into silviculture policy and practice in South Africa at the 1935 conference of forest research officers.

A series of decisions based on Craib's and O'Connor's research from the mid-1930s fundamentally transformed South Africa's pine silviculture. Detailed surveys provided the information needed to match species to site (and later, genotypes within the species). Stands were now to be established by comparatively wide espacement of transplants. Regimes were to be adjusted according to site quality. Foresters planted a single species of pine in a location, spaced the trees widely, thinned early and heavily in order to forestall the competition that set in early with the rapid rates of growth achieved. Successive pruning would prevent loss of strength due to accumulation of knots, and clearfelling harvests were to be on relatively short rotations of around 30 years, to ensure a

---

40   For a compendium of the findings from the program, see the three volumes by Richard Poynton.

41   Toumey's supervision of South African researchers probably had a key role to play in developing the intellectual capital needed for South Africa's novel forestry problems. Whereas with E. J. Neethling the work focused on light as a determinant of forest seedling survival, with Craib the work examined soil-moisture controls on tree growth. Experience of this kind would have informed the discussions with Wicht about the design of forest hydrological research. See Chapter 7.

42   I. J. Craib, 'The Place of Thinning in Wattle Silviculture and its Bearing on the Management of Exotic Conifers', *Zeitschrift für Weltforstwirtshaft*, 1 (1931): 77–108. This publication included observations on the growth and silviculture of pines in South Africa.

43   See A. J. O'Connor's preface in I. J. Craib, *Thinning, Pruning and Management Studies on the Main Exotic Conifers Grown in South Africa*, Forest Research Bulletin (Pretoria: Department of Forestry, 1939).

profitable regime. Instead of deciding when to thin based on visual assessment and judgement of tree condition, they used a quantified system based on formulas from O'Connor's CCT experiments to estimate growth and future yields for given species, site conditions, current densities, and growth rates, and set thinnings accordingly. South African foresters knew that these ideas broke from European orthodoxy. O'Connor, who studied at Tokai and rose to prominence as the Deputy Conservator in the late 1930s, argued that it 'requires courage to abandon the sanction of tradition'. Yet this was necessary in South Africa, where 'our adherence to tradition [in silvicultural methods] has not yielded the expected results'. He reasoned that this perhaps occurred 'because traditional principles meet the needs of the countries in which they were evolved so that there has been no strong urge to examine them'.[44]

This new approach to plantation silviculture reflected the maturation of the department's plantation program. Following Hutchins's direction, every new forest station had its own raingauge installed, and climatic knowledge expanded quickly. By 1930, foresters had indicative experimental results about which pine grew best in different locations, beginning with the results of pre-1910s trial plantations.[45] In 1931, the department published the first silvicultural map of South Africa. The map divided the country into distinct zones according to temperature and rainfall averages, and assigned species for afforestation to zones according to their expected performance.[46] This map provided the means to design and interpret research, and to plan afforestation and silviculture according to the climatic zones of the country. At the same time, foresters kept detailed records of the topography, soil type, and climate at existing planting trials. When combined, the geographical and local data allowed foresters to generate a more specific profile of what species to plant, in any region or site in South Africa.

Afforestation, labour and poor-white resettlement

Afforestation proceeded hand in hand with trial and research, both to overcome obstacles as these arose, as well as to find new ways. Aside from technique and method, finding forest workers for the intensive manual labour involved in afforestation was a vexed problem for South Africa's foresters, with no immediately available solution. In the Cape, some early plantations (such as Tokai, Kluitjes Kraal, and Fort Cunningham—see below) were established by using the labour of probationary prisoners and the inmates of rehabilitation

---

44   O'Connor, ibid.
45   Legat, 'The Cultivation of Exotic Conifers in South Africa', 5–6.
46   Union of South Africa, *Annual Report of the Department of Forestry for the Year Ended 31st March 1931* (U.G. 11/1932), 28.

centres, but mostly foresters relied on a mix of white, coloured, and African workers to do the unskilled work required. Workers tended nurseries, prepared the land (ploughing with oxen, or making planting pits by hand using mattocks), planted trees, weeded, and thinned. The opportunity to hire cheap labour aided public, and most especially, private tree-planting efforts for most of the twentieth century. Compared to other equivalent colonies, such as Australia, South African foresters probably benefitted from having to pay less for labour.[47] South Africa's leading expert on wattle, Ian J. Craib, noted in 1935, 'South African plantations of wattle were only profitable because of cheap labour, indeed, these plantations would not exist without it'.[48]

The employment of forest workers eventually settled into the apartheid system where most employees were contract workers, migrants from diverse 'homelands'. Turbulence in the labour market, following the labour demand and high wages during the South African War and the reconstruction afterward, caused an 'acute labour shortage' in the rural economy of the Transvaal, but the passage of the 1913 *Natives Land Act* soon began to effect a flow of contract labour. In the Sabie area, population increase in the Lowveld, partly caused by evictions from state forest land, caused a gradual shift from the situation where migrant wage earnings supplemented the dominant agricultural income, to a predominance of wage income over agricultural income by the late 1930s.[49] Contract workers for afforestation in the Sabie area would have come from sources such as the 'released' areas under governance of the chiefs or their equivalents, and, later, the communities on Trust land.

Though the department preferred to hire African or coloured labour to establish and manage plantations, various parliamentary decisions from 1916 onward increasingly required foresters to employ poor whites—an onus which bedevilled the afforestation program until foresters succeeded in incorporating a

---

47   See R. N. Parker, *Eucalyptus Trials in the Simla Hills* (Calcutta: Government Printer, 1925), 1. Australia suffered from high labour costs owing to its high-wage White Australia policy. India had 'cheap' labour, but it was felt to be more expensive than that in South Africa.

48   *Fourth British Empire Forestry Conference*, 15.

49   I. Niehaus, *Witchcraft, Power and Politics: Exploring the Occult in the South African Lowveld* (Cape Town, London, and Sterling, VA: David Philip and Pluto Press, 2001), 21–3. Each rural district in South Africa would have had a somewhat different historical response to the developing political economy. See, for example, C. Bundy, *The Rise and Fall of the South African Peasantry* (Berkeley and Los Angeles: University of California Press, 1979). For a history of labour in a Transkei district marked less by coercion than opportunism, see H. de Klerk, 'The Mutual Embodiment of Landscape and Livelihoods: An Environmental History of Nqabara' (PhD thesis, Rhodes University, 2007), 27–9. Also R. Palmer, H. Timmermans, and D. Fay, *From Conflict to Negotiation: Nature-Based Development on South Africa's Wild Coast* (Pretoria: Human Sciences Research Council, 2002). Apartheid measures as well as their 'racial capitalism' progenitors seem to have begun to have their desired effect: a variety of measures caused the freeing up of African labourers for hire in the market as a whole, including the enforcement of the hut tax, the agreement with Mozambique for, among other things, the free recruitment of workers from there (at times, the source of the majority of mine workers) and the 1913 *Land Act*. See, for example, B. Worsfold, *The Reconstruction of the Colonies under Lord Milner*, Vol. II (London: Kegan Paul, 1913), 429, 386–8; Bundy, *The Rise and Fall of the South African Peasantry*, 276.

proper scheme of white settlements into state forestry. All labourers, irrespective of race or ethnicity, faced the reality that forestry was one of the hardest types of work. The white labour settlement scheme drew upon a diversity of destitute people, mainly Afrikaners. Though many succeeded in progressing to a better life,[50] white employees were often poorly endowed. At an inquiry, Legat was asked by the member of parliament Mr Waterston, 'At the present time the Railway and Forests Dept [sic] are dependent upon the class of man who fails in every occupation?' Legat replied, 'Yes, the man who has been unfortunate'.[51]

After the South African War and into the first decade of Union, at the start of the national afforestation program, the market for labour in South Africa was tight and very fluid, affected firstly by the aftermath of high wartime wages and then by the demands of the construction boom that followed.[52] Department officials in the Cape Province and Transvaal tried to use prison labour (probationers from reformatories) in some instances, in order to not compete in localities with farmers in finite labour markets, but the Prisons Department did not always assure sufficient probationers for the department's needs. When the Prisons Department did not give foresters the numbers they requested for Kluitjes Kraal (which was established and had been maintained since 1884 with prison labour and 'inebriate' labourers—alcoholics in rehabilitation—as well as ordinary hire from the labour pool), farmers at Tulbagh protested the department's use of free labour, because the wages of the department, 2s/6d, were much greater than the

---

50 For example, J. C. Heyneke, who had served as conscript with the armed forces during the 1922 Rand Revolt, came from a farming family forced to sell their farm in 1927 at the onset of the Depression; penniless after trying diamond digging, he joined the white labour settlement at Spitskop near Sabie in 1930, set up his own business in 1931, established his own sawmill in 1939, and became a leading figure in civil society. See, *Monumente Langs die Pad van J C Heyneke van Sabie Oos-Transvaal*, Sabie Doprsraad, Sabie Bosboumuseum, Pamphlet No. 51 (26 April 1984), 2. Transcript of interviews with Heyneke conducted by Curator of Sabie Forestry Museum. For a comparable story from Jonkershoek: E. J. Borchardt was quickly promoted to research assistant with C. L. Wicht. J. P. Kleynhans, the forester in charge of Berlin State Forest from 1931 to 1936, regarded the white forest labourers to be generally 'good types', forced into the settlements by the Depression, and names several who progressed to positions of forest and sawmill foremen. '*Herinneringe van die afgelope halwe eeu*', J. P. Kleynhans Sabie Forestry Museum, *Bosboubaanbrekers* Pamphlet No. B 18 (30 January 1981), www.routesmp.co.za/history.html?view=wbsdirmgr&option=com_wbsdirmgr&task=open&fo=1%2520-%2520Mpumalanga%2520Historical%2520Interest%2520Group%252F1%2520-%2520Latest%2520Newsletters%252F (accessed 9 December 2013). The Forestry Schools at Tokai and, later, Saasveld also served as springboards for Afrikaner opportunity (see Chapter 3).

51 Evidence given by Chief Conservator Before Public Accounts Committee, 24 April 1925, E/40-1, FOR 364, NASA-P.

52 For discussion of 'double demand' for labour and labour shortfalls, see Worsfold, *The Reconstruction of the Colonies under Lord Milner*, Vol. I, 376, 264–71; Bundy, *The Rise and Fall of the South African Peasantry*, 276, 208.

1/- plus food and wine (two bottles per day) that local farmers offered.[53] Similar complaints came from national and district agricultural lobbies that worried about the state crowding out the private sector by paying more for workers.[54]

The Union government's rural white labour policy picked up from where the Milner administration had left the problem of relocating Afrikaner farmers and their families from the ruins of war, 200,000 in all, and of reactivating farming.[55] As early as 1907, the Cape government had appropriated funds for employment of woodcutters on afforestation, but the Forestry Department was not ready and little came of this.[56] The first forestry white labour schemes started in 1916, when poor white woodcutters from Knysna were settled at Jonkersberg and Franschhoek, and in 1918, the government made an 'increased provision of funds for afforestation and for employment of white relief labour'.[57] But progress with these two settlements was slow.[58] Legat challenged the plan by pointing out that white labourers worked less efficiently, and for more money, than African or coloured workers.[59] He asked parliament (unsuccessfully) to make special dispensations for employing white labourers because white labourers cost approximately two to three times more than African, coloured, or penal labourers, while their work output was often lower.[60] Concerns about the higher cost of white labour continued throughout the 1920s.

In the early 1920s, the department's afforestation program became firmly enmeshed with government attempts to solve the 'poor white' problem. This issue had deep roots, as Anthony Minnaar describes:

> The unemployed rural poor, who accounted for the largest portion of the so-called 'Poor Whites', had been a problem for many years ... This problem was

---

53   Legat to Minister for Agriculture, 9 January 1922, A98/5, FOR 112, NASA-P. Twenty-four farmers from the Tulbagh area petitioned the Administrator of the Cape of Good Hope on 15 July 1910 against the department's employing 'free' workers; A98/5, FOR 112, NASA-P.

54   Legat to Private Secretary for the Minister of Agriculture, 29 February 1928, E/40-1, FOR 364, NASA-P.

55   Louw, *The Rise, Fall, and Legacy of Apartheid*, 12.

56   Keet, 'Historical Review of the Development of Forestry in South Africa', 63.

57   Ibid., 67.

58   Ibid., 69.

59   Ibid., 67. In 1917, Legat noted that in the Cape, 'The Coloured community, who live side by side with the whites, and are to a very large extent debarred from work in the forests, except on occasions when sufficient white labour is not forthcoming, and who in consequence labour under very heavy disabilities, are steadily rising as compared with many of their poor white neighbours. It may be that the disabilities under which the Coloured population suffer are bringing to the surface the best qualities in the individual. The fact, however, remains that notwithstanding their disabilities they are slowly creeping into posts requiring more or less skilled labour, and that they are far more ardent cultivators of their lands and appear to be generally more successful in their agricultural operations than the poor white who still continues to look to the forest for his main source of income, and to the Government for doles'. See Union of South Africa, *Annual Report of the Forest Department For the Year Ending 31 March 1919* (Pretoria: Government Printer, 1919), 1.

60   Ibid., 2.

tied to that of a rural exodus to the urban centres, and in fact during the Great Depression the Poor White problem became more one of white unemployment and the two problems were often regarded as one.

... Besides the drift to the urban areas of the rural poor the ranks of the unemployed were swelled during the Great Depression years by the thousands of small white farmers and their labourers driven off the land and into acute poverty by the combined economic effects of the Great Depression with its prevailing low prices for agricultural products. It was also a period of prolonged drought (the so-called 'Great Drought' of 1931-33), largescale foot-and-mouth disease outbreaks, countrywide locust infestations and in 1934 widespread floods—all of which impacted negatively on the levels of employment in South Africa. In addition, the almost total collapse of the diamond market led to an influx of large numbers of small independent diamond diggers and their families ...

The stream of those leaving the farms to find other employment, mostly without the requisite training, had, in the course of the Great Depression, become so large that the urban areas found it difficult to absorb them ... This movement to the towns had swelled as the countrywide drought worsened ... There were persistent calls on the government to discourage this urban drift.[61]

In 1922, the department committed to planting 10,000 acres per year, a target that required a large labour force.[62] The expansion of plantations fitted neatly into efforts to solve the 'poor white question', an issue that became an important political platform of Afrikaner politicians in the 1920s and 1930s as they campaigned in the interests of poor white voters. The program also reflected wider concerns about racial mixing and the abject poverty of many of the poorest Afrikaners. Vocal Afrikaner nationalists expressed concern that racial mixing, especially in Knysna, undermined white rule.[63] Politicians turned to poor white resettlement and labour schemes as a way to pull whites out of poverty and to demarcate racial boundaries which they believed had become blurred by impoverished economic conditions, while foresters—leading figures among whom were sympathetic to the plight of the woodcutters—began to see settlement programs as helping to solve the Knysna woodcutter problem. The government pressed foresters to take on white labour settlements the moment new land had been purchased for afforestation, and afforestation plans and procedures suffered as a result.[64]

---

61   A. Minnaar, 'Unemployment and relief measures during the Great Depression (1929–1934)', *Kleio*, 26 (1994): 45–85, 46, 48.
62   Union of South Africa, *Annual Report of the Forest Department For the Year Ending 31 March 1925* (Pretoria: Government Printer, 1925), 1.
63   S. Dubow, *Scientific Racism in Modern South Africa* (Cambridge: Cambridge University Press, 1995), 171–2; Freund, 'South Africa: The Union Years, 1910–1948—Political and Economic Foundations', 227.
64   Roach, 'The White Labour Forest Settlement Program in South Africa 1917–1938', 53–4.

Whereas prior to 1922, most white forest settlers came from the George–Knysna and the agrarian Orange Free State and Transvaal, an expansive program of white resettlement and employment preferences developed in the early 1920s, as a result of a series of mining strikes in the Rand that threatened national political stability.[65] At the heart was a conflict between higher-paid, skilled white miners and the unskilled, impoverished Afrikaners drawn to the mines from a rural economy devastated by drought, the Rinderpest and war, having to accept 'very low' wages or remained unemployed; a 'situation rotten to the core', in Jan Smuts' words. Also, skilled white workers resented being undercut by Africans working for lower wages as the mining houses reduced production costs. South Africa's mining industry at the time was in a deep recession brought on by the fall of gold prices. Union-led agitation culminated on 2 January 1922 in a miners' strike, led mainly by English speakers but involving many Afrikaners, and which, though quickly suppressed by the authorities, had profound consequences. When negotiations between strikers and employers broke down in February 1922, strikers proclaimed a general strike on 6 March. Revolt by 10,000 armed and organised strikers followed soon after. Revolutionaries among the strikers shut down the city and besieged key infrastructure, such as police stations. Smuts called out the army and air force, and applied martial law to put down the insurrection. Over 200 people were killed before the strike was lifted. In the aftermath, 15,000 lost their jobs on the mines.[66]

As a way to alleviate further conflict (and the problem was soon aggravated by the Depression), the Smuts government created a white and coloured (but not African) labour bargaining system and developed a policy of preferential white labour employment in the rural sector. The rural program would employ whites, who included woodcutters but also many from recent migrants to the cities, and returned soldiers. Smuts and the South African Party lost at the 1924 election to the Nationalist-Labour Pact, which placed General J. B. M. Hertzog in power as South Africa's Prime Minister from 1924 to 1933. The Hertzog government pursued the white agrarian program with greater intent, with the initial program aimed at including forest stations as training grounds to move poor whites onto permanent agricultural settlements.[67]

---

65 Giliomee, *The Afrikaners*, 328. Though the broad outlines are well known, the specific details of the causes and consequences of revolt are the subject of an ongoing historiographical debate. See J. Krikler, *White Rising: The 1922 Insurrection and Racial Killing in South Africa* (Manchester and New York: Manchester University Press, 2005); J. Krikler, *The Rand Revolt: The 1922 Insurrection and Racial Killing in South Africa* (Johannesburg: Jonathan Ball, 2005); K. Breckenridge, 'Fighting for a White South Africa: White Working-Class Racism and the 1922 Rand Revolt', *South African Historical Journal*, 57 (2007), pages; P. Maylam, 'The Rand Revolt', *Journal of African History*, 47 (2006), pages.

66 Freund, 'South Africa: The Union Years, 1910–1948—Political and Economic Foundations', 225ff.

67 Union of South Africa, *Annual Report of the Forest Department For the Year Ending 31 March 1925*, 1.

Although white settlers were never the majority of unskilled employees in the department (in 1934, at its height, the program employed 1,354 white forestry settlers, 18 per cent of the establishment of 7,577 unskilled employees),[68] foresters and other technical experts chafed at being burdened with a political mandate that strained finances as well as their operational capacity. The Department of Forestry, the Department of Railways and the Department of Irrigation expressed concerns about the inefficiency of Hertzog's scheme. Legat supported white settlement in principle, but argued that it ran contrary to the original terms of afforestation policy, namely that plantations should return a profit. The costs of white labour meant that afforestation effectively became a social rather than an economic issue.[69] Legat wanted the government to make this an explicit policy, so as to relieve the department from criticism for failing to make its plantations profitable when they could not. The attempt by English-speaking officials did little to convince Hertzog or other Afrikaner politicians who sought to uplift their constituents. The argument that white labour cost too much bore resemblance to the policies used by English-speaking mining magnates before the Rand Strike of 1922.

In 1923, the department transferred Johan Diederik Möhr Keet from his post as District Forest Officer for the Knysna District, to Pretoria in the post of Conservator of Forests for Transvaal and Orange Free State, and from this time on, Keet played a central role in the afforestation program. Keet was born in 1882 in the Afrikaner community of the small agricultural town of Ceres in the Cape Colony, into a family descended from German settlers at the Cape of 1742. Little is known of his youth; he matriculated at the age of 18, and somehow offended the authorities during the South African War, since he was interned by the British authorities for a year during the war (probably in the camp at Tokai, where recalcitrant Cape Afrikaner civilians were held). After his release, he worked for a time on a farm and then entered the South African School of Forestry at Tokai in 1906, receiving his diploma *cum laude* two years later; his record there includes commendation for his interests in nature, and especially trees. (Keet's interest in botany and ecology was lifelong; he discovered species new to science, had a genus in the coffee family, *Keetia*, as well as a wood-rotting fungus, *Trametes keetii*, named after him, and he had many tales to tell of traditional knowledge of plants and timber.) He took up an appointment with the Cape colonial administration in 1908. In 1931, he was appointed Chief of

---

68   Keet, 'Historical Review of the Development of Forestry in South Africa', 84. By 1949, there were just 58 white labourers in the settlements, while daily-paid employees numbered 17,453. See ibid., 58. By 1933, the employment level of 1,340 was just 5 per cent of the total in government unemployment relief schemes. See Minnaar, 'Unemployment and Relief Measures During the Great Depression (1929–1934)', 54, 65.

69   Legat to Minister of Agriculture, 19 September 1925, E/40-1, FOR 364, NASA-P. Also see Reekie, 'The Wood from the Trees', 81.

the Division of Forest Management under the direction of F. C. Geldenhuys, and then as Director of Forestry in 1934. From 1923 he was at the centre of all the strands in the development of forest policy, science, and practice in South Africa, until his retirement in 1942. This included the drafting of the *Forest Act* of 1941, and after retirement, the *Soil Conservation Act* of 1946. As Conservator of Forests and after, he was charged with the largest public afforestation program in the country's history, and his energy and endurance in the field and at his desk was legendary.[70] Keet, closely acquainted with the woodcutter communities (see below), involved himself directly in the development of the white labour settlements.

During 1929, Keet worked with a team including the architect of the Public Works Department, the government health officer, and the welfare officer, touring the Sabie projects, planning the new settlements. 'This was the first time in the history of the program that individuals from these Departments had met before the establishment of a settlement, examined … and together, chosen a site'. Keet reviewed the plan, making adjustments to fit local conditions, and from this came an integrated village plan for the Tweefontein settlement, with a tree-lined avenue, water supply plan, and the forester to live within the community. The Tweefontein design was then copied for other later settlements.[71]

From 1925 to the early 1930s, the forest settlement program expanded rapidly. The poor whites who worked for the Forestry Department came primarily from rural rather than urban areas. These included failed farmers and displaced woodcutters from the Knysna forests, whose numbers with access to the forests for timber extraction had been steadily reduced after decades of conflict, with the last leaving Knysna in 1939.[72] By 1925, 1,074 families had been settled, with accompanying settlement support staff.[73] They lived in villages

---

70   See 'Die eregraad M.Sc. in Bosbou aan Mnr. J.D.M. Keet' and Anon, 'Voorstel om Toekenning van die Grad D.SC. in Bosbou *Honoris Causa* aan Mnr. J.D.M. Keet', attachments to the *Minutes of the Senate of the University of Stellenbosch*, 52, 21 October 1964 – 26 March 1964, in the case of the latter document, unclassified box for the former, University of Stellenbosch Archives; 'Forestry loses one of its greats', *Forestry News*, 4/76 (1976): 1–2. Anon, 'A tribute to the late Dr J.D.M. Keet', *South African Forestry Journal*, 101 (1977): 1–2; accounts of tradition told to Fred Kruger, c. 1964–1965.

71   Roach, 'The White Labour Forest Settlement Program in South Africa 1917–1938', 175–6, 178. These settlements schemes provided social conditions and opportunity for advancement that stood in stark contrast to those encountered by most poor whites in the Transvaal at the time who, frequently displaced in large numbers by more efficient and thrifty African peasants, were forced onto Lowveld farms, often living in 'hovels' and beaten down by poverty and humiliation. See R. Packard, '"Malaria Blocks Development" Revisited: The Role of Disease in the History of Agricultural Development in the Eastern and Northern Transvaal Lowveld, 1890–1960', *Journal of Southern African Studies*, 27 (2001): 591–612, 599–603; Bundy, *The Rise and Fall of the South African Peasantry*, 209–10.

72   Grundlingh, '"God Het Ons Arm Mense Die Houtjies Gegee"'; Keet, 'Historical Review of the Development of Forestry in South Africa', 264: reports a maximum of 2,000 woodcutters in the historical record, 1,500 by 1900, 1,269 registered in 1913 in terms of the new *Forest Act*, and 302 at de-registration in 1939 under the *Woodcutters Annuity Act*.

73   Ibid., 71, 73.

that included provision for staff, schooling, nursery, hygiene, recreation, water supplies, and other facilities. District foresters found themselves caught up in an accelerating project of establishing new forest villages, while having to manage expensive and often unfit and unwilling white workers in difficult terrain, without infrastructure and with as yet unproven forestry techniques. Land shortages occurred in forestry regions of the Transvaal in 1931, so that for a period the department lacked adequate land for afforestation and for the settlements,[74] but during the Depression land did become available, which the department could acquire 'at very reasonable prices'.[75]

Afforestation expanded at a rapid pace from the mid-1920s to the early 1930s in parallel with the white settlement scheme.[76] The rate hovered between 12,000 and 14,000 acres per year, with a peak of 15,642 in 1927.[77] Suitable afforestation sites had to be purchased because most of the state forest land in the Cape and Drakensberg mountains was in unsuitable locations. Foresters looked primarily to the uplands of the eastern and northern Transvaal, especially centred on the town of Sabie. The department built several successful settlements in this area between 1929 and 1932, establishing the equivalent of a small town in the space of four years (Table 1). There they purchased land for afforestation that was freely available on the open market in this region, since the properties in areas suited to afforestation were not well suited to farming, often found no private buyers, and were offered to the state.[78] This included land that had been abandoned by returned soldier settlers.[79] The department was pressed to provide for settlement as soon as it acquired land for afforestation.

74   Memorandum on Employment in Forest Work, 15 July 1931, E120, FOR 364, NASA-P.

75   Union of South Africa, *Proceedings of the Fourth British Empire Forestry Conference*, 27. Quoted from Keet.

76   Union of South Africa, *Annual Report of the Forest Department For the Year Ending 31 March 1931*, 9.

77   The rate of afforestation in acres per year was 13,232 (1925), 15,642 (1927), 14,420 (1928), 12,866 (1929), 14,231 (1930).

78   Roach, 'The White Labour Forest Settlement Program in South Africa 1917–1938', 53–4.

79   *The Annual Report of the Division of Forest Management* records an inspection of the forestry operations in the Sabie area by the Minister, General J. G. Kemp, after which he instructed the department to purchase farms on the eastern escarpment, previously held by returned servicemen, for the protection of indigenous forests and conservation of their water resources (including sources for the Kruger National Park). This instruction was followed by the acquisition of the properties, afforestation, and land rehabilitation programs. See Keet, 'Historical Review of the Development of Forestry in South Africa', 83. Kemp soon retired ill, to be succeeded by Colonel Deneys Reitz, who, in his prior role as Minister of Lands, had promoted the establishment of the Kruger National Park in 1926.

Table 1. White labour forestry settlements in the state forests in the region of Sabie, eastern Transvaal.

Location of settlement	Date established	Number of family units
Coetzeestroom	1929	100
Tweefontein	1929	125
Ceylon	1929	120
Bergplaats Extension	1931	50
Blyde	1931	50
Spitskop	1931	100
Swartfontein	1931	100
Witklip	1931	25
Brooklands	1932	100
Bergvliet	1932	80
Malieveld	1932	19
Total		869

Source: From Roach, 'The White Labour Forest Settlement Programme in South Africa 1917–1938', 147ff.

Conclusion

From the early 1900s, foresters in South Africa were willing, if sometimes critical, partners of a developmental program driven by the minerals revolution, race capitalism, social engineering, and environmental idealism. Forestry was a cauldron of ideas, but also disciplined by stringent budgets, the intellectual and pragmatic demands of innovation, and the drive to achieve the forestry statutory mandate. By 1935, leading foresters had learnt confidence from their successes over the preceding three decades. Establishment of the new Forestry Department in the Union government had gained a secure start by having Cape foresters fill the leading positions, and the 1913 *Union Forest Act*, modelled on the 1888 *Cape Forest Act*, soon provided a coherent legal framework for forest management across the country. However, the wider policy environment, largely given substance by the Milner administration, was at a deeper level characterised by the disrupted rural economy, poverty and migration in the aftermath of the Rinderpest, droughts, and the catastrophe of the South African War. Subsequent developments proceeded in a large degree subject to the direct and indirect demands of the Union's mineral economy, but also in response to government social policies. The Forestry Department had almost immediately had a policy of afforestation approved, while extending its program of protection and management of indigenous forests to reach across the previously underserved provinces. Within a few years, a national drive to set aside and protect mountain catchment areas gained momentum, and the department received responsibility to manage a larger percentage of the country's mountainous areas.

Forestry was drawn rapidly into the politics of rural development and the politics of Afrikaner identity. The most visible manifestation of this, the white labour forestry settlement scheme, while leading to only about 20 per cent of the department's employees being drawn from resettled whites, nevertheless absorbed a disproportionate part of managers' attention. On the other hand, foresters with partners in government succeeded in creating social conditions in forestry settlements highly favourable for settlers' own development. This was at a time when foresters confronted a complex array of intermeshed problems coming from the ambitious program to develop extensive sawtimber plantations, and from all the gaps in knowledge revealed by the domestication of tree species not previously cultivated. Earlier and current investment in advanced education, often at leading institutions overseas, had equipped the department with a leadership group that had the enterprise and determination required to endure a difficult period and to research the innovations needed for progress, while competent field managers trained in South Africa could see to on-the-ground development.

Chapter 5

Competing Agendas? Afforestation, Catchment Management and Indigenous Forests, c. 1910–1935

Afforestation took top priority in the minds of leading South African politicians because it helped to address two problems at once: the negative balance of trade in non-mineral commodities, and the social and economic pressures related to the mining industry, while also offering potential mitigation of the 'poor white' problem. National forestry strategies also focused on two other key areas: catchment management and the conservation of indigenous forests. In the minds of most South African foresters, these three policies had a coherence based on nineteenth-century ideas about the climatic importance of trees. Planting trees produced timber and supposedly moderated climate and conserved water. Catchments should be protected from fire, overgrazing, and other disturbances that would lead to erosion and variable streams.

Yet a growing number of individuals and groups—ranging from Afrikaans- and English-speaking returned-soldier settlers, farmers, Africans, ecologists, botanists, and leading government ministers—began to worry that afforestation was hindering efforts to conserve water and protect indigenous forests. Concerns emerged early in the Cape Colony, where in the 1890s the municipal officials of Cape Town questioned afforestation, substituting plantations on the upper slopes of Table Mountain with water reservoirs.[1] But between 1910 and 1935, concerns became widespread, and these diverse groups questioned the assumption that exotic timber plantations helped to conserve water and improved efforts to conserve soil. They tried to unravel the scientific justification that foresters had

---

1   Pooley, *Burning Table Mountain*, 44.

built up since the middle of the nineteenth century. This involved challenging the authority and science of foresters, a difficult task given the public esteem of the profession internationally at the time. In 1935, tensions over these three policies led to a heated public debate at the 1935 British Empire Forestry Conference, a seminal meeting in the history of South African environmental policy.

This chapter examines the evolution of policies for managing catchments and indigenous forests as part of the Forestry Department's extensive remit. It argues that two distinct criticisms of afforestation arose from specific concerns relating to the hydrology of catchments and the ecology of indigenous forests. Together, these criticisms formed the basis of an argument that evolved throughout the rest of the twentieth century, and form the basis of current criticism of exotic trees in South Africa today. At the time, the hydrological argument arose as the most important, followed second by fears about the destruction of indigenous vegetation. These concerns merged into a coherent argument about the dangers associated with exotic tree planting, first in the Cape, and then later throughout South Africa.[2]

Forests and water

The Forestry Department along with the Irrigation Department (within the Department of Lands) were two key branches directing government water conservation policy after Union. South Africa's water law and policy from 1912 until the 1950s sought to encourage the use of existing water supplies for economic development, particularly in rural areas.[3] The key advance in this field was the Union's *Irrigation and Conservation of Water Act (no. 8) 1912*, which instituted a uniform system of water law for the country, with the main object 'to assist riparian owners to use the water of public streams to achieve [irrigation] development'.[4] The state claimed no ownership of water rights per se, but could exercise 'supervisory control'. Where a stream arose

---

2   See Bennett, 'Model Invasions and the Development of National Concerns over Invasive Introduced Trees', 499–512.

3   South Africa had a tiny manufacturing industry, and water policy was guided by concerns for rural, not urban, usage. See Union of South Africa, *Report of the Commission of Enquiry Concerning the Water Laws of the Union* (Pretoria: Government Printer, 1952).

4   *Irrigation and Conservation of Water Act (no. 8) 1912*; H. Thompson, *Water Law: A Practical Approach to Resource Management and the Provision of Services* (Cape Town: Juta, 2006), 55. A public stream was defined as one flowing in a known and defined channel through more than one property with water available for use by two or more riparian users, as opposed to private water, which was water that arose and flowed over the original property. For an accurate, succinct historical overview, see M. Muller, 'Lessons from South Africa on the Management and Development of Water Resources for Inclusive and Sustainable Growth', in *European Report on Development 2011/2012: Confronting Scarcity: Managing Water, Energy and Land for Inclusive and Sustainable Growth*, 2012, www.die-gdi.de/fileadmin/user_upload/pdfs/dauerthemen_spezial/European-Commission_European-Report-on-Development-2011-2012.pdf., 43.

on a given piece of land the owner had the right to the use of its water as a 'private stream'. Where a stream flowed through land with more than one riparian owner, it was a 'public stream', and riparian owners shared the 'normal flow' (the flow that could be relied upon for irrigation without storage) pro rata to their riparian extent. The Act codified the rules governing the use of public water, and allowed the state, through the Water Court, to adjudicate between competing claims on public water, following a set of principles regarding the sharing of normal water flows among such owners. This statute enabled a rapid growth in irrigated agriculture.

Irrigated agriculture became a key element of the government's rural development policy, just as afforestation was. Prior to Union, forestry and irrigation officials exchanged views on catchment protection and forests; Robertson, recently back in South Africa from his studies at Yale, promoted in a paper on 'Influence of Forests on Climate and Moisture Conditions' to the Philosophical Society of the Orange River Colony (ORC) in June 1907 the ideas of 'climatic forestry' traceable to George Perkins Marsh (and later, in 1889, cautiously by Henry Fourcade in South Africa). Leading experts in forestry and irrigation agreed broadly on the need to protect the headwater catchments of rivers, by conserving the vegetation there.[5] In 1909, C. Dimond H. Braine declared that forestry 'is of vital importance for maintaining the permanence of streams; and the forests that create natural reservoirs on every square yard of their surface, and form the chief source of water-supply, should be preserved at all costs'.[6] Braine, who had worked in India before coming to the Transvaal, echoed the views of most scientists who began their careers during the 1880s – 1900s, and believed deeply in the key tenets of orthodox 'climatic forestry'. These scientists used their positions to disseminate their views to the public. In a letter Braine wrote congratulating Robertson on his presentation on the climatic influence of forests to the Philosophical Society of the ORC, he pronounced favourably that 'with Kanthack [then director of Irrigation for the Cape Colony] … in the Cape and you in the O.R.C. public opinion will gradually be drawn to the importance of the question'.[7]

Francis Kanthack, the Director of Irrigation for the Union of South Africa from 1911 to 1920, saw foresters as the natural guardians of the headwaters of rivers and streams. He drew on examples from France, India, Australia, and the United States to justify his view. In a lecture to the South African Association

5   Beinart notes that foresters and irrigation experts had somewhat different concerns with managing mountain catchments; they nonetheless agreed on the policy. Beinart, *The Rise of Conservation in South Africa*, 182.

6   C. D. H. Braine, 'The Influence of Forests on Natural Water Supply', *South African Journal of Science*, 6 (1909): 111–33, 133.

7   Braine to Robertson, 3 February 1910, Wicht Private Papers, CSIR, Stellenbosch; Beinart, *The Rise of Conservation in South Africa*, 180.

for the Advancement of Science in Grahamstown in 1908, he proposed that 'the [Forestry] Department should have control of the land wherever the physical conditions are such that the removal of the protection afforded by vegetation must result ... in the destruction or deterioration of agricultural conditions'.[8] Kanthack emphasised the importance of the department's ensuring the restoration and conservation of the indigenous vegetation in catchments as well as of its afforesting suitable locations. He argued that '[w]e must learn to clothe the word "forest" with a far wider meaning than is customary. It should stand for the veld generally and the mountain and forest-clad veld in particular'. He noted how the Cape Forestry Department was engaged in the demarcation of watersheds and the protection of their vegetation. In fact, by around 1900, the Cape Department was administering nearly 490,000 ha of forest estate, of which just less than 160,000 ha was under forest, and the rest mainly 'veld'.[9] Subsequently, the 1914 report of the Senate Select Committee on Droughts, Rainfall and Soil Erosion recommended that the state should purchase important catchment land, and place this under care of the Forestry Department.[10] By 1935, when the extent of the public forest estate was nearly 1.5 million ha, over 1 million was catchment, under natural vegetation.

Kanthack's ideas anticipated a policy coherent with earlier Cape policy, and which took effect during the first three decades of the twentieth century: foresters received control over mountainous Crown land under indigenous grassland and fynbos, for the purpose of catchment protection rather than afforestation. These lands included large mountainous regions such as the Cederberg, the Kouga mountains, and the Drakensberg, where the Cathkin Peak Crown Forest was demarcated in 1922.[11] Foresters at the time saw catchment conservation as being the suppression of fire, prevention of overgrazing, and control of extractive activities by limiting access to land; for example, the 1927/28 Annual Plan of Operations for the Monk's Cowl Crown Forest in the Drakensberg focused solely on these objects.[12] This often put foresters in conflict with black and white graziers who desired grazing land for livestock, especially during droughts, and having this responsibility was sometimes seen as a burden to financially strapped foresters. In support of 'enlightened' forest

---

8   F. E. Kanthack, 'The Destruction of Mountain Vegetation: Its Effects upon Agricultural Conditions in the Valleys', *The Agricultural Journal of the Cape of Good Hope*, 33 (1908): 194–204, 196; Beinart, *The Rise of Conservation in South Africa*, 180.

9   Sim, *The Forests and Forest Flora of the Cape Colony*, 17.

10   Beinart, *The Rise of Conservation in South Africa*, 249; Keet, 'Historical Review of the Development of Forestry in South Africa', 146.

11   C. W. Marwick, *Green Shadows*, Pamphlet 372 (Pretoria: Forestry Branch, Department of Environment Affairs, n.d.), 56.

12   Ibid., 56.

officers, Kanthack had in his speech criticised 'the public … [who] looks upon the Forestry Department as an unsatiable [sic] ogre, which is ever seeking to grab more and more land, much of which is unsuitable for profitable tree growing'.[13]

While recognising that much of the forest estate was best maintained under natural veld, Kanthack, Braine and many forest officers continued to promote the notion that afforestation, where feasible, would benefit water supplies. But people from rural and agricultural backgrounds challenged official policy by giving anecdotal testimony that exotic trees dried out rivers, soil and air. Complaints about plantations diminishing streamflow arose on the Cape Peninsula within the first decades following afforestation at Tokai, which began in 1884.[14] In 1923, Alfred Paetzold, a nurseryman in Stutterheim, told readers of *The Farmer's Weekly*, 'Of course, the popular belief is that trees attract moisture, and the Forestry Department, for apparently very selfish reasons, has by silence encouraged this very popular but very erroneous belief'.[15] He pointed to diminished streamflow in his municipality of Stutterheim as an example of how afforestation dried up natural streamflow. Other critics drew on local claims of the desiccating effect of eucalypts and wattles seeming to have dried out vleis or parched the nearby soil. Recurring fears about the progressive desertification of South Africa aggravated concerns about exotic trees. Thomas Sim, Conservator for Natal from 1902–1905, expressed the views of many forest realists when he wrote: 'Africa as a whole has shown a marked advance of desiccation during the past twenty years … the planting of Eucalyptus and Acacias, however useful in other respects, does nothing towards checking this drying out'.[16]

Foresters (such as the aforementioned Sim) did reason that some genera and species tended to exhaust soil moisture rather than to replenish it. Legat had little doubt that eucalypts decreased streamflow (see page 161). Hutchins told forestry sceptics that 'trees have a double effect on water supply, a water-exhausting and a water conserving action, and it will depend on circumstances which is the more powerful'.[17] He warned that the 'general effect [i.e. positive climatic influence] must not be confounded with the local drying effect of quick-growing trees such as Eucalypts and Wattles'.[18]

13   F. E. Kanthack, 'The Destruction of Mountain Vegetation', 194; Beinart, *The Rise of Conservation in South Africa*, 180.

14   J. D. M. Keet, 'Rainfall and Streamflow at the Cape', *Journal of the South African Forestry Association*, 4 (1940), 15–20, 18–9; Pooley, *Burning Table Mountain*, 44.

15   A. Paetzold, 'Mountain Erosion', *The Farmer's Weekly*, 14 March 1923: 48.

16   Sim, *Tree Planting in Natal*, 10–1.

17   Hutchins, 'Extra-Tropical Forestry', 27.

18   Ibid.

These questions about the relationship between plantations and water conservation prompted forestry and irrigation officials to start a scientific investigation of how planted forests affected streamflow and climate. The Union Irrigation Department initiated hydrological research at a freshly established plantation at Jessievale in the east of the Transvaal in October 1910.[19] The purpose was, in Chief Conservator Legat's words, to study 'what influence, if any, afforestation may have on precipitation and on waterflow'.[20] There was a sense of urgency about the study, with Legat enquiring in 1913 from Robertson whether 'the necessary observations are now being taken to obtain the fullest and most reliable data to be collected', and if not, what further could be arranged.[21] Ten years later, Legat inquired from the Transvaal's Conservator, J. D. M. Keet, about whether the results of the experiment had been collected and analysed. Keet suggested waiting until the 900 ha of trees, mostly conifers, had fully grown. He noted that the oldest stands were just 15 years old, still 'unthinned and seriously overstocked'. He wanted to wait until 'the plantations have reached an older stage and forest-like conditions are attained'.[22] But by 1924, it was clear from the failure of gauging techniques and the loss of records that an investigation incidental to routine forest and hydrographical activities was not yielding reliable information.[23] The experiment proved to be a failure: even though it ran until 1963, it yielded no published results.[24]

Leading agricultural experts, irrigation engineers, and foresters during the 1920s maintained a unified front in the face of those who complained that trees were leading to the progressive drying out of Southern Africa. South Africa suffered droughts in 1916 and 1919–1920 that sparked Union-wide concern about climate change and the negative impact of human land uses, such as overstocking and veld burning.[25] A Union Drought Investigation Commission was appointed in 1920 to follow up on the 1914 report of the Senate Select Committee on Droughts, Rainfall and Soil Erosion, and investigate the question of whether South Africa

---

19   Hydrographic Surveyor N. von Warmelo, Irrigation Department to Chief Conservator of Forests C. E. Legat, 'Jessievale Plantation Gauging of Spruit', 22 January 1918, R7649/711, South African Forestry Research Institute [SAFRI] Archives, Council on Science and Industrial Research [CSIR], Pretoria.

20   Chief Conservator of Forests C. E. Legat to the District Forest Officer, Research, C. C. Robertson, 17 January 1913; Hydrographic Surveyor to Chief Conservator of Forests, 22 January 1913. R7649/711, SAFRI Archives, CSIR, Pretoria.

21   Chief Conservator of Forests C. E. Legat to the District Forest Officer: Research, C. C. Robertson, 17 January 1913; Hydrographic Surveyor to Chief Conservator of Forests, 22 January 1913. R7649/711, SAFRI Archives, CSIR, Pretoria.

22   J. D. Keet, Conservator of Forests Transvaal and OFS to Chief Conservator of Forests C. E. Legat, 8 December 1923, R7649/711, SAFRI Archives, CSIR, Pretoria.

23   Conservator of Forests Transvaal and Orange Free State Conservancy J. D. Keet to Chief Conservator of Forests, 7 May 1924, R7649/711, SAFRI, CSIR, Pretoria.

24   The Jessievale study subsequently grew to incorporate three gauged streams, and ran until 1963, but it yielded no scientific results. Email 14 June 2013 from Brian Jackson, Manager: River operations & Data Management, Inkomati Catchment Management Agency.

25   Beinart, The Rise of Conservation in South Africa, 214.

was becoming drier, and if so, why, and how drought losses may be avoided. The commission, led by agricultural officials and scientists, reviewed existing evidence for rainfall declines, desertification, and the relationship between trees and erosion and water conservation.[26] The commission's final report, published in 1923, testified that the country was drying up because 'enormous tracts have been denuded of their original vegetation [since the arrival of the white man] with the result that rivers, vleis and water-holes … have dried up or disappeared'.[27] To halt this process, the commission, though sceptical of 'forestry enthusiasts', recommended, among other things, the large-scale planting of exotic trees in catchment areas to protect water supplies and climate because it would lead to 'decreased soil erosion and a more economic use of the rainfall; the preservation and improvement of mountain catchment areas; the regulation of the flow of rivers, and the clarification of the waters thereof'.[28]

The report of the Drought Commission did little to convince farmers who worried that the expansion of plantations in the eastern and northern Transvaal was diminishing streamflow. By the late 1920s the proximity between freshly created timber plantations and irrigated farm settlements pitted foresters against farmers and engineers. One of the most significant conflicts occurred near White River.[29] The area between White River and Sabie lies on the eastward slopes of the Escarpment, which separates South Africa's Highveld (reaching about 2,000 m above sea level) and the Lowveld (about 300–500 m). Higher elevation sites have a cooler climate and rainfall of around 1,000 mm per year, ideal for plantations. In the foothills below, the warmer tropical climate suited tropical fruit production, but irrigation was required because these sites receive less rain. Soils on these slopes, derived from ancient granites of the original continental crust, are deeply weathered and highly erodible. The upper half of this landscape, from the time of Union onward, became the focal region for investment by both government and mining capital in pine sawlog and wattle and eucalyptus mining-timber forests. Veterans of the South African War and, later, World War I received the opportunity to acquire land for farming in the foothills below this belt. Many of these settlements failed, but government intervention in irrigation allowed some to succeed.

---

26   The committee was dominated by agricultural researchers, officials, and farmers. They included Henrych Sebastian Davel-du Toit, Selby Montague Gadd, George August Kolbe, Arthur Stead, and Reenen Jacob von Reenen.

27   Union of South Africa, *Final report of the Drought Investigation Commission October 1923* (Cape Town: Cape Times Limited, Government Printers, 1923), 8.

28   Ibid., 65.

29   Another significant conflict occurred in the south-western region of the Cape Province. This is discussed in the section following.

One such settlement was what later became the White River Valley Farmers' Association. As part of Alfred Milner's aim of combining rural reconstruction with 'the creation of a British-minded majority',[30] European agricultural settlement began on the White River in 1905 with the so-called 'Milner Settlement' of mainly British immigrants, when water was first supplied by diversion from the White River through a 25-km earth canal serving a series of 100-acre farms. Though heavily subsidised until 1907, most settlers left at the end of their contracts in 1909. The Union government sold the farms in 1911 at a huge loss to a syndicate financed by mining capital, on the condition that the enterprise convert from cash crops to citrus orchards.[31] The scheme expanded to accommodate many settlers after World War I when the state sold land at a discounted rate to demobilised soldiers as part of government efforts to settle ex-servicemen on the land, part of the immigration of 350,000 Europeans, mainly British, in the period to 1936.[32] By 1927, there were five separate irrigation boards in the White River development. The historical record shows that expansion quickly outpaced the availability of water, leading 'to a shortage of water and inequitable distribution of the limited supplies available', with major vested interests facing a looming conflict.[33] Competition was among irrigators, but the conditions were set for conflict about water supply from upstream.

Foresters often found the rivers silt-laden and the land on the mountain slopes to be 'degraded'. This they attributed to sheep grazing, cultivation and excessive veld burning, but it was especially due to mining: the area around Sabie was an early centre of gold mining following the discovery of alluvial gold in 1873, and the rush that began with numerous alluvial diggings ultimately led to 11 shaft-mine operations in this area.[34] Foresters saw afforestation as the way to repair the soil, and argued that creating a belt of pines on the escarpment 3 to 4 miles wide and 40 miles long would enhance rainfall.[35] They reasoned that the land was so degraded by erosion that it did not function effectively as

---

30   Worsfold, *The Reconstruction of the Colonies*, 7; Louw, *The Rise, Fall, and Legacy of Apartheid*, 18.

31   L. V. Praagh (ed.), *The Transvaal and Its Mines (The Encyclopedic History of the Transvaal)* (London and Johannesburg: Praagh and Lloyd, 1906), 157–8.

32   Louw, *The Rise, Fall, and Legacy of Apartheid*, 21; H. T. Glynn (ed.), *Game and Gold: Memories of Over 50 Years in the Lydenburg District, Transvaal* (London: The Dolman Printing Company Ltd, 1926), 221; Schirmer, 'Enterprise and Exploitation in the 20th Century', 303.

33   A. R. Turton et. al., 'A Hydropolitical History of South Africa's International River Basins', *WRC Report* 1220/1/04, Pretoria, Water Research Commission (2004): 331–7.

34   H. S. Webb, 'The Goldfields of Sabie and Pilgrims Rest and Transport Development', in *A Survey of the Resources and Development of the Southern Region of the Eastern Transvaal Lowveld* (Barberton: Lowveld Regional Development Association, 1954): 24–31, 24–5; E. T. E. Andrews et al., 'Minerals and Mining', in *A Survey of the Resources and Development of the Southern Region of the Eastern Transvaal Lowveld*: 58–74, map opposite p. 68.

35   'Invloed van bos op stand van water', J. D. Keet to Chief Forestry Research, 21 October 1933, R9110, Wicht Papers, SAFRI Archives, CSIR, Pretoria; J. D. Keet, C. 203 of 29 June 1934, 'Complaints re: streams drying up on Bultfontein', Wicht Papers, SAFRI Archives, CSIR, Pretoria.

a water catchment because water was not retained in the soil.[36] District foresters and the Transvaal's Conservator, Keet, investigated local water supplies in their planning of the forestry infrastructure, also recognising the potential conflict between afforestation and agriculture. Among others of his inquiries, Keet drew on an analysis of streamflow records in the catchment of the White River from 1922 before embarking on large-scale afforestation in 1927. The conclusion was that, even before afforestation began, the average flows of the White River were too low to support irrigation without the construction of storage dams, a view confirmed by the history of the development of the White River Irrigation Association, which required a succession of new dams to be built to create the water supply needed.

A confluence of environmental and social pressures boiled over in the early 1930s. Prior to afforesting the eastern escarpment, annual rainfall was approximately 25 per cent above normal. Drought set in at the height of planting in the early 1930s, lowering normal rainfall by 25 per cent, or a 50 per cent decline from the pre-afforestation peak in rainfalls (see Figure 10). Complaints from the White River area poured into the Secretary for Agriculture and the Chief Conservator in 1934.[37] Even strong forestry advocates began to worry that tree planting in the region was causing a serious decline in river flows. The Administrator of the Transvaal, S. P. Bekker, himself a farmer and wattle grower in the eastern Transvaal on the edge of the escarpment, laid out the concerns of his farming constituents in 1935: 'I know that farmers in that part of the Province are very much perturbed over the effects of planting these pines and gums on the sources of these streams and rivers which have to irrigate the fertile land at the bottom of these ranges'.[38] Particular concern was laid on the fact that the plantations were young, and would not provide the claimed broader climatic and hydrological benefits for some time.

---

36   J. D. Keet, 'Complaints re streams drying up on Bultfontein'.
37   White River Valley Farmers' Association to Private Secretary, Minister of Agriculture, 18 May 1934, Wicht Papers, SAFRI Archives, CSIR, Pretoria; Captain C. V. Palmer to J. D. Keet, 18 June 1934, Wicht Papers, SAFRI Archives, CSIR, Pretoria; White River Estates Irrigation Board to Director of Forestry, 24 June 1936, Wicht Papers, SAFRI Archives, CSIR, Pretoria; Memo C. 203 of 24 June 1934, J. D. Keet 'Complaints re streams drying up on Bultfontein', Wicht Papers, SAFRI Archives, CSIR, Pretoria; A. O'Connor for Director of Forestry to White River Estates Irrigation Board, 11 August 1936, Wicht Papers, SAFRI Archives, CSIR, Pretoria.
38   *Proceedings of the Fourth British Empire Forestry Conference*, 77.

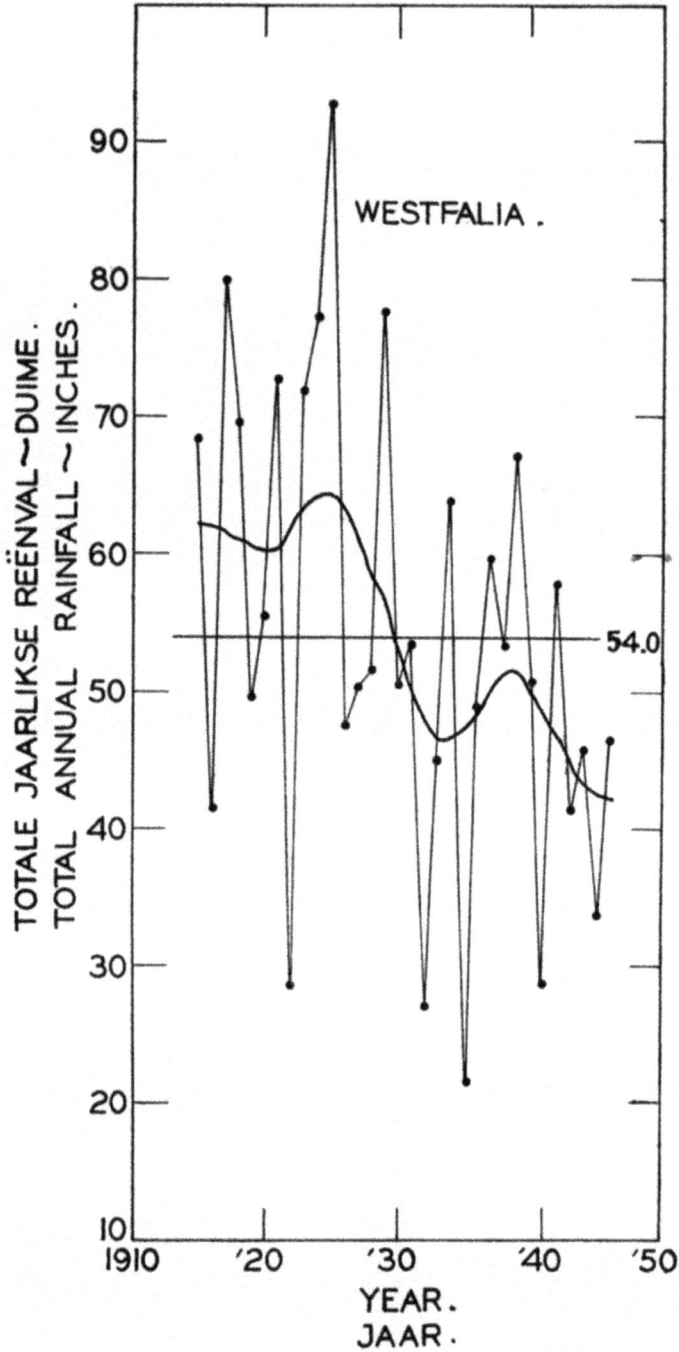

Figure 10. Rainfall variation during the period 1915 to 1946 for Westfalia, a station representative of the Transvaal escarpment.

Source: From C. L. Wicht, *Forestry and Water Supplies in South Africa* (Pretoria: Department of Forestry, Union of South Africa, 1949).

Catchments or forests? Keet's integrated forestry paradigm

By the 1930s, foresters found themselves in a policy dilemma regarding their roles as developers of new plantation resources and protectors of the mountain catchments that formed a growing part of their charge. Legat noted in the first paragraph of his 1930 Annual Report that 'there would seem to be much to be said in favour of all such lands [i.e. mountain catchment areas] being placed in charge of the Irrigation Department, which is the department with interests most directly bound up with the protection of water supplies'.[39] Having to manage un-afforestable land made the department the subject of criticism from farmers and others who wanted to use the land. Controlling such large swaths of land limited foresters' ability to make arguments to Parliament about the need to purchase more land for plantations. Legat protested: 'Of the millions of odd morgen[40] shown as forest reserves on the books of the department, the department is saddled with hundreds of thousands of morgen of this particular type—the rugged ranges of the Western Cape—and Parliament and the public are apt to be under the impression that the department is well endowed with all the land it requires for its purposes'.[41] In spite of this situation and the wider concerns about the hydrological effect of plantations, the department received more, not less, responsibility for catchments during the early 1930s. In 1930, the Lands Department transferred to the forest estate 25,000 morgen (21,420 ha) of land, located primarily in the Cape Province, for the purposes of water conservation, and in 1934, a further 80,000 morgen (68,540 ha).[42]

Against the background of the social and political disaster of the Great Depression, aggravated by the intense hardships of a 'crippling' drought that began in Natal around 1929,[43] extending countrywide, Parliament in 1934 confirmed the assignment of protection of mountain catchments to the department and further acquisition of catchment land. Parliament required the department to devise a scheme to achieve this, and an 'analytical survey' as a basis was to begin immediately. For instance, much of the Drakensberg Mountains became subject to forestry management.[44] Facing the joint challenge of managing catchments as well as indigenous forests and the expanding plantations, together with

---

39   Union of South Africa, *Annual Report of the Forest Department For the Year Ending 31 March 1930* (Pretoria: Government Printer, 1931), 1.

40   One Cape morgen equals 0.856,532 hectares.

41   Union of South Africa, *Annual Report of the Forest Department For the Year Ending 31 March 1930*, 1.

42   Keet, 'Historical Review of the Development of Forestry in South Africa', 200.

43   D. W. M. Edley, 'The Years of Red Dust: Aspects of the Effects of the Great Depression on Natal, 1929–1933' (PhD Thesis, University of Natal, Durban, 1994), 256–67; Pooley, *Burning Table Mountain*, 67.

44   Union of South Africa, *Annual Report of the Secretary for Agriculture and Forestry for the Year ended 31 August 1935* (Pretoria: Government Printer, 1935), 501; Pooley, *Burning Table Mountain*, 67.

growing complaints about the impact of afforestation in mountain catchments, foresters sought a more coherent policy. They chose to harmonise plantation forest management and indigenous forest and catchment management, sharing the costs of infrastructure and in effect subsidising conservation management costs through the provisions for afforestation,[45] a policy that was communicated during the 1935 British Empire Forestry Conference (see Chapter 6).[46] Keet illustrated the reasoning for the new policy in his response to his head of department in answer to intense criticism from Illtyd Buller Pole-Evans, head of the Division of Plant Industry:

> ... The question of forests versus flowers, as with the question of agriculture versus flora and fauna is obviously an economic one. I, for one, stand back to no man in South Africa in my love for its flora; and the forest service has done more than any other, or public body or private person, to preserve the flora. But the nation will not countenance its Forest Service growing only flowers and no timber and thus compel it to continue the importation of some £2,000,000 of timber annually, and incidentally thereby denying a considerable proportion of the population an outlet for healthy and beneficial occupations. It has already been shown ... that we hold, in the Division of Caledon, upwards of 58,000 morgen of land of which only about 5,000 morgen have been afforested. There is thus room for both flowers and forests, with heavy advantage to the former ... Now, in regard to the preservation of water supplies at Elgin, I would observe that there are perennial streams, each of which ... has its origin many miles above the plantations held and managed by the Department ... practically no afforestation has been done in the mountains ... the plantations actually lie well below the sources of the streams.

Forestry officials made internal policy decisions in the early 1930s to forestall problems that might arise from the planting of trees in catchment areas. In 1932, the Department of Forestry adopted the policy of keeping a 20-metre buffer zone between a stream and a plantation to mitigate effects of their plantations on streamflow.[47] That same year they decided to develop a major hydrological research site in the Jonkershoek Valley and began to acquire the necessary land for it.[48] In 1934, soon after his appointment as Chief of the Division of Forestry, Keet put the issue of forest hydrology at the top of his priorities. He travelled personally to meet the White River farmers. He pursued a broader series of investigations about complaints about the negative hydrological effects of

---

45   *Proceedings of the Fourth British Empire Forestry Conference*, 48. Quote from Keet.

46   J. D. M. Keet, 'Memorandum on Afforestation in Relation to Natural Flora and the Effect of Afforestation on Water Supplies', 15 July 1935, from Chief Division of Forest Management to All Professional Officers, Division of Forest Management, 1054/7/2, FOR 330, NASA-P. This was the memorandum briefing South African delegates prior to the 1935 British Empire Forestry Conference.

47   Department of Forestry, *Report of the Interdepartmental Committee of Investigation into Afforestation and Water Supplies in South Africa* (Cape Town: Republic of South Africa, 1968), 13.

48   See J. J. Kotze's comments in *Fourth British Empire Forestry Conference Proceedings*, 293.

afforestation at Jessievale, at other locations around Sabie, in the mountains north of Sabie, and on the Cape Peninsula.[49] As discussed in the next chapter, the question of forest hydrology became a major political issue in 1934–1935 leading up to the fourth British Empire Forestry Conference.

Officials in the Agriculture Department continued to publicly support foresters at the same time that they internally questioned the soundness of existing policy. By the mid-1930s, complaints about the effects of afforestation reached the Minister from farmers in the northern, eastern and south-western parts of South Africa, and from professional botanists in Pretoria and Johannesburg. Yet officials felt compelled to defend afforestation even to the point of absurdity. Dr Philip Viljoen, Secretary for Agriculture and Forestry, who himself was unsure about the climatic impact of forests, felt compelled to tell members of the Wild Flowers Protection Conference Committee that 'no country can be assured of the permanence of its climate with less than 30–50% of its area under forests … [because forests were] the only known and proved way of [ensuring] the amelioration of climate and the conservation of water supplies'.[50] This was a far-fetched statement considering that South Africa had less than 1 per cent of its land surface clothed by forests. Under the force of the need to create a new domestic timber resource, pro-afforestation propaganda was clearly reaching the limits of credulity as efforts began to inform the issue through scientific research.

Discovering ecology in the indigenous forests

The commitment to develop plantations, coupled at the same time with the need to conserve indigenous forests as well as mountain catchments, created the policy and intellectual space for the ongoing debates about the meaning of the concept 'forest', about conservation, about the emerging science of ecology, and about the nature of forest management in a disparate society. This room for dissent would play a central part in the development of ideas about forest hydrological science and ecosystems management in South Africa. As much as the success and failures in the trials of plantation species were an important source of learning for South Africa's foresters, so their experience in the study and management of indigenous forests, woodlands and mountains yielded insights that informed science, policy and practice in the following decades.

---

49   J. D. M. Keet to Chief Forest Research Officer, 21 October 1933, R. 9110, SAFRI Archives, CSIR, Pretoria. The 1935 Memorandum itemises such investigations from all over the forestry regions, including the Cape Peninsula. See, Keet, 'Memorandum on Afforestation in Relation to Natural Flora and the Effect of Afforestation on Water Supplies', SAFRI Archives, CSIR, Pretoria, 12–23.
50   Memorandum on Afforestation in Relation to Natural Flora and the Effect of Afforestation on Water Supplies. Keet likely drafted Viljoen's letter.

As the economy of the Southern African region developed, market forces exposed the natural resources under foresters' care to depletion, at best, and a ransacking, at worst. Whether it was wagon wood or firewood for Kimberley from the Campbell Commonage in the arid acacia savannas; bush tea, buchu, and protea tanbark from the Cederberg; or timber from the forests, the lesson was the same: the vast tracts of often remote and sparsely peopled land were effectively open to access by extractors.[51] Existing colonial and indigenous institutions that protected indigenous forests would never have been sufficient to regulate rising demand for scarce forest products given Southern Africa's limited resource endowments. Quite simply, there were too many demands for the country's finite resources given the onset of capitalism and modern globalisation. Forest conservation would not work without taking sufficient account of these local and global forces.[52]

In 1910, the Union Department of Forestry inherited from the colonial administrations numerous, scattered indigenous forest resources divergent in tenure, condition and governance; a situation determined by the different histories of the four provinces that constituted the Union. Small in overall extent as well as being patchily distributed, these forests had by that time been degraded to a greater or lesser extent almost everywhere. Foresters reported severe depletion by indigenous communities for construction materials, fuel, household commodities, and swidden agriculture—the practice of exploiting the enriched soils of the forests, or the 'forest rent', repeatedly cited by diverse observers.[53] But it was especially white settlers, and the bands of woodcutters who sprung up wherever there was extractable timber, who accelerated forest degradation.[54]

---

51   See Keet's summaries from Cape Superintendent of Forests Reports: J. D. M. Keet, 'Historical Review of the Development of Forestry in South Africa', (Pretoria, c. 1970), MS available online, www2.dwaf.gov. za/webapp/resourcecentre/Documents/Publications_And_Media/Keet_Forestry_History_page_41-66.pdf, 51–3. *Acacia (Vachellia) erioloba* woodlands on Crown Forests were 'denuded' for firewood and timber for the Kimberley diamond fields; *Agathosma betulina* was intensely harvested in the Cederberg for the pharmaceutical and perfume trade, as were bush teas of different species of *Aspalathus*, and *Protea nitida* and *Heeria argentea* for tanbark, for domestic as well as export markets.

52   Laughton, *The Sylviculture of the Indigenous Forests of the Union of South Africa*, 96.

53   Austin defines the forest rent as 'the difference in the cost of producing a unit of beans on a farm that has been replanted with cocoa, compared to one freshly cleared from forest'. See Austin, 'Resources, Techniques and Strategies South of the Sahara', 599. On this practice in South Africa, Fourcade, in the *Report on the Natal Forests*, recounts many instances of forests exploited this way, as does Keet in J. D. M. Keet, 'Management of the Indigenous Forests of the Transkei', 18 January 1934, R1814, SAFRI Archives, CSIR, Pretoria.

54   Fourcade, *Report on the Natal Forests*, 4; Sim, *The Forests and Forest Flora of the Colony of the Cape of Good Hope*, 3–9, 22, 48–9, 528; Laughton, *The Sylviculture of the Indigenous Forests*, 20–34; for the Transvaal, see Tempelhoff, 'Die Ontginning van Noord-Transvaal se Houtbronne', 67–74; for Natal see D. McCracken, 'Dependence, Destruction and Development: a History of Indigenous Timber Use in South Africa', in *Indigenous Forests and Woodlands in Southern Africa: Policy, People, and Practice* (Pietermaritzburg: University of KwaZulu-Natal Press, 2004): 277–308, 278, 280–1.

In the Cape, the department controlled most of the Knysna and Tsitsikamma forests—about 65,000 acres (26,000 ha)—and smaller forests to the west, as well as the forests in the eastern territories conquered during the Frontier Wars of the nineteenth century, and afterward, forests in the annexed territories of the Transkei. By around 1910 the Union forest estate as a whole included nearly 500,000 ha, of which less than one-third had indigenous forest cover: most reservation was for catchment protection, and some for the establishment of plantations.[55] In the Transkei (including Pondoland), the process of gaining control of indigenous forests was well advanced, but encountered significant conflict with the traditional leaders, the magistrates who administered the territories, and commoners who used the lands and the forests: the program had been informed by bitter lessons from the period of free access. In the Transkei, the indigenous forest area of around 100,000 ha was distributed over about 1,300 forests, scattered through most of the territory, hundreds being very small—100 ha and less in area. Here (as in Zululand in Natal), forest reservation intruded on communal tenure systems, creating a source of continuous conflict. The Forestry Department came to control scores of the demarcated forests there, as well as attempting, by working through traditional institutions, to protect many more, smaller 'Headman's' forests.[56] In the province of Natal, Henry Fourcade's survey during the 1880s had provided an inventory of indigenous resources and an assessment of their tenure and exploitation, although not all major forests had been surveyed, and the authorities had vacillated about instituting an effective administration; few forests had state protection at that time, and many had been severely depleted by settler farmers or woodcutters supplying buyers for the interior housing and wagon-making markets.[57] In the ORC, there were virtually no indigenous forests and, during the last decades of the nineteenth century, much of the demand for timber was met by bringing it in over the Drakensberg passes, mostly by illegal extraction.[58] Finally, official knowledge of indigenous forests in the Transvaal was poor, though concern for the future of timber resources was clear from repeated submissions from Boers in the north-east to the South African Republic (ZAR) government. But, clearly, early legislation by the ZAR government had been ineffective.[59] The only official

---

55   Sim, *The Forests and Forest Flora of the Colony of the Cape of Good Hope*, 17–8. Sim's data are drawn from reports of the Conservators of Forests.

56   Tropp, *Natures of Colonial Change*, 8–9.

57   D. P. McCracken, 'The Indigenous Forests of Colonial Natal and Zululand', *Natalia*, 16 (1986): 19–38, 24, 28ff.

58   Ibid., 25; Fourcade, *Report on the Natal Forests*, 17.

59   Tempelhoff, 'Die Ontginning van Noord-Transvaal se Houtbronne', 67, 71–2.

source at the time was Hutchins's *Transvaal Forest Report*, which contained information only on the Woodbush complex of forests, estimated at 20,000 ha, and overlooked many other forest complexes in the region.[60]

Charged by the 1913 *Forest Act* to protect and manage these forests, foresters found this to be one of the department's most contentious assignments. Because, in the majority, each was so small, conserving the forests in effect meant controlling access by many groups of people to the resources and places that they had used for years, decades and even centuries. As a result, intense 'negotiations' ensued over access to resources in indigenous forests throughout the country.[61] Though there was no uniform experience, foresters gradually gained control over most of the remaining indigenous forests. In doing so, they used legislation, policing, market forces and resettlement to regulate the use of the forests, and especially sought to substitute the timber from these forests with timber from plantations. Despite their small contribution to the timber economy, management of the indigenous forests figured large because of their political importance.

Three principal phases characterised the department's forest protection and management. Reservation, as defined in the *Forest Act*, involved an initial step of acquiring the forest and accessioning it to the forest estate as a forest reserve. The second was the demarcation of the forest reserve, which involved the survey of the forest boundaries followed by statutory demarcation—a demarcated forest could not be alienated without two-thirds parliamentary majority approval. And the third aspect was the regulation of access to and use of forest resources, of whatever kind: timber, minor forest produce, or recreation.[62]

Once reserved, forest management and use was regulated through the provisions of the Act, mainly through fee-based licensing, but also allowing free access and use—only for headloads of dry firewood in territories with communal tenure, the *teza*. Colonial foresters restructured access to indigenous forests through a variety of means, including policing, fines, and negotiation with elected, and later appointed, headmen, while in the Transkei and other regions of communal

---

60   Hutchins, *Transvaal Forest Report*, 136, 14–21. Hutchins's estimate of 20,000 hectares was a major exaggeration; the entire area of Northern Mistbelt Forests, of which the Woodbush complex is a part, is about 19,000 ha; G. Von Maltitz et al., *Classification System for South African Indigenous Forests: An Objective Classification for the Department of Water Affairs and Forestry*, Environmentek report ENV-P-C 2003-017 CSIR Pretoria, planet.uwc.ac.za/nisl/biodiversity/Attachments/Final%20document%20Full%20forest%20 classification%20report.pdf (accessed 13 December 2013), 140, 279.

61   See Tropp, *Natures of Colonial Change*, 8–9; McCracken, 'The Indigenous Forests of Colonial Natal and Zululand', 33.

62   Fourcade, *Report on the Natal Forests*, chapter on 'Survey and Demarcation of the Crown Forests'.

tenure, permitting the *teza* and access to other resources by licence.[63] Foresters, however, found white and 'coloured' sawyers entering the territories from the Cape to present as great or an even greater problem; eventually, access to indigenous forest resources was limited to residents of the territories, and sale of timber to outsiders prohibited.[64] Regulations to control access to forests and to prevent fires led to the apprehension and fining of both interlopers from the Cape as well as Africans on the frontier boundaries of what is now the Eastern Cape. After 1910, foresters in the then Transkei worked with Native Affairs officials to regulate access to remaining indigenous forests, and to 'wean' Africans off pre-existing communal forest use by the planting of fast-growing trees, usually wattle—a project that gradually succeeded in supplying the needed substitutes.[65]

Throughout the forest regions, bands of woodcutters, 'of unrecorded descent and origins', extracted timber from the indigenous forests.[66] Their exemplar was the Knysna woodcutter, member of a series of loose-knit communities, some originating from the 1777 VOC timber colony in the Tsitsikamma. They made a bare living extracting timber for sale to merchants. These poor white (and coloured) woodcutters had sole rights to timber extractions, and

---

63   *Teza* applied to Transkei forests, and the *Forest Act* provided for it. Keet sets out the *teza* as follows: 'All the forests, except those in the Territory of Port St. Johns, are subject to the servitude of "Teza", that is the collection, except by means of axe, saw, or other cutting implements, and removal, without permit, free of charge, of dry wood for fuel, for his own domestic requirements only by each native resident of a location from forests situated in such locations'. 'Teza-wood cannot be sold, at least not to Europeans', J. D. M. Keet, 'Management of the Indigenous Forests of the Transkei', 18 January 1934, R1814, SAFRI Archives, CSIR, Pretoria, 4, 12. Tropp traces the origin of the institution to regulations formulated by Magistrate Walter Stanford in the 1880s. See Tropp, *Natures of Colonial Change*, 81. See also F. S. Laughton, *The Sylviculture of the Indigenous Forests*, 31.
64   By direction of Prime Minister J. B. M. Hertzog. See Keet, 'Historical Review of the Development of Forestry in South Africa', 88
65   Tropp, *Natures of Colonial Change*, Chapter 3.
66   Woodcutter groups sought a living almost throughout South Africa where timber could be extracted. See, for example, Laughton, McCracken, and Tempelhoff, cited earlier. Also see Henkel in Sim, *The Forests and Forest Flora of the Cape Colony*, 46. Though commonly described as 'poor white', these groups were more likely in various degrees 'creolised' frontier bands of mixed origins, like the raiding mountain bands described in S. Challis, 'Creolisation on the Nineteenth-Century Frontiers of Southern Africa: A Case Study of the AmaTola "Bushmen" in the Maloti-Drakensberg', *Journal of Southern African Studies*, 38 (2012): 265–80, 273–5. See the account of the arrest of 'large numbers' of British Army deserters extracting timber from the Alexandria forest between Port Elizabeth and Grahamstown, where they lived in huts dispersed through the forest, with their Khoi-Khoi common-law wives and children, in Gordon-Brown (ed.), *The Narrative of Private Buck Adams, 7th (Princess Royal's) Dragoon Guards, on the Eastern Frontier of the Cape of Good Hope, 1843–1848* (Cape Town: The Van Riebeek Society, 1941), 99–104. J. D. M. Keet knew and worked with the woodcutters while District Forest Officer at Knysna. He was able to converse with them in their Afrikaans. He wrote of 'woodcutters of unrecorded descent and origins', that the 'Whites … must be understood to include people of different races', and expressed his sympathy in quoting Gray's *Elegy to a Country Churchyard*: 'the short and simple annals of the poor' in Keet, 'Historical Review of the Development of Forestry in South Africa', 264–5. Reporting on the Transkei, Keet refers to woodcutters 'of a semi-nomadic type' in Keet, 'Management of the Indigenous Forests of the Transkei' R1814, 18 January 1934, SAFRI Archives, CSIR, Pretoria, 4. F. S. Laughton provides a perceptive and sympathetic account of the Knysna woodcutter society. See Laughton, *The Sylviculture of the Indigenous Forests*, 27–8.

received consistent political support and public attention in their constant appeals against restrictive regulation. Foresters began exercising progressively stricter control over their felling from 1874 onward, leading to registration in 1912 (soon ratified under the 1913 *Forest Act*) of all woodcutters with extractive rights, i.e. any who had taken out timber in the preceding two years; the list was closed, and thereafter only registered woodcutters could take timber from state forests, according to what became a quota system.[67] In the beginning, approximately 1,200 woodcutters registered, but by 1929, numbers had fallen to 557 (many having left for the white forestry settlements). However, it was only with the passage of the 1939 *Woodcutter's Annuity Act*, a social support measure, that the woodcutters finally left the forests. But, 'giving the sole right to the working of timber to the registered wood-cutters has linked the question of forest management and the alleviation of distress', so that it was often 'necessary to sell very much more timber from the forests than what was considered permissible'.[68]

The largest continuous extent of indigenous forests, 60,500 hectares, grew in an east to west line at around 34°S along the southern slopes of the Outeniqua and Tsitsikamma mountains and inland of the coastal town of Knysna.[69] In these forests, de Vasselot had left an important legacy. His 1882 tour of inspection, informed by experienced conservators, especially Captain Christopher Harison, gave him the intelligence for a rational forest management plan for timber-yielding indigenous forests. Harison had established a form of selection silviculture when he was appointed Conservator in the Tsitsikamma in 1866. De Vasselot introduced regulations in 1883 which had the objects of governing fellings so that 'the forest can yield in perpetuity so that each year the quantity felled can be replaced', and of assuring adequate regeneration within the forests. With this, he advanced the selection system, building on French silviculture and Brandis's system for the teak forests of Burma.[70] He compiled, from his own appraisal and the know-how of his Conservators, a fair forest resource assessment and from this, for the Knysna forests, estimated a sustainable annual timber yield of 500,000 cubic feet (about 14,000 cubic metres). De Vasselot's system involved the selective removal of mature trees, within the limits of the allowable

---

67   Laughton, *The Sylviculture of the Indigenous Forests*, 28–9.
68   Grundlingh, '"God Het Ons Arm Mense Die Houtjies Gegee"', 40–56, 44, 54; Keet, 'Historical Review of the Development of Forestry in South Africa', 264–9.
69   The modern designation is the Southern Cape Afrotemperate Forests, and these form the largest forest complex in Southern Africa measured at approximately 60,500 hectares. See Von Maltitz et al., *Classification System for South African Indigenous Forests: An Objective Classification for the Department of Water Affairs and Forestry*, 88.
70   Laughton, *The Sylviculture of the Indigenous Forests*, 96–7, citing C. B. McNaughton's 1898 policy memorandum for the working plan for the Sourflats (*Zuurvlakte*) Forest near Knysna. Fourcade recommended an analogous management system in his chapter 'Management of the Forests', using the French term *tire et aire* for the selection system. See Fourcade, *Report on the Natal Forests*, 55ff.

yield for any given block of forest, rather than clearfelling or other alternative systems. This was the system to regulate extraction by the woodcutters—who enjoyed sole rights to the timber—initially in the Knysna forests, but soon after, with Hutchins's arrival, in the forests of the Amatolas and, later, the few timber-yielding forests of the Transkei. Each block was divided into sections, each section was methodically inspected and trees for felling identified, measured and marked, and then woodcutters were assigned trees for felling by lots, to distribute the allowable cut fairly. Once the allowable cut had been completed for a block, the block was closed to felling for 40 years to allow regrowth. In their marking for felling, foresters took care to distribute the removals so as to achieve the desired composition of the forest, to prevent excessive opening of the canopy, and to protect seedlings from excessive light, desiccation, and competing growth from undergrowth species. Steps to measure tree growth in order to inform estimates of allowable fellings began immediately on de Vasselot's institution of the system, though the initial extent of measurement was small.[71]

De Vasselot's system, progressively refined and improved,[72] was applied for several decades afterward. But foresters found that they could never achieve the norms set for the management of each forest or block in the forest: their personnel were too few, the woodcutters persisted in felling according to what was easiest sold, and political forces invariably intervened.[73] With the decline in demand following extension of the railways, especially the loss of the wagon-wood markets, woodcutters selectively felled high-value timber trees for the local markets, undermining the desired forest composition—threatening to render the forests 'ironwood wastes'.[74] During World War I, private forests were depleted; woodcutters focused greater demands onto the state forests, increasing the extractions from them, while foresters, lacking timber inventories for the forests, could not convince their fellow authorities and the public to contain the fellings, while knowing that the forests were being overexploited. Again, still lacking figures to prove permissible, sustainable harvest, foresters were obliged to considerably increase the allowed cut from the forests from 1924 to 1930, to substantially more than the guide figure of 500,000 cubic feet per year (and, to make matters worse, the recorded volumes did not account for

---

71   See Cape of Good Hope, *Report of the Superintendent of Woods and Forests for the Year 1889 (Part 1)*; Laughton, *The Sylviculture of the Indigenous Forests*, 25–7, 90–4; Keet, 'Historical Review of the Development of Forestry in South Africa', 48–51. Keet cites an attempt to extend the selection system to the Kalahari woodlands, too late to stop the denudation. See Keet, 'Management of the Indigenous Forests of the Transkei'.
72   A. H. W. Seydack et al., 'An Unconventional Approach to Timber Yield Regulation for Multi-Aged, Multispecies Forests. II. Application to a South African Forest', *Forest Ecology and Management*, 77 (1995): 155–68.
73   Cape of Good Hope, *Report of the Superintendent of Woods and Forests for the Year 1889 (Part 1)*, 9, provides an illustrative case. Vasselot reports having marked 130,524 cubic feet of ironwood for felling, whereas just 10,111 was sold. Of 644,405 cubic feet marked for thinning that year, just over 20 per cent was sold. Similar bias existed later, for different reasons, and still later, the quantities felled exceeded the allowable cut.
74   Laughton, *The Sylviculture of the Indigenous Forests*, 25.

timber wasted in the forests, when woodcutters sawed and removed the prime timber and abandoned the rest). By 1930, best estimates were that the forests could yield 183,000 cubic feet per year, sustainably, not the 500,000 or so being harvested. But in sympathy for the plight of the woodcutters, the authorities could not admit to reducing the felling quotas. Instead of imposing an allowable cut they resolved to allocate a quota of 700 cubic feet per woodcutter per year, and watch the extraction diminish as the number of entitled woodcutters declined with time. It was only with the pensioning of the last woodcutters in 1939, in terms of the *Woodcutter's Annuity Act* (no. 11) of 1939, that the Knysna forests could be managed by the standards reckoned by the foresters.[75]

Figure 11. Woodcutters participating in the allotment of trees for their extraction, Knysna Forest, 1926.

Seymour Laughton describes the procedure: each woodcutter draws a numbered disc from a hat, '[t]he holder of number one has the first choice. He shouts out the number of the tree he wants and its species and his name … They proceed thus until the man holding the highest number has his turn … they work their way back in reverse order … until all the trees have been allotted.' The process is supervised so that each allotment for the year approximates 700 cubic feet, but not more.

Source: George Museum; photographer unknown.

---

75   Ibid., 87–93; Laughton cites Legat's 1930 report as Conservator for much of his account. Keet, 'Historical Review of the Development of Forestry in South Africa', 92; Grundlingh, '"God Het Ons Arm Mense Die Houtjies Gegee"'.

Foresters lacked the scientific information to justify the policies they promoted. They lacked necessary data on the growth rates of the different species of tree, and they needed to adapt their silviculture better to the different types of forest encountered, while assuring successful regeneration in and rehabilitation of the forest after over a century of misuse. Efforts to regenerate valuable species, such as Yellowwood (*Afrocarpus falcatus* and *Podocarpus latifolius*), Stinkwood (*Ocotea bullata*), and Ironwood (*Olea laurifolia*) had proved difficult. Cleared or overworked parts of the forests—'the wounds caused by … heavy exploitation'—were slow to recover, becoming infested with shrubby weeds.[76] Attempts to rehabilitate these by transplanting indigenous trees and by planting Tasmanian Blackwood (*Acacia melanoxylon*) had yielded little success.

Agricultural, botanical and forestry researchers in South Africa agreed in 1920 on the need to pursue research across departments to solve fundamental problems relating to indigenous tree management.[77] Attempts to foster closer cooperation between Pole-Evans's Plant Industry Division and the Forestry Department began with joint meetings to discuss these issues,[78] and in 1922, the department appointed John Phillips to study the Knysna Forest, at the Diepwalle Forest Research Station. Phillips's subsequent experiences in Knysna shaped his criticisms of forestry in the 1930s, and are valuable to study more closely because he became one of the strongest critics of prevailing forest policy and practice, and played a key role challenging forestry ideas at the 1935 British Empire Forestry Conference.

John Phillips grew up outside of Grahamstown in what is now the Eastern Cape Province. He became interested in forestry after participating in the pro-conservation Boy Scouts. While employed as a pupil forester by the Forestry Department in the 'magic mountain country' of the Pirie and Amatola forests near King Williams' Town, his aptitude, and the encouragement of a mentor, R. W. Rose Innes, encouraged him to study botany and forestry at Edinburgh University, as a student sponsored by the department.[79] At Edinburgh, he devoted himself to botany, studying under the tutelage of the then influential morphologists Isaac Bayley Balfour and Frederick Orpen Brower. Phillips was influenced by their approach to botany, which focused on studying a plant's evolutionary history to understand its attributes and distribution. Phillips also engaged with the nascent field of ecology, and was especially attracted to ideas of the Nebraska ecologist Fredric Clements. Clements famously developed the theory of the climatically determined climax community and ecological succession as being the dynamic

---

76   Laughton, *The Sylviculture of the Indigenous Forests*, 109.
77   *Journal of the Department of Agriculture*, 1 (1920), 180.
78   'Notes', *Journal of the Department of Agriculture*, 2 (1921), 3.
79   J. Phillips, *Kwame Nkrumah and the Future of Africa* (London: Faber and Faber, 1960), 16. For a brief outline of his career at Knysna see Anon, 'J.F.V. Phillips Revisits Knynsna Forest', *Bosbounuus/Forestry News* (1984): 12–4; B. W. van Wilgen, 'Introduction to John F. V. Phillips' Article', *Fire Ecology*, 8 (2012): 1–2. Bennett and Kruger, 'Ecology, Forestry and the Debate over Exotic Trees in South Africa', 100–9.

'unswervable' processes of orderly, predictable seral stages culminated in this stable, 'organismic' climax: an evolutionary model that saw the formation of climax communities—organic entities—as the culmination of progressions of a series of increasingly complex and stable communities.[80] The idea that ecological communities always developed progressively towards a climax community, which was an organism in equilibrium, drove much of Phillips's ecological research, and led to his later conflict with the Oxford ecologist Arthur Tansley.

Figure 12. John F. V. Phillips and Jeannie Phillips, Edinburgh, 1920.
Source: Department of Agriculture, Forestry and Fisheries, Pretoria; photographer unknown.

---

80   See D. Worster, *Nature's Economy: A History of Ecological Ideas*, 2nd ed. (Cambridge University Press, 1994), 209–11; Pooley, 'Pressed Flowers', 603. The quote is from Worster. Little is known of Phillips's time in Edinburgh, though his later work bears the imprint of Balfour and Brower. See Phillips, *Kwame Nkrumah*, 17 (note 30).

Robertson, head of the Forestry Department's forestry research branch, appointed Phillips to improve the scientific foundation for the selection system of silviculture.[81] His duties included improving and maintaining the monitoring of forest growth and increment, resuming the measurements on the 23 sample plots established by McNaughton in 1902, but neglected during the succeeding 12 years for want of staff.[82] His duty was to conduct experimental research to address questions about the regeneration of the forest: how regeneration varied between moist and dry sites, and how this was determined by the effects of silvicultural operations on light near the ground, and soil moisture.

Phillips believed that foresters should use ecology to study the *whole* forest rather than focusing only on economically valuable *trees*. This view frequently put him at odds with many of South Africa's first- and second-generation foresters. Leading colonial foresters, such as David Ernest Hutchins, had emphasised that ecology was a tool to understand the 'relation of trees to their environment'.[83] For Hutchins, ecology was not a study of relationships merely for the sake of relationships—it was for the effective manipulation of valuable trees. The differences in outlook between Phillips and the first generation of colonial foresters became apparent on his arrival. In a 1924 letter written to the former Cape forest surveyor and botanical expert of Knysna's flora, Henry Fourcade, Phillips included a draft publication of his own work and a copy of Fredric Clements's 'wonderful book', *Plant Succession*.[84] Fourcade demurred by writing to Phillips, 'I do not quite share your enthusiasm over Clements work. To my mind it is undoubtedly valuable, but seriously marred by his ... mania for coining new words',[85] echoing the critique of Clementsian ecology as 'a vocation for turning out glossaries', in Worster's words, or Keet's comment at the fourth British Empire Forestry Conference, expressing the hope that ecology 'is not merely a case of adding new words to our vocabulary'.[86] This perception bedevilled his relations with colleagues, as much as any other issue (see Chapter 6). Still, Phillips received the freedom to design his own research agenda at Diepwalle, with support from Robertson. Robertson's sympathy perhaps related to his having been a key contributor to Illtyd Buller Pole-Evans's ecologically informed, national botanical survey for South Africa;[87] he saw the botanical survey as the key to unlocking the 'secrets of the reasons for the distribution of our forest and of the species comprising them'.[88]

---

81   C. C. Robertson to all conservators, 8 June 1923, circular minute 1360 R. 10, FBT 1/3, NASA-CT.
82   Bennett and Kruger, 'Ecology, Forestry and the Debate over Exotic Trees in South Africa'.
83   D. E. Hutchins to Charles Curry, 24 July 1901, B 872/10, AGR 723, NASA-CT.
84   J. Phillips to H. Fourcade, 4 July 1924, Fourcade Bequest BC 246, UCT.
85   H. Fourcade to J. Phillips, 24 July 1924, Fourcade Bequest BC 246, UCT.
86   Worster, *Nature's Economy: A History of Ecological Ideas*.
87   Anker, *Imperial Ecology*, 65–6.
88   Botanical Survey, 27 May 1921, Minute No. F 4226/R 3500, FBT 1/3, NASA-CT.

Diepwalle was an extensive experimental forest with good nursery facilities and many silvicultural trials underway, a natural laboratory where Phillips developed many of his most important ideas about ecology. Here, his research exposed imported European and North American ecological theories to the facts from detailed local fieldwork. His program proved enormously fruitful, yielding over 19 publications in influential journals, covering a wide range of topics. Forest growth and increment measurements were put on a sound footing, and continued later by F. S. Laughton and his successors. His important studies focused on the natural and artificial reproduction of tree species, tree mortality, tree diseases, the ecological impact of exotic species on indigenous tree species reproduction, and descriptions of the microclimate, soil moisture regimes, and a variety of other important forest conditions. He studied the ecological interaction between plants and mammals and birds, the effect of the naturalisation of Tasmanian Blackwood on forest regeneration, and wrote theoretical papers on experimental ecology.[89] In 1909, in an attempt to encourage reproduction, the Forestry Department itself had begun to plant the faster-growing Blackwood to provide shade for light-sensitive indigenous species. Others saw Blackwood as a more economic replacement of the shrinking forests. Some even asked the government to harvest the remaining ancient, 'over-mature' trees and plant exotic forests in their place.[90] Phillips noted that '[r]egeneration of native species was not only very, very rare under stands of "Blackwood", but also appeared moribund'.[91] He designed tests to measure how Blackwood influenced soil moisture and light intensity, which in turn affected the growth of indigenous tree species. The results of the experiment demonstrated that Blackwood lowered the soil moisture and light intensity, making it nearly impossible for native tree species to grow. Yet Phillips thought that, without human disturbance, Blackwood would not become 'uncontrollable' throughout Knysna because

89   J. F. V. Phillips, '*Platyophus trifoliatus* Don: A Contribution to Its Ecology', *South African Journal of Science*, 22 (1925): 44–60; J. F. V. Phillips, 'The Propagation of "Stinkwood" (*Ocotea bullata* E. Mey.) by Vegetative Means', *South African Journal of Science*, 23 (1926): 418–43; J. F. V. Phillips, 'Fossil *Widdringtonia* in Lignite of the Knysna Series with a Note on Fossil Leaves of Several Other Species', *South African Journal of Science*, 24 (1927): 188–97; J. F. V. Phillips, 'Dendrographic Experiments: *Ocotea bullata* E. Mey. (Stinkwood)', *South African Journal of Science*, 24 (1927): 227–43; J. F. V. Phillips, '*Faurea Macnaughtonii* Phill. ("Terblanz"): A Note on Its Ecology and Distribution', *Transactions of the Royal Society of South Africa*, 14 (1926): 317–36; J. F. V. Phillips, 'Experimental Vegetation: A Second Experiment', *South African Journal of Science*, 24 (1927): 259–68; J. F. V. Phillips, 'Plant Indicators in the Knysna Region', *South African Journal of Science*, 25 (1928): 202–24; J. F. V. Phillips, 'Mortality in the Flowers, Fruits and Young Regeneration of Trees in the Knysna Forests of South Africa', *Ecology*, 8 (1927): 435–44; J. F. V. Phillips, '*Olea laurifolia* Lam. ("Ironwood"): An Introduction to Its Ecology', *Transactions of the Royal Society of South Africa*, 16 (1928): 169–90; J. F. V. Phillips, 'The Principal Forest Types in the Knysna Region—an Outline', *South African Journal of Science*, 25 (1928): 188–201; J. F. V. Phillips, '*Curtisia faginea* Ait. ("Assegaai"): An Ecological Note', *Transactions of the Royal Society of South Africa*, 17 (1930): 29–41; J. F. V. Phillips, 'The Behaviour of *Acacia melanoxylon* R. Br. ("Tasmanian Blackwood") in the Knysna Forests: An Ecological Study', *Transactions of the Royal Society of South Africa*, 16 (1928): 31–43; J. F. V. Phillips, 'The Influence of *Usnea* Sp. (near *barbata* Fr.) upon the Supporting Tree', *Transactions of the Royal Society of South Africa*, 17 (1929): 101–7.
90   See file, working of Knysna and other Midlands Forests, A225, FOR 158, NASA-P.
91   Phillips, 'The Behaviour of Acacia Melanoxylon', 3.

they could not grow under the dense forest canopy unless humans or elephants made openings.[92] Through his research on Blackwood, Phillips gained important insights that informed the study of forest regeneration in general.

The recalcitrance of indigenous trees proved almost impossible to overcome. For example, attempts to multiply Stinkwood artificially proved so difficult that Phillips wrote that he was 'forced to conclude that the propagation of *Ocotea bullata* by vegetative means is not possible on a practical scale'.[93] While transplanting to rehabilitate degraded forest would continue, matching silvicultural treatment to site remained a priority concern.

To address the problem of matching silvicultural treatment to site conditions, Phillips following the ideas of A. K. Cajander, settling on the use of plant indicator species in the ground layer for the classification of forest types. Cajander was a forester and Finnish nationalist, whose main contribution to forest science was his theory of 'forest types'. He argued that nature produced distinct ecological forest types that could be understood by studying 'indicator species' found among the ground flora, not the dominant trees.[94] Cajander's theoretical ideas also developed as a response to growing northern European, particularly German, concerns about the sustainability of monoculture timber plantations and clear-cutting.[95] Cajander took part in a northern European movement that advocated a more organic model of forestry based upon an intricate knowledge and replication of the ecological and biological processes of natural forests.[96] The view perhaps found sympathy in Phillips, while the fact that Cajander dealt with a wide range of site types, from bog forest to dry, very likely promised solutions to Phillips's classification problem. From the indicator species approach, Phillips generated a set of broad categories that distinguished soil moisture regimes, from moist to dry, which correlated with the response of the forest, especially the ground flora and soil moisture regime, to the degree of opening of the canopy of the forest through silvicultural treatment, one of the principal concerns to foresters in Knysna.

Phillips's emphasis on studying ground species indicators led him to become dissatisfied with the silvicultural paradigm of forestry then being pursued in South Africa. In his 1928 article devoted to Cajander's method, 'Plant Indicators in the Knysna Region' in the *South African Journal of Science*, he reproved foresters as 'they consistently failed to formulate a system of indicators, because their real attentions were directed towards the growth and silviculture of stands

---

92   Ibid., 42.
93   Phillips, 'The Propagation of "Stinkwood" (*Ocotea bullata* E. Mey.) by Vegetative Means', 433.
94   A. K. Cajander, 'Über Walden Typen', *Acta Forestalia Fennica*, 1 (1909).
95   J. Raumolin, 'The Formation of the Sustained Yield Forestry System in Finland', in H. K. Steen (ed.), *History of Sustained-Yield Forestry: A Symposium* (Santa Cruz: Forest History Society, 1983), 159–60.
96   M. L. Anderson, 'Review of 'Forest Types and Their Significance', *Forestry*, 24 (1951): 72–3.

of trees, and not towards the prime factors of the habitats occupied by such'.[97] Whereas Cajander used his theory to show an ecological difference between Russia and Finland, Phillips used his ideas to critique exotic afforestation in South Africa.

Phillips drew his entire research program together through the framework of Clementsian succession theory. He used this scheme to make sense of the overall complexity of the Knysna forest through the model of succession to the climax, tracing the conjectural pathways of forest change, from the 'hydrosere' at the wet end of the ecological spectrum, to the 'xerosere' at the dry. This again was the theoretical frame for linking silviculture to site type. His research culminated in a PhD in Ecology from Edinburgh University in 1927, 'Forest-Succession and Ecology in the Knysna Region', published as a monograph in Pole-Evans's Botanical Survey of South Africa series.[98]

In 1925, he fell under the influence of Jan Smuts, the former war general, prime minister and amateur botanist, after meeting and hiking the mountains with Smuts at the South African Association for the Advancement of Science meeting at Oudtshoorn in the Cape.[99] Smuts convinced Phillips of the importance of Holism, explained in his 1926 publication *Holism and Evolution*. Smuts argued that individuals in nature (such as cells, individual animals or plants) tended to cooperate in a specific direction, leading towards 'wholes'. He suggested 'that groups, societies, nations, and Nature are *organic without being organisms*'.[100] Phillips took these ideas to his next post.

Smuts's holistic concept strongly influenced Phillips's thinking on ecology during this period; soon after, Phillips formulated his idea of the 'biotic community', the ecological assemblage as something 'more than the sum of its parts', in a series of three papers in the *Journal of Ecology*.[101] The biotic community he saw as comprised of interdependent species that had progressively evolved into a holistic, organismic system. Phillips's writings did not stand criticism from A. G. Tansley, who invalidated the ideas on logical as well as empirical grounds.[102] But Phillips's observations of forest ecology, free of philosophical elaboration, did provide a valid analysis of much of the dynamics of these forests.

---

97   Phillips, 'Plant Indicators in the Knysna Region'; Phillips, 'The Principal Forest Types in the Knysna Region—an Outline'.

98   Later published as J. F. V. Phillips, 'Fossil *Widdringtonia* in Lignite of the Knysna Series with a Note on Fossil Leaves of Several Other Species'.

99   See Anker's careful study of their relationship. Anker, *Imperial Ecology Environmental Order in the British Empire, 1895–1945*, Chapter 2.

100   J. F. V. Phillips, 'The Biotic Community', *Journal of Ecology*, 19 (1931): 1–24.

101   J. F. V. Phillips, 'Succession, Development, and the Complex Organism I: An Analysis of Concepts', *Journal of Ecology*, 22 (1934): 554–71; J. F. V. Phillips, 'Succession, Development, and the Complex Organism II: Development and the Climax', *Journal of Ecology*, 23 (1935): 210–36.

102   A. G. Tansley, 'The Use and Abuse of Vegetational Concepts and Terms', *Ecology*, 16 (1935): 282–307.

Figure 13. Members of the South African delegation entraining for Cape Town to embark for London and the September 1931 Centenary Meeting of the British Association for the Advancement of Science.

General Jan Smuts, centre, Jeannie Phillips fourth from right, John F. V. Phillips third from right, Colonel Deneys Reitz, second from right. The forestry network was close: Deneys Reitz served as President of the South African Forestry Association from 1934, while J. D. M. Keet was Chairman.

Source: From a print possessed by James Paterson, grandson of John Phillips, and reproduced with his permission.

In 1931, Phillips offered a stinging ecological critique of exotic tree planting. His ecological criticisms, based on his idea of the 'biotic community' and his previous work in forestry, differed fundamentally from previous criticisms in South Africa. He did not justify his critique based upon their unsightliness or the utilitarian argument that exotic trees used more water than indigenous vegetation. Rather, he blamed exotic trees for inhibiting the functioning of indigenous biotic communities.[103] He blamed *Eucalyptus* for reducing the bee population in Knysna (a requirement for reproduction of some indigenous trees) and criticised Australian *Acacia* for inhibiting the reproduction of native trees by decreasing the overall soil moisture and depleting soil fertility.[104]

Phillips's research gave weight to growing criticisms of the difference in ecological and hydrological properties between indigenous forests and plantations. Scientific experts and members of the public in the Cape Province

103   J. F. V. Phillips, 'Ecology the Foundation of Forestry', *Empire Forestry Journal*, 10 (1931): 86–105. This article was also reprinted as a pamphlet by the South African government. We cite the reprint.
104   Phillips, 'The Behaviour of *Acacia melanoxylon*', 14.

began to challenge the argument that all trees increased streamflow and had positive climatic influences. Alfred Paetzel noted in his *Farmer's Weekly* letter that 'the most striking feature of difference [between indigenous and exotic forests] is the total absence of any undergrowth—that most valuable vegetation for the conservation of moisture'.[105] Arguments about the nature of the forest floor and its soil played a key role in later debates about whether plantations stopped or encouraged soil erosion.

From the point of view of his campaign to influence forest policy and practice, however, John Phillips had organised his ideas, and his policy views, in three key papers, which took his conceptual analysis into the realm of practice. In his 1928 paper in *Nature*, 'Influence of Forest Formation on Soil Moisture', Phillips reasons, from a Clementsian perspective, that the development of forest, natural or planted, on sites at the wet end of the soil-moisture spectrum (the hydrosere) will progressively dry out the soil, whereas development on the drier sites— the lithosere and the psammosere—will, stage by stage, add soil moisture, a view that echoed Hutchins's concern about the 'double effect' of forests on the water balance (see earlier). He buttressed his argument with evidence from his own observations and experiments involving eucalypts and pines, as well as indigenous forest treatment, including streamflow observations. He also concluded that 'Water-voracious indigenous plants not natural to the particular sere, and demanding exotics, draw strongly upon the soil moisture no matter what the successional history of the site'. So he was addressing the same questions as his forestry colleagues, with a little more science.[106]

His 1931 paper in the *Journal of Ecology*, 'Ecological Investigation in South, Central and East Africa: An Outline of a Progressive Scheme', is a broad schedule of ideas explicitly stated as the framework for a regional program in ecology that gives prominence to the biotic formation as the category for planning, anticipating the biome-based mapping to come in South Africa, by John Bews in 1936[107] and John Acocks in 1953.[108] He argues that the natural class of the biotic community would be the 'main indicator significance' from which to infer aspects of 'practical significance', such as 'the development of progressive forest policies in the sphere of conservation, re-forestation, afforestation, silvicultural

---

105   Paetzold, 'Mountain Erosion', 48.

106   J. F. V. Phillips, 'Influence of Forest Formation on Soil Moisture', *Nature*, 122 (1928): 53–4, 53.

107   Bews had anticipated this form of ecological classification in his 1916 work 'An Account of the Chief Types of Vegetation in South Africa, with Notes on Plant Succession'. See Pooley, *Burning Table Mountain*, 51.

108   J. F. V. Phillips, 'Ecological Investigation in South, Central and East Africa: An Outline of a Progressive Scheme', *Journal of Ecology*, 19 (1931): 474–82; J. P. H. Acocks, *Veld Types of South Africa*, Botanical Research Institute, Pretoria (1953); Rutherford and Mucina, 'Introduction', in *The Vegetation of South Africa, Lesotho and Swaziland*, 3–11; Rutherford and Mucina do not acknowledge Phillips.

management, and forest protection', and 'the conservation of moisture in natural water-catchment regions. In this paper he foreshadows much of what was to come in South Africa's natural resource management approaches.

The third paper, 'Ecology the Foundation of Forestry', in the 1931 *Empire Forestry Journal*, argues that 'forestry virtually is applied ecology'. He promotes land classification and land policies on Clementsian principles as in his *Journal of Ecology* paper; the classification of forest types, both for 'natural or artificially-raised stands of trees'; and silvicultural treatment 'prescribed to meet the particular requirements of each type'.

Taken together, the ideas in these three papers mirror closely the concerns of Phillips's other colleagues in forestry; leading figures such as Ian Craib, J. J. Kotzé, J. D. M. Keet all shared a common interest with Phillips. They were in a large degree wrestling with the same concepts and problems, though the latter worked outside the Clementsian philosophical framework. But with the background of his substantial body of ecological research, and the practice analysed in these three papers, Phillips would have felt ready and confident to engage in the public forest policy debate anticipated in the forthcoming fourth British Empire Forestry Conference. His paper, 'Ecology the Foundation of Forestry', concluded determinedly that 'possibly some of the points raised may be considered as provocative! If indeed these have the effect of persuading a few more foresters that all is not so clear cut as some text-books at least aver, then my writing of this note shall not have been altogether in vain'.[109]

While Phillips had developed a scientific critique of plantation forest programs, J. D. M. Keet was in effect the official voice of forest policy, but as the author of this policy, not merely the messenger. His views derived not from prolific research within a single forest, but from extensive practice, detailed field investigations, and the lessons learned from dealings with workers, field managers, politicians, and his official peers. He had succeeded in harmonising policy for indigenous forests, plantations, and catchment management, had played the larger role in resolving the question of woodcutters, stabilised the white forestry settlement program, and facilitated progress in silvicultural and wood properties innovation. He was sure of his direction.

---

109   Phillips, 'Ecology the Foundation of Forestry', 13.

Figure 14. Johan D. M. Keet at the foot of an enormous Stinkwood (*Ocotea bullata*), Diepwalle Forest, c. 1920.

Keet served as District Forest Officer with responsibility for the management of the Knysna Forest from about 1918 to 1923, coinciding with part of John Phillips's period of research at Diepwalle.

Source: Department of Agriculture, Forestry and Fisheries; photographer unknown.

Conclusion

Forces creating the overarching field of change were the politics of rural development, and the politics of Afrikaner identity. Against this backdrop, the afforestation program emerged as a focal point for conflict among rural sectors about water supplies, and later, about the protection of natural flora and landscapes. Vigorous arguments against enthusiastic claims about the water resource benefits of plantations led early on to foresters paying serious attention to water supplies in forestry regions, the influence of the cycles of drought and rain, and whether or not claimed afforestation effects were real. Early steps in an experimental program were taken. A pioneering program of ecology in the indigenous forests of Knysna provided an intellectual counter to 'forestry enthusiasts', providing a necessary counter-position in the dissent about forests. Strong personalities, well informed and passionate about the discipline, confident in their success, stood ready for their parts in the major policy forum provided by the forthcoming fourth British Empire Forestry Conference.

Chapter 6

1935: The Fourth British Empire Forestry Conference in South Africa and the Origins of a Consensus Science Program

The significance of the mighty words of Humboldt that 'by felling trees, which are adapted to the slopes and summits of mountains, men in every climate prepare for future ages at once two calamities: want of wood [sic] and scarcity of water seem no longer to be realised or heeded.[1]

—J. D. M. Keet, then Chief of the Division of Forestry in the Department of Agriculture, Union of South Africa, in his presentation to the fourth British Empire Forestry Conference.

Prelude to the fourth British Empire Forestry Conference, 1931–1935

The first British Empire Forestry Conference was held in London in 1920 in response to the British government's establishment of a Forestry Commission after World War I. British Empire Forestry Conferences held in 1920, 1923 (Canada), and 1928 (Australia) brought together foresters and imperial officials to share information, encourage inter-colonial trade, develop important statistics and coordinate central management problems.[2] Each conference had a significant local context and outcome. The Australian 1928 meeting, for instance,

---

1   Union of South Africa, *Proceedings of the Fourth British Empire Forestry Conference*, 116.
2   Rajan, *Modernizing Nature*, 113–29; J. M. Powell, 'Dominion over Palm and Pine: The British Empire Forestry Conferences 1920–47', *Journal of Historical Geography*, 33 (2007): 852–77.

was instrumental in saving the Australian Forestry School in Canberra.[3] So too would the fourth British Empire Forestry Conference have a lasting influence on South African forest history.

South Africa was to have hosted the fourth British Empire Forestry Conference in 1932. The conference was delayed by three years because of the Depression and South Africa's decision to postpone the meeting until 1935 because of financial problems. The three-year delay may have fundamentally changed the nature of the conference. In 1934, the government placed the control of the Forestry Department (which became the Forestry Division) under the Agricultural Secretary, creating a joint Agriculture and Forestry Department, with Colonel Deneys Reitz as Minister of Agriculture and Forestry. Reitz strongly sympathised with farmers who worried about the negative effects of tree planting in catchments, while also being sympathetic to foresters in his role as President of the South African Forestry Association from 1935 to 1939.[4]

By 1935, plantation forests in South Africa extended to about 400,000 hectares, with around 90,000 planted for sawlog production.[5] The Union government foresaw around 400,000 hectares of new sawlog plantations to substitute for sawtimber imports. Foresters, increasingly confident in their ability to profitably grow exotic trees because of new scientific innovations, saw a bright future for plantation forestry. This enthusiasm was tempered by shifting political winds and a growing scientific and agricultural critique of exotic timber plantations. Afforestation had accelerated by 1935 and would see an increase in area by 1 million hectares over the following 60 years.[6] The intense contest over water resources, and indeed the protection of the country's flora and fauna, may easily have headed off the trajectory.

In the 1930s, in South Africa and elsewhere, professionals and scientists in the fields of forestry and hydrology had no scientific evidence available to them of the relationships between forests, climate and hydrology adequate to the questions that they faced. Instead, they had a set of disparate observations, mainly from Europe, sundry strands of thought, usually tendentious, many myths from historically influential figures, and ancient and recent anecdotes, all loosely and variously woven together to create a polemicised story about forests, water and floods that could suit the teller and the audience. Such experimental work that had been done mostly had political, rather than scientific, purpose.

---

3   Bennett, 'An Imperial, National, and State Debate'.
4   See PSC 4/6, SDK 37, NASA-P.
5   Union of South Africa, *Proceedings of the Fourth British Empire Forestry Conference*, 98; Keet, 'Historical Review of the Development of Forestry in South Africa'.
6   D. W. van der Zel, 'Sustainable Industrial Afforestation in South Africa under Water and Other Environmental Pressures', in *Sustainability of Water Resources under Increasing Uncertainly (Proceedings of the Rabat Symposium S1, April 1997)*, IAHS Publication No. 24 (1997): 217–25, 220.

The orthodox view traceable to books by Alexander von Humboldt and Jean-Baptiste Boussingault in the early 1800s following their enquiry into the shrinking of Lake Valencia in Venezuela, and the apparent desiccation of the western aspect of Peru, was the notion of 'climatic forestry', with several standard rubrics: deforestation dries up streams, forests 'economize and regulate their flow', deforestation causes the diminution of rainfall, and so on.[7] This paradigm was in turn set within the conviction of a nature in steady-state equilibrium, and the 'pristine myth'—where nature is found to be in disequilibrium, this owes to human action; ideas which continue today to 'haunt the sciences of climatology, ecology, and conservation'.[8] And in a large degree through the force of the American George Perkins Marsh's authorship, these Humboldtian ideas underlie much of modern environmentalism, as they did in the 1930s.

Although in South Africa John Croumbie Brown had written extensively on the topic of climatic forestry,[9] in professional forestry circles it was Henry Fourcade who first paid careful attention to the authoritative reports on this question, and detailed his summary in his 1889 *Report on the Natal Forests*. In a chapter titled 'On the Utility of the Forests', Fourcade deals with forests and rainfall, forests in relation to climate, erosion, catchment degradation following deforestation, and the benefits of forests in flood control. He cites observations from the 1860s and later in France, Germany, Italy and Russia, as well as his own acute observations on local weather and climate. His authorities include Alexandre Surell, George Perkins Marsh, Ernst Ebermayer, Robert Hartig, John Tyndall, and D. E. Hutchins on rainfall fluctuations. On rainfall, he concludes that 'forests do not materially, [but] they unquestionably regulate it, promote the frequency of showers, control the flow of water, which, on the whole, is a preferable effect', and emphasises the reports from Europe of the decline in river flow attributed to deforestation. Fourcade describes the rainfall, topography and erosion in Natal in detail, drawing the analogy with the south of France, and concludes '[b]esides being cheaper, it is also better to preserve natural forests than to destroy them to substitute plantations which take centuries to acquire the deep layer of humus and the surface growth which give such a peculiar climatic value to the natural forest', citing B. E. Fernow in the United States on the value of forests in the 'rational management of the water capital'.

---

7   Cushman, 'Humboldtian Science, Creole Meteorology, and the Discovery of Human-Caused Climate Change in South America', 38. The quote is from J. B. Boussinghaut in 1845. V. Andréassian, 'Review: Waters and Forests: From Historical Controversy to Scientific Debate', *Journal of Hydrology*, 291 (2004): 1–27.

8   Cushman, 'Humboldtian Science, Creole Meteorology, and the Discovery of Human-Caused Climate Change in South America', 41, 44.

9   See Beinart, *The Rise of Conservation in South Africa*.

When characterising the ideal humus and soil beneath the forest, Fourcade echoed Marsh's vision of the ideal forest floor condition—the characteristic of the 'true forest'. Marsh summarised his views on the general consequences of the destruction of the forest:

> With the disappearance of the forest, all is changed ... the melting snows and vernal rains, no longer absorbed by a loose and bibulous vegetable mould, rush over the frozen surface, and pour down the valleys seaward, instead of filling a retentive bed of absorbent earth, and storing up a supply of moisture to feed perennial springs ... The face of the earth is no longer a sponge, but a dust heap, and the floods which the waters of the sky pour over it hurry swiftly among its slopes, carrying in suspension vast quantities of earthy particles ... The rivulets, wanting their former regularity of supply and deprived of the protecting shade of the woods, are heated, evaporated, and thus reduced in their summer currents, but swollen to raging torrents in autumn and in spring.[10]

Ideas of what a forest is circled around these characterisations of climatic forestry, and much of the debate about 'forest influences' at the forthcoming conference reduced the question to whether plantations were *true* forests, and whether the state of the forest floor and its humus was similar with a natural forest. Clearly forewarned, Jan Smuts took care in his speech to emphasise the fact that South Africa was 'not a forest country', that unlike 'the other parts of the world' the delegates came from—'the climax there is forest'. In South Africa, he noted that 'the climax is something quite different'. Smuts believed that '[s]oil development is a complex phenomenon in South Africa, where the rate of decomposition of organic matter is very rapid, and where formation of the ideal humus rarely takes place'.[11]

George Perkins Marsh helped politicise forestry in the United States, inspiring a series of developments leading to new forest law in the US, and the acquisition and reservation there of millions of hectares as national forest under the banner of protecting the 'water capital'.[12] Bernhard Fernow, the first professionally trained forester to head the then Division of Forestry in the US Agriculture Department, from 1886 to 1898,[13] took up the polemics of forest management in favour of the water capital. In 1893, he published a book called *Forest Influences*, in which he reviewed the ideas of the time on the effects of forests

---

10   George Perkins Marsh, *Man and Nature, Or Physical Geography as Modified By Human Action*. Edited by David Lowenthal (The Belknap Press of the Harvard University Press, 1965), 186–7.

11   Union of South Africa, *Proceedings of the Fourth British Empire Forestry Conference*, 79–81.

12   K. J. McGuire and G. E. Likens, 'Historical Roots of Forest Hydrology and Biogeochemistry', in D. F. Levia, D. Carlyle-Moses, and T. Tanaki (eds), *Forest Hydrology and Biogeochemistry: Synthesis of Past Research and Future Directions* (Berlin: Springer-Verlag, 2011): 3–26.

13   A. D. Rodgers, *Bernhard Eduard Fernow: A Story of North American Forestry* (Princeton: Princeton University Press, 1954); Fernow would later employ the idea of climatic forestry to motivate the planting of a vast shelterbelt of trees to prevent the spread of the dustbowl in the American Mid-west, 100 miles wide and 1,150 miles long.

on climate and rainfall, effects which he acknowledged were difficult to prove.[14] Reasoning from first principles, as well as empirical evidence from Europe and India, he argued: 'without forest management no rational water management is possible'.[15] Gifford Pinchot, the first Chief of the US Forest Service (1905–1911), knew France well since he had studied at the forestry school of Nancy and was familiar with the early forest hydrological work there.[16] Writing in 1905, he emphasised: 'It is unfortunate that so much of the writing and talking upon this branch of forestry has had little definite fact or trustworthy observation behind it. The friends and the enemies of the forest have both said more than they could prove'.[17] (And, by 1912, Zon also was warning that the US Forest Service had overstated the significance of forests to flood control.)[18]

These ideas characterised what might be termed the 'propaganda period' in forestry, a period when forestry advocates used anecdotal evidence and professional authority to justify the setting aside of vast swaths of forest for climatic and hydrological purposes. Some of the most vivid illustrations come from the USA, where foresters such as Gifford Pinchot argued strenuously for the expansion of US Forest Service powers to manage catchments. The conflict over the Hetch Hetchy Valley, where it was proposed to build a dam to provide San Francisco with water, was one of a series of conflicts regarding the role of forestry and dams in water hydrology management.[19]

At their height of power in the 1910s, US Forest Service officials sought to have Congress authorise the federal purchase of large areas in the east to restore forests, and argued for this on the basis of flood-mitigation claims and the benefits to navigable waters, seeking power and fiscal provision for this purpose under the Weeks Law of 1 March 1911. This law required that forest land be acquired by the Forest Service only in those cases where 'the purchased lands would be permanently maintained as federal forest reserves as a way to protect navigable waterways'.[20] To meet the requirements of Congress for evidence of forest benefits to navigable waterways, investigators executed a dubious experiment in

---

14   B. E. Fernow, *Forest Influences*, US Department of Agriculture, Forestry Division. Bulletin no. 7 (Washington: Government Printing Office, 1893).
15   Ibid., 12.
16   Andréassian, 'Review: Waters and Forests', 7.
17   G. Pinchot, *A Primer of Forestry Part II: Practical Forestry* (Washington: Government Printing Office, 1905), 56.
18   R. Zon, *Forests and Water in the Light of Scientific Investigation*, Reprinted with Revised Bibliography, from Appendix V of the Final Report of the National Waterways Commission, 1912. (Senate Document No. 469, 62nd Congress, 2nd Session), 1912.
19   Conflicts raged throughout the US as foresters tried to have vast catchments set aside for hydrological purposes. One of the most well-known examples—then and now—was the fight over the Hetch Hetchy Valley. See R. W. Righter, *The Battle over Hetch Hetchy: America's Most Controversial Dam and the Birth of Modern Environmentalism* (New York: Oxford University Press, 2005).
20   M. Williams, *Americans and Their Forests: A Historical Geography* (New York: Cambridge University Press, 1989), 454.

the White Mountains of New Hampshire. The US Geological Survey conducted a brief study of 11 watersheds during the spring and summer of 1911–1912, compared two catchments—one that had been cut-over and another that had not—during a spateflow, and without any prior data of the previous streamflow of the two catchments, concluded that the forested area had less runoff, and hence less damaging floods, than the deforested: 'a given amount of precipitation on a deforested or burned area causes a sharper flood wave and a greater flood flow than does the same precipitation on a tree-covered area; further, that the streams draining the former are not flow-sustaining during long rainless periods as well as those draining the latter'.[21] This evidence was submitted to Congress, which used it to justify the application of the Weeks Law, an action that led to the subsequent purchase of 6 million acres of privately owned land in the American east.[22]

Foresters in South Africa who thought about forests and water were connected to the ideas of Humboldt, George Perkins Marsh and Bernhard Fernow through the work of Henry Fourcade as well as through their education, and used the idea of climatic forestry to motivate the public interest in plantation forests. Their colleagues in the irrigation sector, such as F. E. Kanthack and C. D. H. Braine, allied with them in this argument.[23] Forestry was an environmentalists' domain in the northern hemisphere, and strongly politicised. But in South Africa, sceptical questions about climatic forestry had led as early as 1910 to the early catchment experiment at Jessievale, while J. D. M. Keet, as much as he promoted plantation forestry, took the sceptics seriously by his careful enquiries into cases of ostensible stream declines caused by afforestation. The fourth British Empire Forestry Conference, populated as it was by articulate proponents of climatic forestry,[24] would prove to be the forum where this orthodoxy would encounter full on the arguments of the sceptics. It was in a large degree owing to the intellectual leadership of Jan Smuts and Deneys Reitz that the dissent at this conference yielded a positive, scientific synthesis of ideas, rather than a one-sided loss.

21   G. O. Smith, Preliminary statement on White Mountains, New Hampshire, US Geological Survey Report No. 13 (1912). This is streamflow study that helped justify the passing of the Weeks Law. Unpublished manuscript available at FHS, Durham, NC.

22   See, for example, J. D. Hewlett, *The Principles of Forest Hydrology* (Athens, OH: University of Georgia Press, 1982), 5; W. G. Hoyt and W. B. Langbein, *Floods* (Princeton: Princeton University Press, 1955), 461, 155, 182–6.

23   Beinart, *The Rise of Conservation in South Africa*, 178.

24   Delegates included R. S. Troup, then of the Imperial Forestry Institute at Oxford. Troup had been a student of Wilhelm Schlich at Cooper's Hill, who in turn had been mentor of Gifford Pinchot. Sir Roy Robinson of Australia had also passed through Schlich's hands. C. G. Trevor quoted the same sources on desiccation in the ancient world as did George Perkins Marsh.

Public and scientific criticism of afforestation reached a crescendo prior to the conference.[25] That year Parliament had confirmed the Forestry Division in its role as manager of the nation's catchments, a controversial decision that shaped hydrological research in South Africa for the rest of the twentieth century. The Professor of Botany at Witwatersrand University, John Phillips, and the director of the South African National Botanical Survey, Illtyd Pole-Evans, tried to use their connections with Jan Smuts and other Afrikaner agricultural officials to prevent afforestation at various sites on the basis of their negative impact on streamflow and on the natural flora.[26] The Minister for Agriculture and Forestry, Colonel Deneys Reitz, confronted by questions in Parliament, deferred making a decision based on these concerns until he had received the advice of the Conference.

Foresters and critics of forestry staked out their positions on afforestation months before the conference began. In 1931, Phillips had returned to South Africa to take up the post of Professor of Botany at Witwatersrand University in Johannesburg, Smuts having lobbied on his behalf for the appointment.[27] Phillips became actively involved in forestry politics. Phillips and Pole-Evans led a behind-the-scenes lobbying against the Forestry Division's policies, taking forward Pole-Evans's campaign.[28] Before the meeting, Phillips and Pole-Evans sought to convince Smuts and Reitz about the negative impacts of exotic plantations. Pole-Evans used his position as the Chief of the Division of Plant Industry in the department to appeal to Reitz to shut down exotic afforestation near streams and rivers. In a memorandum dated 4 March 1935, he declared that afforestation 'will inevitably destroy ... the streams' and the flora. Reitz did halt the afforestation in one instance, though the area concerned was small.[29] At the same time, Pole-Evans wrote to Smuts about his letter, telling him that foresters made 'a great mistake in planting these exotic trees' and that the effects on water 'were becoming apparent all over the country'.[30] Pole-Evans also began an exchange of memoranda across the desks of their joint head of department, P. R. Viljoen, Reitz as Minister, and Keet, the Chief of the Division of Forest Management in South Africa, regarding his concerns about the desiccating effects of exotic trees planted near water sources.

---

25   For examples see debate between Keet and Pole-Evans in the memorandum and correspondence, Keet, 'Memorandum on Afforestation in Relation to Natural Flora and the Effect of Afforestation on Water Supplies', 15 July 1935, FOR 330, NASA-P.

26   Also see Phillips to Smuts, 3 March 1935, Smuts Collection, 238/53, NASA-P; Phillips to Smuts, 4 August 1935, Smuts Collection, 238/53, NASA-P.

27   Anker, *Imperial Ecology*, 144.

28   Keet, 'Memorandum on Afforestation in Relation to Natural Flora and the Effect of Afforestation on Water Supplies', 15 July 1935, FOR 330, NASA-P.

29   Ibid.

30   Pole-Evans to Smuts, 8 March 1935, Smuts Collection, 238/53, NASA-P.

Pole-Evans was not able to attend the conference, and he proposed Phillips as his substitute. Phillips also corresponded privately with Smuts during the conference. Keet had adamantly opposed this substitution, because Phillips was a strong critic of exotic afforestation and an ally of Smuts. Phillips was aware of the heated debate between Keet and Pole-Evans. He wanted to represent Smuts and Pole-Evans but sought a compromise by asking Smuts to help solve 'water-supply controversy that has been raging between Dr Pole-Evans and Keet'.[31] Keet was not convinced by Pole-Evans's arguments: his own investigations of several cases of complaint in the field persuaded him that the current claims of afforestation effects were groundless, because streamflow decline could be ascribed to the current droughts, or because apparent shortage was owing to overestimates by would-be irrigators of the streamflow available for their new farms, or because effects could not be attributed to the plantations, their being too small in extent, or too young, to have decreased streamflow.

Despite the public concerns and behind-the-scenes moves on forestry and water, the approach by South African delegates and participants to the conference was reasoned and measured. Jan Smuts had prepared a speech emphasising the economic necessity of plantation forestry in South Africa and hence the necessity of a comprehensive research program, while Deneys Reitz provided a cool appraisal of farmers' concerns and his own observations about forests and water, while respecting, and challenging, the opinions of his forestry officials. Two months before the meeting, Keet distributed an official memorandum to all his professional staff and all concerned parties, including Phillips, briefing them on 'the whole question' of exotic trees and their impact, which he identified as 'a major issue for discussion' at the conference.[32] This memorandum included excerpts from official correspondence exchanged between Keet and Pole-Evans during 1935, as well as passages from parliamentary debates, extracts from letters from conservation bodies and material from Smuts's speeches. Keet's memo highlighted all the points that would arise in the conference on these issues.

The South African organisers prepared not only for the regular agenda and program of the conference, but also for special deliberations on the particular question of the 'water-supply controversy'.

---

31   Phillips to Smuts, 3 March 1935, Smuts Collection, 238/53, NASA-P; Phillips to Smuts, 4 August 1935, Smuts Collection, 238/53, NASA-P.

32   Keet, 'Memorandum on Afforestation in Relation to Natural Flora and the Effect of Afforestation on Water Supplies', 15 July 1935, FOR 330, NASA-P.

The conference

The conference brought together foresters from around the British Empire. At that time, the fourth British Empire Forestry Conference was the major international forum for the discussion of forest science, policy, and practice. Its delegates 'managed environmentally every major forest type in the world ... Fifty separate forest services protected not only trees but also soil, water, and— so foresters believed—the climate of entire continents and regions'.[33] (The other forum, now IUFRO, the International Union of Forest Research Organizations, at that time represented only a number of forest research stations in Europe.) The conference assembled in South Africa the leading foresters in the world at that time, outside Europe and the USA.

In September 1935, delegates to the fourth British Empire Forestry Conference travelled by train and steamship to Durban, South Africa, for the opening, with Deneys Reitz as host. They met in Durban, Pretoria and Cape Town and toured as a group throughout the forestry regions of the Transvaal, Natal, and Cape provinces. They enjoyed excursions to indigenous and plantation forest stations, as well as a tour of the Kruger National Park. The conference began on 2 September 1935 in Durban.[34] The opening speeches by Colonel Deneys Reitz and the Secretary for the department, Dr P. R. Viljoen, laid the foundations for a searching debate. Reitz, elected President of the conference, challenged delegates to address questions about afforestation effects on water supplies and climate. He said that there was a growing belief among farmers that exotic trees were desiccating South Africa:

> For more than a century we in South Africa have been planting trees, chiefly pines and eucalypts, under the impression that such plantations were valuable for the conservation of water. It has now been put to me that in this way we are decreasing the humidity and drying the soil ... I think we may possibly be in the position that while we are now spending millions on planting trees, we may ultimately have to spend more millions in uprooting them.[35]

Reitz related that 'irate farmers' had accosted him recently to complain about the dangerous desiccating effects of exotic trees.[36] He noted that Dr Viljoen had also raised the troubling issue of exotics and their influence on water and soil erosion.[37] Reitz requested the conference, and delegates agreed to form

---

33   Barton, *Empire Forestry and the Origins of Environmentalism*, 1.
34   See Department of Agriculture and Forestry, *Fourth British Empire Forestry Conference, General Program and Notes on Tours to Be Undertaken* (Pretoria: Government Printer, 1935).
35   Anon, 'Forest Planting in South Africa: Colonel Reitz's Alarm at the Effect of Erosion', *Natal Mercury*, 3 September 1935.
36   Ibid.
37   Ibid.

a committee to investigate the issue.[38] This became the Committee on Forests in Relation to Climate, Water Conservation, and Erosion (the Committee on Forest Influences). Professional foresters led the committee, although other scientists and attendees offered comments, which were published in the proceedings. The committee discussed the relationship of forests and climate in its broadest sense. In particular, South African participants debated the effect of forests on stream runoff;[39] 'the drying up of streams'; choice of different forest species and management regimes, and the effects on catchment hydrology, including water physiology;[40] soil conservation and effects of plantations on soil processes compared with natural vegetation;[41] 'the question of veld burning' and veldfire management;[42] the need for long-term catchment experiments;[43] and aesthetic and economic trade-offs with afforestation.[44]

Conference attendees spent most of their time in Pretoria, the administrative capital of South Africa. General Jan Smuts, the present and future prime minister of the Union of South Africa, welcomed foresters to Pretoria on 8 September. He explained the necessity of afforestation with exotic species of trees, being 'pessimistic about the prospects of extending the country's indigenous forests', but criticised the ecological and aesthetic impact of exotics.[45] Smuts hated exotic pines being planted on his beloved Table Mountain, although he had been a strong advocate of the Forestry Department throughout his former prime ministership (September 1919 – June 1924).[46] Again, Smuts reminded the foresters at the conference about widespread popular criticisms of exotic trees:

> There is no doubt that a popular feeling is arising in South Africa that afforestation is causing the drying up of springs and water sources. Although this has not been proved it can be said that a sufficient case has been made out for thorough research.[47]

Jan Smuts urged delegates to 'frankly admit that South Africa is not a forest country', and that 'we are ... thrown for afforestation purposes to the planting of exotics'.[48] But he then raised all the issues about the use of exotic species

---

38   Union of South Africa, *Forests in Relation to Climate, Water Conservation and Erosion Pretoria* (Pretoria: Department of Agriculture and Forestry, Division of Forestry, 1935), preface.

39   *Fourth British Empire Forestry South Africa 1935 Proceedings*. For Smuts, 81; Keet, 114–6.

40   Ibid., Smuts, 81.

41   Ibid., Viljoen, 6; Smuts, 81; Keet, 114–6.

42   Ibid., Smuts, 82.

43   Ibid., 82; Phillips, 127.

44   Ibid., Smuts, 82.

45   Ibid., Smuts, 79–82; Anon, 'Forestry Research Essential: General Smuts Stresses Need for Scientific Development', *Rand Daily Mail*, 9 September 1935.

46   W. K. Hancock, *Smuts: The Fields of Force, 1919–1950* (Cambridge: Cambridge University Press, 1962), 411.

47   Anon, 'Forestry Research Essential'.

48   *Fourth British Empire Forestry South Africa 1935 Proceedings*, Smuts, 79.

of trees—the concern about alien invasive plants, the protection of flora, and the management of fire in vegetation—that pervaded the debates in the formal meetings, as well as during the extensive field tours that the delegates enjoyed. He challenged the assembled forest scientists on all fronts: 'We know little about our own forests, and practically nothing about the strange forms which we have been importing into South Africa. Our forestry problem is therefore one of research, and endeavour to get at the facts'. He advocated forest policy research: 'where the best national policy would dictate afforestation or the conservation of the natural vegetation'; 'careful research into the water question'; 'the soil question', 'comparative study of the water and soil building and conserving characters of natural forest, *fijnbos*, … and other vegetation, and of exotic plantations, a necessity', 'long range experiments upon watershed areas', and 'aesthetic and economic issues'.[49] This speech contained all the ideas that later influenced the research agenda that emerged with the establishment of Jonkershoek.

The work of the Committee on Forest Influences

On Wednesday, 11 September, the Committee on Forest Influences met to debate the influence of exotic trees on climate, the conservation of water and soil erosion. The panellists comprised a wide variety of attendees, though it was chaired by C. G. Trevor from India and dominated by foresters who supported afforestation.

After hearing reasons in favour of afforestation from an Indian and Australian forester, Keet spoke.[50] He strongly defended the use of exotic trees, especially against those whom he called '[f]orest alarmists … who regard grass and veld generally, as more efficient agents in conserving the water supplies of this country'.[51] These forest alarmists—probably specifically with Phillips and Pole-Evans in mind—criticised exotic timber plantations for using more water than the native plants of South Africa, and viewed exotic trees, especially *Acacia* and *Eucalyptus*, as helping to cause the decline of streamflow and water tables in the country. To the forest alarmists:

> It is our plantations, especially, that stand suspect. They are accused of being ecologically foreigners to our climate, and South African foresters are accused of confusing natural forest conditions with exotic forest conditions, and that generally we have ignored the ecological outlook.[52]

49   Ibid., 79–82. See also Powell, 'Dominion over Palm and Pine: The British Empire Forestry Conferences 1920–47'.
50   Union of South Africa, *Forests in Relation to Climate, Water Conservation and Erosion*, 17–22.
51   Ibid., 25.
52   Ibid.

Keet challenged the argument that forestry as practised in South Africa was not 'ecological' by responding: 'The forester lives with nature, studies nature, and follows nature's law, and if that is not ecology, I fail to see what ecology can be'.[53] Keet disagreed that exotics used more water, pointing out that they often acted as pioneers for native species of 'moisture-loving' trees, eventually leading to the creation of indigenous forests. He ended his speech by noting that 'we are on perfectly right lines with our exotic plantations in South Africa'.[54] The foresters in the audience applauded, 'Hear, Hear'.[55]

The strongest critic at the committee was John Phillips, who was then at the height of his career and had the ear of Jan Smuts.[56] Phillips challenged the views of foresters by pointing out the erosion and local desiccating impacts caused by exotic trees that he witnessed throughout South Africa. He argued that *Eucalyptus* and *Acacia* often encouraged greater erosion than caused by native grasses and used more water:

> [T]hose studies that we have carried out in this country most definitely show that the water requirements and the rate of water usage and transpiration are considerably greater in the gums, acacias, and the Blackwood [*Acacia melanoxylon*] … as compared with those native shrubs and trees which have been investigated at the same time.[57]

Phillips challenged foresters to provide empirical data to support their conclusions, rather than relying—as did Phillips himself and other foresters in attendance—on anecdotal claims. He challenged foresters:

> I ask you again to forget traditional concepts. Here we need long range research— we want to select certain catchment areas which will have a number of streams. We must investigate the regime and life-history, as it were, of these streams over a sufficient period. We must do this quantitatively, and must work most critically and with a sufficient number of these, and treat them in different ways.[58]

Going beyond the matter of water supplies, Phillips noted the negative ecological impacts of exotic trees. Phillips asserted that Keet's example of exotics pioneering indigenous trees was wrong. Indigenous trees rarely grew in areas planted with exotic trees.[59] He provided anecdotal evidence of gums and wattles killing off the growth or grasses nearby. Phillips implored attendees to look at the problem scientifically, through rigorous studies that examined the actual

---

53 Ibid.
54 Ibid., 25–6.
55 Ibid., 26.
56 Anker, *Imperial Ecology*, 129–223.
57 Union of South Africa, *Forests in Relation to Climate, Water Conservation and Erosion*, 32–3.
58 *Fourth British Empire Forestry South Africa 1935 Proceedings*, 127.
59 Union of South Africa, *Forests in Relation to Climate, Water Conservation and Erosion*, 35.

water usage and transpiration of native flora compared with exotic plantations.[60] He called for nothing less than an expansive (and expensive) research program into the hydrological and ecological impact of exotic trees. Phillips suggested the research would require no less than a 20- to 30-year program to ultimately resolve the debate about the influence of exotic trees on South Africa's climate and water cycle.

South African foresters consistently rebutted the claims that the exotic plantations had caused the drying up of streams and increased erosion in South Africa, compared with indigenous vegetation types. They admitted that forests, along with certain types of vegetation, did use water resources. The question, as posited by J. D. M. Keet, during the fourth British Empire Forestry Conference, was rather what type of land use provided the maximum benefit for South Africa's water supply: 'It is not the maximum quantity of water, however, but the maximum beneficial yield of water that we are after. We cannot cater for the demands of those reservoir engineers, who realising that all plant growth, like all living matter, must use up water to live and grow, demand catchment areas paved with bare rock'.[61] Foresters emphasised that forestry provided, in certain contexts, the best use of land and water at the same time that it offered climatic and ecological benefits.

In the committee's final report, foresters exonerated South Africa's forest policy. The report noted: 'We cannot do otherwise than commend any and all efforts which are being made, for which can be made, to bring under forest cover a greater proportion of the land area of the Union'.[62] It saw no evidence to indicate that exotic trees reduced soil moisture and streamflow: 'we do not consider that the afforestation policy of the Government of the Union of South Africa has been detrimental to the general water supply of the country'.[63] The report noted that 'diminishing rainfall' was the primary culprit.[64] To allay the fears of the 'farming community', the report noted that the belief that exotic trees caused desiccation 'is at variance with the generally accepted conclusions regarding the effects of forests on streamflow'. The report dismissed Phillips but recommended a program of research: 'All things considered, we are not disposed to place much credence in the complaints instanced above … but we suggest that a comprehensive scientific investigation on the effects of tree-planting upon local water supplies would be of value', 'which would be of inestimable advantage to it and the world at large, and which might allay certain fears which

---

60   Ibid., 34.
61   *Fourth British Empire Forestry South Africa 1935 Proceedings*, 115.
62   Union of South Africa, *Forests in Relation to Climate, Water Conservation and Erosion*, 9.
63   Ibid., 13.
64   Ibid., 12.

the public have expressed'.[65] While it would take many years before results would emerge, 'in order to allay public anxiety', fast-growing exotics should not be planted 'in the actual sources of streams and the eyes of springs; in such places the natural vegetation should be carefully protected', affirming that 'the Forest Department is already acting on these lines'.[66]

Common purpose

Thus, in spite of disagreements, South African foresters were receptive to research. Keet agreed with Phillips on the need to pursue empirical research into forest hydrology. And in fact, three years prior to the meeting, the Forestry Department had acquired land in the Jonkershoek Valley near Stellenbosch to pursue a major hydrological experiment studying whether exotic timber plantations used more water than indigenous vegetation.[67] South African foresters had also acknowledged the need to balance forest development with the protection of water catchments. That is why Keet had emphasised in 1935 that the department would afforest only one-fifth of the state forest–managed land, while protecting catchments and their indigenous vegetation on the other four-fifths. Thus instead of talking past each other, foresters and their critics found common ground on the need to afforest conservatively and to study the impact of exotic trees on water supplies. This agreement between foresters and ecologists formed the basis of a new research program that lasted for over 50 years.

The 1935 British Empire Forestry Conference should be understood as the event that helped to crystallise political support for a major research program that was already underway. Participation by leading politicians such as Jan Smuts and a broad array of scientists and agricultural officials ensured that foresters had widespread support to pursue what would be a long and expensive research program. The conference also helped to enshrine a basic underlying belief that would govern research and policies regarding forests and water from the mid-1930s until the early 1990s. Politicians and critics of forestry demanded policies that were evidence-based, practical, and would be constantly reviewed.[68] Foresters in turn agreed that '[in] the reciprocal relationship between theory and experience the latter must hold sway, until at least the former has been proved in practice'.[69]

---

65   Ibid., 13, 16.
66   Ibid, 13.
67   *Fourth British Empire Forestry Conference Proceedings*, 293.
68   *Fourth British Empire Forestry Conference Proceedings South Africa, 1935*, 80.
69   Ibid., 155.

Early foundations of a consensus agenda

The debate about hydrological research at the 1935 conference must be understood as the culmination of a discussion that began at least as early as the Union in 1910, when the first hydrological research began, not as the origins of Jonkershoek.[70] Informed opinions on the question of forests and water supply were not orthodox, and not all foresters agreed that afforestation would benefit water supply. Charles Legat, for example, did not believe that eucalypts, at least, favoured increased streamflow. Writing in 1920 in response to an enquiry from one of his officials, he states that 'Eucalypts have frequently been planted near springs in this country and ... little doubt that their effect is prejudicial' [to water supply]. In support of his argument he cited a report by D. E. Hutchins, 'A Discussion on Australian Forestry', as well as the direct arguments of W. E. Abbott, who in a paper, 'Forest Destruction in New South Wales', read before the Royal Society of New South Wales in July 1880, claimed to 'predict with absolute certainty that ... the effect of destroying the natural eucalypt forests will be to cause a permanent increase of water in such water-courses and produce springs where there were none before', Legat thus inferring the reverse with afforestation.[71]

Soon after Union, questions about plantation forests and water prompted officials to attempt measurements of effects on streamflow. Notwithstanding the views of Kanthack and Braine, expressed just a few years before, the Union Irrigation Department initiated hydrological research at a newly planted state forest at Jessievale in the east of the Transvaal beginning in October 1910.[72]

But by 1924, it was clear from the failure of gauging techniques and the loss of records that an investigation incidental to routine forest and hydrographic activities was not yielding reliable information.[73] The Jessievale study subsequently grew to incorporate three gauged streams, and ran until 1963; but it yielded no scientific results,[74] probably because scientists did not trust the experiment to the extent of investing in its analysis.

---

70   Showers, 'Prehistory of Southern African Forestry', 311.

71   C. E. Legat, Chief Conservator of Forests, to the Acting District Forest Officer Doornboom, 26 February 1920, 'The Effects of Eucalypts on Water Supply', digi.nrf.ac.za/dspace/bitstream/handle/10624/419/CLW_Effeuc1920001.pdf?sequence=1 (accessed on 12 August 2013).

72   Hydrographic Surveyor N. von Warmelo, Irrigation Department to Chief Conservator of Forests C. E. Legat, 'Jessievale Plantation Gauging of Spruit', 22 January 1918; R7649/711, SAFRI Archives, CSIR, Pretoria.

73   J. D. Keet, Conservator of Forests Transvaal and Orange Free State Conservancy to Chief Conservator of Forests, 7 May 1924, R7649/711, SAFRI Archives, CSIR, Pretoria.

74   Email 14 June 2013 from Brian Jackson, Manager: River operations & Data Management, Inkomati Catchment Management Agency.

In parallel with this research effort, foresters had made strenuous efforts to anticipate and respond to concerns from downstream water users. At the fourth British Empire Forestry Conference, the then Administrator of the Transvaal, S. P. Bekker (speaking some years later, himself a farmer in the east of the Transvaal), laid out a succinct statement of the problem in his constituency:

> we have a belt a couple of hundred miles long and 30 miles broad, which has an assured rainfall, and where the only perennial streams and rivers are to be found in the Transvaal Province ... I know that farmers in that part of the Province are very much perturbed over the effects of planting these pines and gums on the sources of these streams and rivers which have to irrigate the fertile land at the bottom of these ranges.[75]

Bekker's belt of country with assured rainfall is the escarpment of South Africa's interior highland, cresting in this part of the country at about 2,000 m above sea level, and falling rapidly to the undulating lowlands of the Lowveld to the east, about 300 m above sea level. The cool climates of the higher elevations, with rainfall of around 1,500 mm per year, are ideal for forestry; in the foothills, the warmer climates suit tropical fruit production, but irrigation is needed under the lower rainfall regime there. The ancient granites that underlie the foothills along the lower half of this gradient yield deeply weathered, highly erodible soils. The upper half of this landscape, from 1910 onward, became the focal region for investment by both government and mining capital in sawlog and mining-timber forests, of pine and eucalyptus; eventually, this investment created what for a time were the largest planted forests in the world. Below this belt, World War I veterans received the opportunity to acquire land for farming.[76] In this context, the exchange between the White River Valley Farmers' Association, irrigators of the fertile land below the escarpment, and forestry officials illustrates the many lengthy engagements arising from concerns about forests and water supply.[77]

Finding the land on these slopes often to be degraded through mining, sheep grazing, cultivation and veld burning, foresters saw afforestation as the way to repair this, and held still to the view that afforestation with pines of the belt

---

75   Union of South Africa, *Fourth British Empire Forestry Conference Proceedings, South Africa*, 77.

76   This was part of the government's hesitant program to promote rural development through immigrant anglicisation. See Fedorowich, 'Anglicisation and the Politicisation of British Immigration to South Africa, 1899–1929'.

77   White River Valley Farmers' Association to Private Secretary, Minister of Agriculture, 18 May 1934, Wicht Papers, SAFRI Archives, CSIR, Pretoria; Captain C. V. Palmer to J. D. Keet 18 June 1934, Wicht Papers, SAFRI Archives, CSIR, Pretoria; White River Estates Irrigation Board to Director of Forestry, 24 June 1936, Wicht Papers, SAFRI Archives, CSIR, Pretoria; Memo C. 203 of 24 June 1934, J. D. Keet 'Complaints re streams drying up on Bultfontein', Wicht Papers, SAFRI Archives, CSIR, Pretoria; A. J. O'Connor for Director of Forestry to White River Estates Irrigation Board, 11 August 1936, Wicht Papers, SAFRI Archives, CSIR, Pretoria.

along the escarpment would enhance water resources.[78] Still, aware of potential concerns regarding afforestation effects on water supplies to downstream irrigators, foresters assessed streamflow in the catchment of the White River in 1922, prior to planting from 1927 onward; they estimated that the flows were in any case too low to support irrigation without the construction of storage dams, a view confirmed by the history of the development of the White River Irrigation Farmers Association. They thought that the land was so degraded by erosion and soil crusting that it did not function effectively as water catchment; afforestation would improve the condition of the land. The severe drought that set in from around the time of afforestation onward confounded attempts to respond to these concerns—in this location, annual rainfall was about 25 per cent above 'normal' immediately prior to afforestation, and 25 per cent less afterward. Drought was accompanied by rising complaints from farmers. Foresters wrestled with their preconceived ideas and conflicting views about forests and climate, their sketchy knowledge of hydrology, the distress of the farming community, and the scepticism of Colonel Deneys Reitz, Minister of Agriculture and Forestry.

Forestry officials, though confident of the wider economic and environmental benefits of exotic tree planting, having learnt from their detailed local case studies, began to take action in the early 1930s to forestall any localised problems arising from the planting of trees in catchment areas. In 1932, the Department of Forestry adopted the policy of keeping a 20-metre buffer zone between a stream and a plantation to mitigate effects of their plantations on streamflow.[79] That same year they decided to start a major hydrological research site in the Jonkershoek Valley and began to acquire the necessary land for it.[80] And in November 1935, J. D. M. Keet issued a circular instruction to all forest officers reflecting the recommendation of the conference committee on 'Effects of Afforestation on Climate, Water Supplies and Erosion', containing instructions confirming the moratorium on afforestation along streams and near springs, but also instructing officers to take measures in afforestation plans to protect 'species of special botanical interest or vistas of peculiar beauty'.[81]

---

78   J. D. Keet, R9110 of 21 October 1933 to Chief, Forestry Research, 'Invloed van bos op stand van water', Wicht Papers, SAFRI Archives, CSIR, Pretoria.

79   *Report of the Interdepartmental Committee of Investigation into Afforestation and Water Supplies in South Africa*, 12.

80   See J. J. Kotze's comments in Union of South Africa, *Fourth British Empire Forestry Conference Proceedings, South Africa*, 293.

81   J. D. M. Keet, 'Water Conservation and Aesthetics in Relation to Afforestation and the Propagation and Preservation of Indigenous Trees and Flora', Department of Agriculture and Forestry, 22 November 1935, Wicht Papers, SAFRI Archives, CSIR, Pretoria.

Conclusion

The fourth British Empire Forestry Conference marked an important shift in the history of environmental sciences in twentieth-century South Africa. At the time of the conference, forestry was still the dominant conservation discipline—ecology having only recently been developed in South Africa by foreign-born and foreign-educated scholars.[82] Yet the fourth British Empire Forestry Conference was a turning point in the power of professional foresters. They had to work within an increasingly complex scientific and policy arena that was often hostile to forest policies. The establishment of the Jonkershoek Forest Research Station was a response to forestry critics that also gave foresters the opportunity to pursue empirical research that could guide forest policy. The decision showed the serious intent on the part of the South African community of foresters to pay meaningful attention to a question that not only was the source of public controversy, but which also vexed their own minds.

---

82   Pooley, 'Pressed Flowers', 599–618, 602–6.

Chapter 7
Jonkershoek as Fulcrum: The Forest Hydrological Research Program

It is possible, and indeed it is all too frequent, for an experiment to be so conducted that no valid estimate of error is available. In such a case an experiment cannot be said, strictly, to be capable of proving anything. Perhaps in this case it should not be called an experiment at all, but be added merely to the body of experience on which for lack of anything better, we have to base our opinions.

—R. A. Fisher, 1937[1]

Place

The dissent over afforestation did not abate after the conclusion of the 1935 Empire Forestry Conference, although open antagonism decreased as scientists and politicians waited to see the results of the experimental program. All parties had agreed that a comprehensive experimental program was required to solve the fundamental question of whether forests transpired more water than indigenous vegetation types, and whether or not the forest helped to equalise streamflow throughout the year. There would be no quick resolution, though the complaints from Eastern Transvaal farmers sustained the pressure to resolve the issue, while forest officers persevered with local studies to answer the complaints.[2] But the centre of attention shifted, to focus on the development of a major program of

---

1   R. A. Fisher, *The Design of Experiments*, 1937, in C. L. Wicht, 'Determination of the Effects of Watershed-Management on Mountain Streams', *Transactions of the American Geophysical Union*, 24 (1943): 594–606.
2   For example, White River Estates Irrigation Board to Director of Forestry, 24 June 1936; R. S. Harriss to the Hon. Deneys Reitz, 22 March 1938; District Forest Officers Sabie E. B. Domisse to Conservator of Forests Transvaal J. D. M. Keet on effect of rainfall on water supplies, 6 June 1939; all in Wicht Papers, SAFRI Collection, CSIR Pretoria.

science in an infant discipline, and its extension from the start in the mountains of the Cape Fold Belt, to the Drakensberg mountains to the east, and ultimately to the north, near the furthermost reaches of the forestry regions.

At the fourth British Empire Forestry Conference, Phillips and Smuts had argued that it could take 20 to 30 years of continuous experimental observation to offer definitive evidence about the impact of trees. These predictions assumed, of course, that the experiment would be designed and executed in a way that assured valid findings, so that parties across the spectrum of opinion could abide by the conclusions, including both the sceptics of afforestation, and the pro-forestry advocates who wished to know whether plantations really did benefit water supplies, or at least, prove to be the more beneficial use of water than alternative uses of catchment land.

While the experiment at Jessievale had been a remote and lonely enterprise, seldom attended by any expert, the choice of Jonkershoek as the new site yielded immediate rewards. Located as it was, close to universities and Cape Town, an emerging major city, Jonkershoek soon became the focus of energetic activity. The annual reports of the Chief Forest Research Officer and Jonkershoek's resident researcher both speak of an immediate start, with a geological, a soil, and a botanical survey, the 'small beginning' of planting 17.8 acres of *Pinus radiata* in 1935/36, the ordering and receipt of '£450 worth of equipment', the building of the foundations of the gauging weirs, and installation of raingauge networks, visits from senior leadership, and lectures at the University of Stellenbosch, all within 24 months.[3]

The department's leadership had chosen not to import someone from abroad to design and run the research, as they had done with the new field of timber technology, probably trusting their own competence, and believing that the forest hydrological problem was uniquely 'South African', and like its other forest research projects, required a South African who understood local environmental and social conditions. The department instead selected Christiaan L. Wicht to lead the country's new research program. While Wicht was not obviously the best-trained person for the job, he was well known to his colleagues, who had followed his overseas studies closely. There was, in any event, no candidate through the whole of South Africa who came fully equipped with the knowledge and expertise required to design a program in forest hydrology, an emergent field of science that lacked internationally agreed-upon concepts and methodologies.

---

3   Annual Report of the Forest Research Officer Jonkershoek, 1935/36, Annual Report DFO Jonkershoek 1935/36, Annual Report DFO Jonkershoek 1936/37, Annual Report RO Jonkershoek 1936/37, all at Jonkershoek archives, CSIR, Stellenbosch and digi.nrf.ac.za/dspace/bitstream/handle/10624/378/CLW_Annrep1937p001. pdf?sequence=1 (accessed 11 August 2013).

In his new role, Wicht worked hard during the late 1930s and early 1940s to design an experiment to his satisfaction and to that of his colleagues, and later in the mid-1940s to early 1960s confronted some of the world's pre-eminent hydrologists, who claimed that his experimental design and methods were too empirical, lacking adequate theoretical foundation. That he cultivated the support of a wide collegial network may help to explain how he maintained his prestige despite strong criticism, but greater credit owes to his close attention to methodology, technique, and relevance to the forest policy issues that directed the program in the first place.

Hydrological research, like other field sciences, was naturally place-based, and the study of field sciences such as hydrology reveal key intersections between the history of science, historical geography and environmental history.[4] Wicht's role from 1935 to 1947 and after was deeply connected with the place of Jonkershoek. For Wicht, and many other foresters after him, Jonkershoek was a 'place' in Kohler's sense of the term: it was not just a geographical location but a locale with a particular ecological and social context with intimate 'connections between doing science and living lives'.[5] Wicht's background and the character he developed during childhood and youth was deeply inspired by the landscape and culture of Jonkershoek, Stellenbosch and the surrounding Cape landscapes. Jonkershoek's close association with Stellenbosch University— an important cultural and academic centre for Afrikaners as well as home to the country's undergraduate and postgraduate program in forestry—made it a ready intellectual focal point for hydrological research in Southern Africa, with proximity to Cape Town an added merit.

Jonkershoek's distinct climate and its topography, however, did not make it the self-evident choice for the new program. It was 2,000 km away from the main centres of afforestation and from Jessievale and White River and the other locations where Keet had recently investigated complaints about afforestation. Its Mediterranean winter rainfall climate differed markedly from the summer rainfall that prevailed through most of the forestry zones in Natal, Zululand and Transvaal. Despite its climatic and geographic differences, Jonkershoek had the benefit of proximity to intellectual centres. And it would prove to be the laboratory for the ideas that emerged to form the methodological basis for the program that later included

---

4   Interest in the study of 'place' and 'locality' are recent historiographical developments that have arisen from an intersection of methods from the history of science, historical geography and environmental history. See, for instance, J. Vetter (ed.), *Knowing Global Environments: New Historical Perspectives on the Field Sciences* (Piscataway: Rutgers University Press, 2011) and the special issue of the *Journal of the History of Biology*, 45 (2012). There has been less work on this subject within the field of South African environmental history, although recent publications by the authors of this work address these questions. See Bennett and Kruger, 'Ecology, Forestry and the Debate over Exotic Trees in South Africa'.

5   R. E. Kohler, 'Practice and Place in Twentieth-Century Field Biology: A Comment', *Journal of the History of Biology*, 45 (2012): 579–86, 581.

South Africa's major forestry regions, taking its methodological shape through a series of carefully situated sites representing the different forestry regions within South Africa. In fact, while the department was establishing Jonkershoek, the reconnaissance to select a second site was already underway.[6]

Christiaan Lodewyk Wicht

Wicht was born at Schoongezicht, a farmstead on the outskirts of the town of Stellenbosch, on 15 August 1908, to Johan Hendrik Wicht and Susanna Johanna Wentzel, the *'natrossie'*—the late bunch (of grapes)—in the family.[7] The next youngest sibling, a brother, was 18 years older. At the age of three years, his father died, and he moved with his mother to Stellenbosch, where she raised him. The young Wicht frequently visited Schoongezicht, where an elder bother, Henry, farmed until 1922. His second oldest brother, Dr William Frederick Wicht, was his mentor, and Christiaan spent a lot of time with him in Cape Town, where William was the Port Doctor at the old harbour.

Wicht's father came from a tradition of Cape Afrikaners who were leading Cape Town 'capitalist landlords' and politicians. His grandfather was a member of the Cape Anti-Convict Association, who succeeded in having the convict ship *Neptune* turned away from the Cape in 1850, and later a member of the Cape Legislative Assembly.[8] His family and their associates were members of what Saul Dubow describes as the 'progressive forms of Afrikanerdom'.[9] They supported the liberal tradition in the Cape and resisted the imperialism of Alfred Milner while maintaining distance from the two Boer republics to the north.

Wicht developed an early affinity with nature as a youth living in the Cape Fold mountains in the south-western Cape. His mother would make annual excursions to the farm, and he more often, to enjoy the wildflowers and to bring bouquets home to her. Visits involved walking expeditions, including his first summit, when at the age of five or six years, his brothers helped him climb a buttress 500 metres high, north of the farm.[10] In 1918, during the influenza epidemic, schools closed

---

6   U. W. Nänni, 'Forest Hydrological Research at the Cathedral Peak Research Station', *Journal of the South African Forestry Association*, 27 (1956): 2–35.

7   Unless otherwise shown, this biographical information is from an autobiographical manuscript, '*Bergmaats*', compiled by Wicht while recuperating in hospital from a heart attack, in 1959, and another darkly lyric essay from the same time reflecting on illness, death, human ecology and art, *Die Swart Kat* (Jonkershoek Archives, CSIR, Stellenbosch), as well as emails from Susan Wicht Clark, 7 September 2010, 16 September 2010, 19 September 2010.

8   H. Giliomee and B. Mbenga, *A New History of South Africa* (Cape Town: Tafelberg, 2007), 141. For the Wicht family in nineteenth-century Cape Town, see N. Worden, E. Van Heyningen, and V. Bickford-Smith, *Cape Town: Making of a City* (Cape Town: New Africa Books, 2004), 170.

9   S. Dubow, *A Commonwealth of Knowledge*, 156.

10   *Die Swart Kat*, 5, MS personal memoir c. 1959, Wicht Papers, SAFRI Archives, CSIR, Pretoria.

and his mother sent him to Schoongezicht, perceived to be out of harm's way. As the only child on the farm, he entertained himself with adventures in the veld and exploration of the hills nearby. These experiences taught him a deep love of the veld and the mountains of Stellenbosch, and a practical botany from an early age.[11] Later, this was reinforced during school excursions into the mountains. Before his assignment, Wicht had walked the Jonkershoek terrain and climbed the peaks many times as a child and youth. He recounted these experiences in his 1959 memoir '*Bergmaats*' ('Mountain Comrades'), recalling his first visit to the valley of Jonkershoek as a child of 12 years, led by his school hostel master, Oom Jannie Krige, a sympathetic lover of the veld.[12]

Mountains played an important social role in his personal and professional life. As a student at the University of Stellenbosch, he expanded the reach of his childhood sallies into Jonkershoek, frequently climbing the Jonkershoek peaks by different routes, often in the company of leading academics from the university, including R. W. Wilcocks, later rector of the university, and professors in geology, history, law and other fields. His growing familiarity with the terrain accompanied the building of a collegial network that prepared him well for his professional career. After marriage in 1935, he and his wife made the enjoyment of nature and the mountains 'the foundation of [their] family life'.[13] His attachment to Jonkershoek as a place was profoundly strong. In the text for his school radio talk entitled 'The Balance of Nature and the Story of a Valley', he wrote: 'I am very attached to this valley … [Except for his period of study overseas] … I have always, as I do today, lived where I could hear the roar of the river over the stones when it comes down in the winter floods'.[14]

In 1925, Wicht matriculated from what is now the Paul Roos Gymnasium in Stellenbosch. This was the school attended by Jan Smuts, and many other prominent South Africans, especially Afrikaners. At school, he enjoyed the close friendship of Uys Krige, the future prominent author and poet, and Uys's brothers Francois and Bokkie—they played in the same rugby team, and enjoyed art and literature together. These companions became important figures in what became the 'Thirties Movement' (*Die Beweging Dertig*), a loyal resistance movement that pursued the alternative to conventional politics and culture in

---

11   Wicht's experience growing up was part of a wider culture shared by other white Afrikaners in the Cape. For instance, Keith Hancock's descriptions of Jan Smuts's childhood in Swartland, a wheat-growing valley along the west coast of South Africa, echoes a similar aesthetic: W.K. Hancock, *Smuts: Sanguine Years 1870–1919* (Cambridge: Cambridge University Press, 1962). For a critical analysis of landscape and whiteness see Jeremy Foster, *Washed with Sun: Landscape and the Making of White South Africa* (Pittsburg: University of Pittsburg Press, 2008).

12   C.L. Wicht, 'Bergmaats' (Stellenbosch, n.d.), Jonkershoek Papers, CSIR, Stellenbosch; copy in author's possession.

13   Ibid.

14   'The Balance of Nature: The Story of a Valley' c. 1947, MS, Wicht Papers, SAFRI Archives, CSIR, Pretoria.

Afrikaans at the time, and was the forerunner to the later *Sestigers*.[15] It was probably in this company that he developed his own taste for art and poetry, collecting the masters of South African painting, and frequently quoting South Africa's poets in Afrikaans and English. He participated keenly in tennis, rugby and saltwater fishing. He and his wife Peggy enjoyed the style of the Arts and Crafts movement, and he designed and built furniture for his home in that style. He later made his own contributions to the modernising of Afrikaans, contributing to two lexicons, *Afrikaanse Bosboomname* and *Voorlopige, Gedeeltelike Hidrologiese Terminologie*.[16]

After matriculating, he applied to the Public Service Commission for a sponsored studentship in forestry, but was rejected on the grounds of his defective eyesight. Persisting, he was authorised to undertake a second eye examination, and then after passing the exam, was accepted.[17] As a sponsored student of the Department of Forestry he completed the study for a degree in science at Stellenbosch University, graduating in 1929 with majors in botany and geology. His teachers at Stellenbosch were eminent people, and would have impressed him strongly. Although in Wicht's time the Professor of Botany was Gert Nel, it was A. V. Duthie who would have led him in this field (he would often speak of her with his own students).

Augusta Duthie, born in 1881 at Belvidere near Knysna, studied under Bertha Stoneman at the Huguenot College in the Cape country town of Wellington and gained her BA from the South African University of the Cape of Good Hope (University of Cape Town) in 1901. In 1902 she became the first South African-trained lecturer in botany when she took a post in in the embryonic department of plant sciences at the Victoria College, later the University of Stellenbosch. She was strongly influenced by Rudolph Marloth, then beginning his 1913–1932 *Flora of South Africa*, and she became a highly productive botanist, publishing regularly. She was head of department until Gert Nel's appointment in 1921 allowed her to concentrate on research, and her 1929 PhD dissertation was an

15   G. Olivier, '"The Dertigers and the Plaasroman": Two Brief Perspectives on Afrikaans Literature', in D. Attwell and D. Attridge (eds), *The Cambridge History of South African Literature* (place: Cambridge University Press, 2012), 313; S. P. van Aardt, 'Uys Krige se familiebriewe uit Frankryk en Spanje 1931 – 1935: teksuitgawe, met historiese oriëntering, teoretiese verantwoording en annotasies' (Master's Thesis, University of Stellenbosch, 2009), 9: 'Krige stel hom in teen die letterkundige status quo en hou vol dat hy nooit deel daarvan sal raak nie ... en 'dat Afrikaans himself op politieke, maatskaplike, filosofiese en kunsterrein buite die eng Afrikaner-sfeer moet ontwikkel'; see also page 10 on Krige's commitment to the Afrikaner culture.

16   See C. L. Wicht, 'Afrikaanse Bosboomname', *Journal of the South African Forestry Association*, 5 (1940): 41–61; C. L. Wicht, 'Preliminary, Partial Hydrological Terminology', with F. J. Snyman (1968), at digi.nrf.ac.za/dspace/bitstream/handle/10624/436/Vooged1968001.pdf?sequence=1 (accessed 12 September 2013).

17   Public Service Commission to Wicht, 11 May 1927; Public Service Commission to Wicht, 22 July 1927; Susan Wicht Clark says that he lost the sight in one eye in an accident playing with a china doll, at the age of three years; later he was refused when volunteering for military service in World War II and, as a result, continued with the work at Jonkershoek while taking on the duties of colleagues in military service; emails from Susan Wicht Clark, 7 and 19 September 2010.

intensive study of the entire ecology of a 10-km^2 area of the Eerste River terrace on the campus.[18] Such attention to botanical detail later became evident in Wicht's study of the effects of fire in fynbos at Jonkershoek.[19]

In geology, the then professor was S. J. Shand, an eminent scientist who later became Professor of Geology at Columbia University in New York City, but it was A. V. Krige who would have had a strong influence on Wicht; again Wicht often referred to Krige and his work in his lectures. Krige built on Shand's earlier work in his study of sea-level fluctuations around the Cape Coast over the preceding 120,000 years, and used the succession of terraces in the valley of the Eerste River for field evidence of the geomorphological relationship between sea-level variation and riverine erosion.[20] These were an excellent combination of knowledge and world views to create a broad mind in a student of nature.

Like other sponsored students, he spent two months per year assigned to work on a plantation or visit other sites, 'acquiring a practical knowledge of Forestry as practised in this country' and of the various forestry environments and species.[21] After graduation, he spent the period from January to August 1930 employed at Tokai on the Cape Peninsula before leaving for Oxford as a sponsored student.[22]

He received from Oxford University the BA Forestry in 1931. There he kept in touch with his seniors in the department through regular reports, and received encouraging correspondence from them. He enjoyed extracurricular activities, made many friends, played university rugby, assisted Harry Champion, then at the India Forest Research Institute at Dehra Dun, by translating a paper on teak plantation growth models from the German, corresponded with forest scientists around the world, and represented South Africa at an IUFRO conference in Munich in 1931 and later, in September 1932, at the IUFRO Congress in Nancy.

---

18   P. G. Jordaan, 'A. V. Duthie en Haar Bydrae tot die Plantkunde', *South African Journal of Botany*, 33 (1967): 47–57; M.R.S. Creese and T.M. Creese, *Ladies in the Laboratory III: South African, Australian, New Zealand, and Canadian Women in Science: Nineteenth and Early Twentieth Centuries* (Lanham, MD: Scarecrow Press, 2010), 14–6; A.V. Duthie, 'Vegetation and Flora of the Stellenbosch Flats', *Annals of the University of Stellenbosch*, 7 Series A, no. 1 (1929), 52. At the Huguenot College, Augusta Duthie was taught by Bertha Stoneman, graduate of Cornell University, a progressive and gifted teacher and author of *Plants and their Ways in South Africa*, the first South African botany textbook (H. Engel and M. Smiley, *Remarkable Women in New York State History* (Charleston: The History Press, 2013), 264).
19   C. L. Wicht, 'A Statistically Designed Experiment to Test the Effects of Burning on a Sclerophyll Scrub Community. I. Preliminary Account', *Transactions of the Royal Society of South Africa*, 31 (1948): 479–501.
20   S. J. Shand, 'The Terraces of the Eerste River at Stellenbosch', *Transactions of the Geological Society of South Africa*, 16 (1913): 147–55; A. V. Krige, 'An Examination of the Tertiary and Quaternary Changes of Sea-Level in South Africa, with Special Stress on the Evidence in Favour of a Recent World-Wide Sinking of Ocean-Level', *Annals of the University of Stellenbosch*, 5, Series A, No. 1 (1927).
21   C. C. Robertson for Chief Conservator of Forests, Memorandum for Nominated Students – Professional Staff, A.57/1/22 29 October 1929, as addressed to C. L. Wicht, Wicht Papers, SAFRI Archives, CSIR, Pretoria.
22   Chief Conservator of Forests to Wicht, 21 December 1929; Wicht to Chief Conservator of Forests, 3 July 1930, Wicht Papers, SAFRI Archives, CSIR, Pretoria.

He then proceeded to the forestry school at the Forest Academy at Tharandt (now the Institute of International Forestry and Forest Products, Dresden University of Technology).[23]

The regular contact between officials and Wicht, as with other sponsored students, ensured that he received the necessary encouragement and direction, while the officials had sufficient opportunity to grasp his capabilities and promise. C. C. Robertson gave Wicht ample advice on his intentions to study at Oxford.[24] Wicht also received the 1931 memo from E. J. Neethling on the choice of speciality direction, and there followed a careful correspondence about Wicht's choice of a doctoral program. Wicht had considered Yale, Zurich and, in Germany, Eberswalde and Tharandt as options; Neethling would have liked Wicht to have worked with Toumey at Yale, but Yale's School of Forestry could not finance Wicht's proposed research. Wicht finally chose Tharandt, to do a methodological dissertation rather than an experimental program, and Neethling concurred, telling him in a letter *'dat ek ten volle saamstem met die Tharandt skema wat u voorgelê het'*—'I agree fully with the Tharandt scheme that you have submitted'.[25] While at Tharandt, J. J. Kotzé, then Chief Forest Research Officer, advised Wicht on the direction of his studies. Kotzé informed Wicht that he would join the department's Research Branch on his return, to work in the area of silviculture. Wicht thanked Kotzé for his advice and for giving Wicht 'a very good idea of what is happening in South Africa' whilst he was in Germany.[26]

At Oxford during the forestry course, Wicht encountered the work of R. A. Fisher in mathematical statistics and experimental design, and developed a close knowledge of the field, which he later employed rigorously in his research career. At Tharandt, his dissertation was a theoretical examination of experimental methodologies for the study of thinning in forest stands.[27] He returned from Germany in 1934, and once back with the department he was assigned to silvicultural research on the then eastern Transvaal, based in Pretoria. But soon after he was posted as Forest Research Officer at the Jonkershoek Forest Influences Station.

---

23   Emails from Susan Wicht Clark 7 September 2010, 16 September 2010, 19 September 2010; H. G. Champion Dehra Dun to Wicht, 22 December 1932; Chief: Division of Forestry Education to Wicht, 12 August 1932, Wicht Papers, SAFRI Archives, CSIR, Pretoria; see also 'Obituary: Dr C.L. Wicht', *South African Forestry Journal*, 106 (1978), 78.

24   C. C. Robertson to Wicht, 17 June 1929.

25   Yale School of Forestry to Wicht, 21 September 1931; E. J. Neethling for Chief Conservator to Wicht in Oxford, 6 September 1931, 3 November 1931, Wicht Papers, SAFRI Archives, CSIR, Pretoria

26   J. J. Kotzé (personally) to Wicht, 7 April 1932; Wicht to Kotzé, 9 January 1933, Wicht Papers, SAFRI Archives, CSIR, Pretoria.

27   C. L. Wicht, *Zur Methodik Des Durchforstungsversuchs. Mitteilung Aus Der Sächsischen Forstlichen Versuchsanstalt, Abteilung Für Ertragskunde.* (Dresden: Bufra, Buchdruckerei Otto Franke, 1934).

Wicht had a strong social and political consciousness, illustrated by his own memories and those of his daughter. He related years after of walking the countryside outside Dresden and encountering a march of the Hitler Brown Shirts brigade, and his fearful apprehension of this menace.[28] His daughter Susan tells that while there he kept his suitcase packed, ready to flee Germany. Susan tells also of his opposing the exclusion of blacks from the University of Stellenbosch and his outrage when a lecturer was rejected from the university for being Jewish. His defence of the rights of coloureds was reported in *The Argus* on Thursday, 5 August 1971 in the article 'Maties, UCT call for Coloured uplift', which reported: 'More than 100 academics of the Universities of Stellenbosch and Cape Town as well as writers, professional men and prominent farmers in the Boland today came out in support of a call for full citizenship for the Coloured people which was made by a group of Transvaal academics last week', and Wicht was listed among these 100 academics.[29]

Wicht became Fellow of the Royal Society of South Africa in 1941 and served as the first Chairman of the Table Mountain Preservation Board during 1952–1957, after chairing the Royal Society of South Africa's investigation into the preservation of vegetation in the south-western Cape in 1945, as well as being the principal author of its report.[30] From 1949, he played a leading role in successive Southern African and Pan-African science initiatives, which eventually lost momentum during decolonisation, though this work continued until the 1970s. Among other things, he had a long service on the Standing Committee for forestry of Southern African Regional Commission for the Conservation and Utilization of Soil, in which role he contributed to many regional initiatives, especially in the field of hydrology.[31]

---

28 Personal comments, Wicht to F. J. Kruger, around 1977.

29 *The Argus*, 5 August 1971.

30 C. L. Wicht, 'Preservation of the Vegetation of the South Western Cape, Part I', *Transactions of the Royal Society of South Africa*, 30 (1943): 7–18; R. S. Adamson, 'II. Characteristics of Vegetation', *Transactions of the Royal Society of South Africa*, 30 (1943): 18–56.

31 SARCCUS, the Southern African Regional Commission for the Conservation and Utilization of Soil, representing 14 member states from Malawi to South Africa, was formed in 1948 at the First Inter-African Conference on Soil held at Goma in the then Belgian Congo, at which Wicht presented a paper on hydrological research; SARCCUS served to promote technical cooperation among its members in the fields of agriculture and forestry, and all other aspects of the utilisation of all related natural resources; A. B. Bridgens, 1987, 'South Africa's Contribution to the Progress and Stability of Southern Africa through Her Membership of SARCCUS, Secretary-General of SARCCUS, June 1987', unpublished report, University of Pretoria library, 4; Anonymous (ed.), Proceedings of the Twenty-Third Ordinary Meeting of SARCCUS, Sanbonani, Hazyview, South Africa, 28–30 April 1992, 114.

In his private capacity, Wicht ardently advocated nature conservation and the protection of the land against erosion and other forms of degradation.[32] Among other things, he wrote and delivered in English and Afrikaans a series of school radio talks, presented around 1947, in which he communicated the concepts of the ecosystem, biome, conservation, and other talks in ecology, hydrology, forestry and land management.[33] His childhood provided the inspiration: 'We had a lovely time and it was a very beautiful farm. In the summer the hills were clothed in green, and in the kloofs the trees grew tall and stately ... There were the great adventures', this being the idyllic condition set in contrast to a degraded land. These texts made up a short book on conservation, erosion and land management in Afrikaans, published in 1952, which contains lucid explanations of ideas such as the nascent concept of biodiversity (*verskeidenheid van diere en plante*, p. 18), the role of biodiversity in sustaining the plant community (*veld*) in unfavourable climatic conditions, and other aspects of ecology and conservation, as well as an account of the history of Jonkershoek.[34]

Jonkershoek as locale

Jonkershoek is a steep fault-valley penetrating the mountains south-east of the town of Stellenbosch, and forming the source catchment of the Eerste River. It aligns on geological faults dating from the time of the breakup of Gondwana and which run from south-east to north-west; faults that directed the downcutting by the Eerste River to expose the granite footslopes. Rugged peaks of sandstone crest the granites, rising above the valley floor to elevations over 1,500 m. A transverse downward block fault closes off the south-eastern corner and forms the Dwarsberg, a wall of resistant sandstone that forces the rain-bearing north-westerly air masses to rise 1,000 m above the floor of the

---

32   For overviews of ideas about erosion and land degradation in South Africa, see W. Beinart, 'Conservation and Ideas about Development: A Southern African Exploration, 1900–1960', *Journal of Southern African Studies*, 11 (1984): 52–83; B. Dodson, 'A Soil Conservation Safari: Hugh Bennett's 1944 Visit to South Africa', *Environment and History*, 11 (2005): 35–53; Wicht met Bennett at the time of his visit and cited him in *Land en Lewe* (see below).

33   The concept of the 'biome' had deep connections with ecology in South Africa. Frederic Clements and Victor Shelford coined the term 'biome' in 1939 first but credited John Phillips's notion of the 'biotic community' as a 'veritable magna carta for future progress' on the study of biomes. See Frederic E. Clements and Victor E. Shelford, *Bio-Ecology* (New York: J. Wiley & Sons, Inc., 1939), 24. Clements and Shelford defined biome as '[t]he biome or plant-animal formation is the basic community unit, that is, two separate communities, plant and animal, do not exist in the same area. The sum of plants in the biome has been known as vegetation, but for animals no similar distinctive term has become current', ibid., 20; the current, detailed vegetation map of South Africa derives from a prior classification of 11 biomes and then their subsequent subdivision into bioregions, each containing several allied vegetation types: Mucina and Rutherford, *The Vegetation of South Africa, Lesotho, and Swaziland*, 11.

34   Texts of radio talks, nine chapters, including 'The Balance of Nature', 'Forests', 'Wild Nature Preserved', Wicht Papers, SAFRI Archives, CSIR, Pretoria; C. L. Wicht, *Land en Lewe: Die Behoud van Grond, Water, Plante, en Diere*, Excelsiorboekies, VI No. 7 (Afrikaanse Pers-Boekhandel, 1952), 18–9.

valley. Beyond this wall lies a plateau with the highest average rainfall in South Africa, at over 3,500 mm per year. At the mouth of the valley, rainfall averages around 1,000 mm per year. Jonkershoek thus houses a complex ecosystem distinguished by strong gradients in its physiography. These gradients create a striking diversity in soil and habitats, with a vegetation that includes over 1,200 species of plants, mainly from the Fynbos Biome.[35] The several tributaries to the main stem of the Eerste River provided the space and opportunity for large experimentation.

Jonkershoek differed greatly from the hills and mountains of south-eastern and north-eastern South Africa, where the Forestry Department's afforestation program was directed. Jonkershoek was located on the outskirts of a major university town and urban area, Stellenbosch (12 km away) and was only a short distance (60 km) from Cape Town. Researchers at Stellenbosch had access to Stellenbosch University's library and new forestry faculty, then under the direction of E. J. Neethling. As a place, Jonkershoek had several attributes lacked by, for example, the Jessievale experiment: Jonkershoek was close to sophisticated urban communities, with as good an intellectual and service environment as could be found in the country at the time, it offered proximity to the good schools and a rewarding social milieu that a professional scientist with a family would need, and could be developed to provide competent on-site supervision.

Prior to the 1935 Empire Forestry Conference, the department purchased some small farms and land from the Municipality of Stellenbosch at the mouth of the valley in 1934 and added this to the adjoining catchments on Crown land to form the 11,000-ha Jonkershoek State (Crown) Forest. This gave enough space for expansive experiments, as well as land suited to *Pinus radiata* plantations (though most of the area was too remote and rugged to accommodate these developments). Wicht's position was not just that of Research Officer, but also District Forest Officer, which meant that he was responsible for the management of the whole state forest, including its *Pinus radiata* plantations. The development formed a complete forestry concern, which in the long run was meant to provide revenue and employment, as well as the facilities for the research experiment. Foresters envisioned that the location, both in terms of its physical and social environment, would provide the means to overcome pre-existing uncertainties that stemmed from earlier case-based investigations. Rather than being the 'centre' of forestry research, it was imagined that the experiment would be the first step of a larger research program that would see

---

35   F. J. Kruger and H. C. Taylor, 'Plant Species Diversity in Cape Fynbos: Gamma and Delta Diversity', *Vegetatio*, 41 (1979): 85–93.

research stations extended to inform the forests and water question throughout South Africa's forestry regions, an intention already expressed in Keet's 1935 policy memorandum, with the then eastern Transvaal in mind.[36]

Figure 15. The view of the Jonkershoek catchments from the mountain ridge that demarcates the south-western divide of the Eertse River catchment, c. 2005.
Source: Greg Forsyth, reproduced with his permission.

Catchment hydrology: Concepts and context

To engage with the history of the program on forests and water requires an acquaintance with certain essential concepts in catchment hydrology.[37] Scientifically, the term *catchment* means the drainage basin that delivers water to the channel of its stream. Catchment hydrology—which is the scientific focus of research on forests and water—is the study of the hydrological cycle in higher-lying, usually steep landscapes that are the source areas for rivers. The typical

---

36   J. D. M. Keet, 'Water Conservation and Aesthetics in Relation to Afforestation and the Propagation and Preservation of Indigenous Trees and Flora' (Department of Agriculture and Forestry, 22 November 1935), Wicht Papers, SAFRI Archives, CSIR, Pretoria.
37   For a complete and lucid account of concepts, principles and terms in forest hydrology, see Hewlett, *The Principles of Forest Hydrology*.

catchment has a perennial stream, which may be fed by smaller tributaries. Landscape elements such as hill slopes, riparian zones and streams predominate; floodplains are usually absent.[38] The scale of catchments tends to range from about 1 to 50 km^2 in area. The streams in such catchments are most frequently bedrock rivers, i.e. non-alluvial rivers. Stormflows are very brief, and the volume of streamflow is principally in the form of baseflow (see below). On sites in the Northern Hemisphere, and in the humid and sub-humid tropics where there is enough precipitation to support perennial streams, these catchments are normally forested. In South Africa, however, these catchments usually have as natural cover a grassland, fynbos, or perhaps savanna vegetation.

The special interest in upland catchments arises from the fact that they are the headwaters and source areas for river systems and downstream water supply, and that downstream effects cannot be understood without knowledge of upstream causes. It is here that land-use change has its immediate, direct effects on river flows. The kinds of questions asked in catchment hydrology relate to the fate of rainfall in the catchment, the routes by which water during and after rain reaches the stream, how long water is retained within the catchment, and how changes in vegetation cover affect these processes.

The hydrological cycle in the watershed results in two outputs, the losses of water as vapour to the atmosphere in evapotranspiration, and the outflow in the stream, i.e. streamflow (as well as through deep seepage below or beyond the stream channel). Hydrologists record streamflow from a given watershed continuously, and the graph of the time series they record is the hydrograph. Though modern hydrology employs isotope tracer techniques to detect the sources of water that constitutes the hydrograph, in Wicht's time much of the science in watershed hydrology focused on analysing the hydrograph. This begins with the discrimination between stormflow (or spateflow), and baseflow (or normal flow), through a procedure called hydrograph separation. This was a key element in the South African research, since the program was driven to a large extent by questions about land management and its effects on floods and on dry season water supplies.

---

38   B. L. McGlynn and J. J. McDonnell, 'Quantifying the Relative Contributions of Riparian and Hillslope Zones to Catchment Runoff', *Water Resources Research*, 39 (2003): 1310, doi:10.1029/2003WR002091; Hoyt and Langbein, *Floods*, 13. A first-order (perennial) stream in the Horton–Strahler system is one with no tributaries, a second-order stream is one formed by the confluence of two tributaries, and so on; D. Knighton, *Fluvial Forms and Processes: A New Perspective* (London: Arnold, 1998), 13.

Clearing the way through the intellectual heritage

Wicht took up his position at Jonkershoek at the beginning of December 1935. Logistics and communication were extremely difficult: there was no telephone, and for transport there were two ox-drawn scotch carts and one trolley wagon. The buildings acquired with the land from the Municipality were old and in disrepair. He was newly married to Margaret Donaldson-Selby, daughter of a Transkei magistrate. Margaret and his family appear to have created a wholesome and caring counterpoint to his constant work, and the outlet for the artistic and humane sides of his character. Margaret in due time wrote several articles for the popular press on fynbos and forests.[39]

Practical and administrative issues absorbed most of Wicht's effort during the first few years of the program. Several years passed before essential resources were in place. Administrative and managerial tasks necessarily used time: procuring equipment, mainly from overseas; constructing and proving devices for streamflow and rainfall management; recruiting and training both lay and professional researchers; and seeing to the afforestation program and forest management overall.[40] His first professional assistant, Mike de Villiers, posted there in April 1938, soon left on assignment to establish the next afforestation experiment, at Cathedral Peak in the east, after training under Wicht's supervision.[41] His next assistant arrived in 1942, and in 1943, H. B. Rycroft, later director of the Kirstenbosch National Botanical Garden, joined the team as a graduate student.

Added to this was the fact that the program was to be funded off the budget of the Forestry Division in the Department of Agriculture and Forestry, in competition with operational budgets. Each of the seven stream gauging stations required at Jonkershoek cost £250 or more,[42] whereas the Research Grants Board, then the

---

39   See, for example, M. L. Wicht, 'Creeping Invasions of the "Green Cancers"', *African Wildlife*, 25 (1971): 11–4.

40   Including direction of and participation in the fighting of a major 11-day fire in the surrounding mountains in December 1942, which among other things threatened the experimental treatments at Jonkershoek; from this Wicht learnt many lessons that would influence his ideas on the ecology and management of wildfire. In his report he gave fulsome praise to the many parties who assisted (copy of Minute M. 600, undated, Jonkershoek Archives, CSIR, Stellenbosch). This was one among many fires that would threaten the hydrology program over time; not long after this he made carefully argued proposals to introduce controlled burning as part of catchment management: District Forest Officer C. L. Wicht to Conservator of Forests Cape Town, M. 600 222 March 1944, Jonkershoek Archives, CSIR, Stellenbosch.

41   H. Söhnge, 'Bosboupionier: AM de Villiers—Die Man van Mike's Pass', *ARBOR*, December 1996: 12–3; de Villiers confronted similar challenges at Cathedral Peak as Wicht had at Jonkershoek, being able to access the research terrain only on foot or by horse, and his first major task in 1945, when resuming the work after World War II, was to supervise the building of the pass through the cliffs of the Clarens Sandstone to provide road access to the research catchments.

42   Annual Report of the District Forest Officer, Jonkershoek 1937/38, Jonkershoek Archives, CSIR, Stellenbosch.

primary science funder in South Africa, dispensed only about £1,300 per annum between 1919 and 1936, mostly in the range of £50 to £100—far too small to be of much use for Wicht's program.[43]

When the department assigned him to Jonkershoek it would have been in the knowledge that his academic record was good, he had an enquiring mind, and that he knew the terrain of the chosen research site, as well as being well connected in the strong intellectual community of the Cape.[44] The thorough theoretical study for his Doctoral dissertation and his exposure to R. A. Fisher's work on experimental design while at Oxford seem to be the key factors in learning the scientific method; he quoted Fisher: 'If the design of an experiment is faulty, any method of interpretation which makes it out to be decisive must be faulty too'.[45] Much of the program at Oxford had been taken up by applied topics, such as completing a working plan for a nearby forest, and this together with his experience at Tokai and elsewhere in South Africa prepared him for managing a forest station. He was also familiar with the emerging field of ecology, having at that time copies of books by Andreas Schimper, Arthur Tansley, Eugenius Warming, A. J. Cajander, and John Weaver and Frederic Clements in his personal library. As well as these personal qualities, he enjoyed the constant support of his colleagues in his department and at the universities and other institutions in the Cape.

Wicht encountered a situation where in 1935, and for some decades afterward, the field of forest hydrology lacked a coherent set of methods and concepts—effectively it was a topic without a professional discipline. The body of knowledge called catchment hydrology, much less forest hydrology, was hardly discernible, and the community of practice just taking their first steps.[46] Basic questions about experimental design and data collection recurred as problems throughout the first half of the twentieth century—as late as 1970, the eminent hydrologist John Hewlett could still say that an observation of hydrological research 'revealed disjointed purposes, terminology and methods, among the workers in

---

43   Dubow, *A Commonwealth of Knowledge*.

44   Senior officials corresponded frequently with Wicht before and while at Oxford and Tharandt; e.g. Chief Conservator of Forests C. C. Robertson to Wicht, 17 June 1929, E. J. Neethling for Chief Conservator circular minute to to Wicht and others, 10 June 1931, Director of Forestry to Wicht on choice of school for doctoral study, 6 September 1931, all in Wicht Papers, SAFRI Archives, CSIR, Pretoria.

45   Fisher, *The Design of Experiments*. The mark-ups in Wicht's copy of the book shows he studied it carefully (copy in the possession of Fred Kruger).

46   M. Winsor, 'The Practitioner of Science: Everyone Her Own Historian', *Journal of the History of Biology*, 34 (2001): 229–31. We refer to catchment hydrology as the field of science that seeks to explain the hydrological cycle in upland catchments, and how vegetation and other aspects of biodiversity influence this cycle.

several countries'.[47] The first principal body of scientific knowledge on forests and water, Joseph Kittredge's 1948 book *Forest Influences*, was still 13 years away from publication when Wicht started in 1935.[48]

In the mid-1930s, the leading theory describing the behaviour of catchments was Robert E. Horton's theory of infiltration-excess overland flow. Horton argued, from laboratory evidence, that in upland catchments, stormflow in the stream originates as flow over the land surface when rain during a storm exceeds the infiltration capacity of the soil, and only then (or at least, stormflow is mostly generated this way). Thus, stormflow does not arise from subsoil seepage; moisture that does infiltrate the soil is stored or returned in the evapotranspiration process, or seeps via groundwater to the stream to feed baseflow; groundwater flow seldom contributes to stormflow. So for that reason, the quantity and timing of stormflow during precipitation events depends on the infiltration capacity of the soil, all else being equal, and this depends among other things on the condition of the humus layer. This idea shaped hydrological science for decades, including the program in South Africa, and still influences engineering hydrology.[49]

Wicht employed the Hortonian concept in analysing Jonkershoek hydrographs, and spent huge effort in generating Hortonian baseflow recession curves, constructs later rejected by Ugs Nänni in his research on baseflow at Cathedral Peak.[50] The Hortonian model stands in contrast to the alternative, the variable source area–subsurface storm flow theory, according to which spateflow originates, first from the rain that falls on the channels and on the shallow groundwater adjoining the channels, and second from the accelerated seepage downslope of soil moisture—'old water' from the upslope soils, displaced by infiltrated rain.[51] Since rainstorm intensity in upland catchments rarely exceeds

47   J. D. Hewlett, 'The Relation of Forests and Forestry to Water Resources. Address to the Annual General Meeting of the South African Institute of Forestry, Stellenbosch, 15 May 1970', *South African Journal of Forestry*, 75 (1970): 4–8.

48   J. S. Kittredge, *Forest Influences: The Effects of Woody Vegetation on Climate, Water, and Soil, with Applications to the Conservation of Water and the Control of Floods and Erosion* (New York, McGraw-Hill Book Co. Ltd, 1948).

49   K. Beven, '"Robert E. Horton's Perceptual Model of Infiltration Processes", Scientific Briefing', *Hydrological Processes*, 18 (2004): 3447–60; T. Burt, 'Valley-Side Slopes and Drainage Basins. Runoff and Erosion', in T. P. Burt, R. J. Chorley, D. Brunsden, N. J. Cox, and A. S. Goudie (eds), *The History of the Study of Landforms or the Development of Geomorphology Volume 4: Quaternary and Recent Processes and Forms (1890–1965) and the Mid-Century Revolutions* (London: Geological Society of London, 2008).

50   C. L. Wicht, 'Depletion of Ground-Water Flow in Jonkershoek Streams', *Journal of the South African Forestry Association*, 8 (1942): 50–63.

51   See, for example, M. G. Sklash and R. N. Farvolden, 'The Role of Groundwater in Storm Runoff', *Journal of Hydrology*, 43 (1979): 45–65.

soil infiltration capacity, and from other lines of evidence (see Chapter 8), the Hortonian model is obviously invalid. This has profound implications for catchment management and land management in general.

Coincidentally, Jonkershoek was started at about the same time that important catchment experiments developed in the US, at San Dimas in the chaparral of California in 1933, and Coweeta in South Carolina in 1934.[52] While those could not inform Wicht's thinking at the outset of his work, the exchange with the scientists in these and later programs in the US (as well as in Africa) was to become a vital part of South Africa's catchment hydrology, and an exchange that contributed to the early stages of the development of the discipline, as we shall see. Wicht did take what lessons were to be found in earlier studies, especially in those in Switzerland, at Wagon Wheel Gap (see below) and on the Wasatch Plateau in Utah (begun in 1912), as well as studies of parts of the hydrological cycle, such as the work on riparian zone evaporation in California reported in 1936.[53] But these were early pieces of the bigger picture that was yet to emerge.

A South African science of forest hydrology

At the time in South Africa, an atmosphere of excitement had taken hold in the science community under Smuts's leadership, which would energise the work at Jonkershoek. Saul Dubow describes Smuts as a 'powerful' champion of South African science, distinguished by his 'key role … as a patron of ecological thinking'. Smuts, in his 1925 address to the South African Association for the Advancement of Science, urged the audience to develop what he termed the South African point of view, to overcome the 'habits of thought and the viewpoints characteristic of its birthplace in the northern hemisphere', and to 'correlate scientific developments across a range of disciplines in new and creative ways, with potentially far-reaching implications for "universal science"'.[54] J. S. Hofmeyr, addressing another joint meeting of the association and the British Association for the Advancement of Science on behalf of Smuts in 1929, emphasised the need to 'Africanise' science, with South Africa as the 'southern

---

52   Paul H. Dunn et al., *The San Dimas Experimental Forest: 50 Years of Research*, General Technical Report PSW-104 (Berkeley, CA: US Department of Agriculture, Pacific Southwest Forest and Range Experiment Station, 1988); K. J. McGuire and G. E. Likens, 'Historical Roots of Forest Hydrology and Biogeochemistry', in D. F. Levia, D. Carlyle-Moses, and T. Tanaki (eds), *Forest Hydrology and Biogeochemistry: Synthesis of Past Research and Future Directions* (Berlin: Springer-Verlag, 2011): 3–26.
53   See C. L. Wicht, 'Determination of the Effects of Watershed-Management on Mountain Streams'; H. C. Troxell, 'The Diurnal Fluctuations in the Ground Water and Flow of the Santa Ana River and Its Meaning', *Transactions of the American Geophysical Union*, 17 (1936): 496–504.
54   Dubow, *A Commonwealth of Knowledge*, 207–8, 212, 235.

gateway' to Africa. The Forestry Department Research Branch, a state entity, sought to realise Smuts's vision.[55] This atmosphere and the urgency from the Empire Forestry Conference gave ample scope for ambition in Wicht's science.

This sense of urgency, excitement and satisfaction is evident in the progress reports that Wicht and others wrote about the work as it unfolded. Developments progressed with the support of key influential people, who with Wicht formed an intellectually powerful cohort. Wicht's boss, the Chief Forest Research Officer (first J. J. Kotzé, and later Ian Craib) visited Jonkershoek from Pretoria on inspection at least twice a year. J. D. M. Keet was a frequent visitor, and on one occasion, in Wicht's absence, when told that there were no funds to complete a gauging weir, he instructed work to proceed, that the funds would be found.[56] The Minister of Agriculture and Forestry, Colonel Collins (Deneys Reitz having moved on), visited to assess progress.[57] Colleagues from other disciplines came too: John Phillips, I. B. Pole-Evans, and South Africa's leading meteorologist T. E. W. Schumann among them.

Scientists and the state in the mid-1930s fully supported Wicht's long-term research experiment. The forest research community was naturally committed to the long view: by 1935, A. J. O'Connor had already set up his 30- to 50-year CCT trials. These he designed to study the relationship between the density of plantation tree populations, the onset of competition among the trees, how this affected productivity, and how competition effects could be mediated by thinnings. Results became available within a decade, but findings were still being analysed and refined 50 years later. The entire basis of forest management in South Africa turned on this work, but forest managers knew that they would need to design and adapt practice progressively over the long run as the new information emerged.[58] Since Jonkershoek catchment treatments involved afforestation, final results could come only in 40 years. This idea did not disconcert the officials in the department, despite the urgency to see progress.

Among all the preparations that Wicht made immediately after arriving at Jonkershoek, he began to recruit support and collaboration from scientists in the Cape, reaching out to the University of Cape Town. Within the first two

---

55   Like the situation in agronomy and veterinerary science: ibid. 178–80.

56   Forest Research Officer C. L. Wicht to Chief Forest Research Officer, Minute R.J. 120/21 of 26 March 1941, Jonkershoek Archives, CSIR, Stellenbosch.

57   Monthly reports of the Forest Research Officer, Jonkershoek, 1936–1945, Jonkershoek Archives, CSIR, Stellenbosch.

58   O'Connor, *Forest Research with Special Reference to Planting Distances and Thinning*; B. V. Bredenkamp, 'The C.C.T. Concept in Spacing Research—a Review', in D. C. Grey, A. P. G. Schönau, and C. J. Schutz (eds), *Proceedings of the IUFRO Symposium on Site and Productivity of Fast-Growing Plantations, 30 April–11 May 1984, Pretoria and Pietermaritzburg, South Africa*, vol. 1 (Pretoria: South African Forest Research Institute, 1984): 313–32.

years, he quickly gained the interest of University of Cape Town botanists R. S. Adamson and Margaret Levyns, who both began a botanical survey of the valley, while Wicht himself worked at building the botanical collection. For advice on stream gauging he consulted experts from the Irrigation Department, the Town Engineer from Stellenbosch (Mr Sandeberg), and the Agricultural Engineer from the Stellenbosch–Elsenburg College of Agriculture. For the precise calibration of the V-notch stream-gauging weirs soon being installed at Jonkershoek, he enlisted Professor Snape of the Civil Engineering Department at the University of Cape Town and his Master's student, J. P. Kriel (later Head of the Department of Water Affairs).

He was also good at mobilising people from within: half the workers at Jonkershoek were uneducated white settlers, mostly from Knysna, and he sought out talent from among these. He spotted the intelligence and initiative of 'Settler E.J. Borchardt' and in 1937 had him appointed Research Clerical Foreman to assist in technical work;[59] throughout his career he succeeded in mentoring young candidates.

Wicht soon created an important international network through correspondence. He exchanged ideas with Walter C. Lowdermilk, H. G. Wilm and other prominent hydrologists in the USA who were beginning to engage in forest hydrology. He noted that he received a 'very friendly letter ... from Dr. W. C. Lowdermilk, Assistant Chief of Soil Conservation Service in the United States ... [he] has offered constructive criticism and is forwarding some publications to me'.[60] Other correspondents included Joseph Kittredge, C. G. Bates (of the Wagon Wheel Gap study), Andrew P. Mazurak at the University of California, Harry F. Blaney, G. W. Musgrave (who offered to publish Wicht's work in the *Transactions of the American Geophysical Union*), C. E. Ramser of the US Soil Conservation Service, E. N. Munns of the US Forest Service, C. F. Brooks of Harvard University, and Robert E. Horton.

Wicht cultivated widespread public understanding of his project. Within the first year, he had granted interviews to all daily newspapers in the Cape. He wrote a popular text on Jonkershoek for *Farming in South Africa* and for the University of Stellenbosch alumni magazine, lectured the Rotarian Club of Cape

---

59   Annual Report of the Forest Research Officer 1937–38, Jonkershoek Archives, CSIR, Stellenbosch, digi. nrf.ac.za/dspace/bitstream/handle/10624/380/CLW_AnnRep1938p001.pdf?sequence=1 (accessed 24 July 2013).

60   Monthly Report, Forest Research Officer, Jonkershoek, December 1940, Jonkershoek Archives, CSIR, Stellenbosch.

Town, and conducted 'numerous visitors' around the Valley.[61] His sociability allowed him to create a supportive coalition of leaders within the department as well as from the South African and international science communities, South African politics and the public in the south-western Cape. Wicht's forceful thoroughness, intellect and powers of persuasion were well supported by his grace and social skills, and his professional networks when his controversial methods and later findings received international criticism.

Rough terrain, erratic rainfall, and necessary precision

Work began immediately in 1936 with the installation and instrumentation of a meteorological station, rainfall stations and streamflow-gauging weirs, as well as studies of local meteorological patterns. Finding accurate ways to measure rainfall and streamflow in the difficult terrain chosen dominated much of Wicht's research agenda from 1935 to 1940. The most urgent technical problems were those of sound estimation of rainfall over the whole experimental site, and the accurate measurement of streamflow from all gauged streams, over the very wide variation in flow levels that characterised the Jonkershoek streams. Resolving these problems taught Wicht key features of the catchment ecosystem and the physiographic trends and secular variations in the climate of the area.[62]

Early experiments in raingauging began to reveal the magnitude and relevance of local variations in rainfall. After resolving raingauge design and orientation,[63] the work moved on to studying the variability of measured rain, in time and space. Fifteen raingauges were installed at increasing distances from the Dwarsberg plateau, so as to represent altitudinal zones throughout the valley.[64] In addition, Wicht began local studies designed to test the requirements for random sampling of rainfall.

---

61   Annual Report to Chief Forestry Research Officer 1936–37, Jonkershoek Archives, CSIR, Stellenbosch, digi.nrf.ac.za/dspace/bitstream/handle/10624/378/CLW_Annrep1937p001.pdf?sequence=1 (accessed 24 July 2013); see digi.nrf.ac.za/dspace/handle/10624/381 for the progress reports that contain further information (accessed 13 September 2013).

62   See C. L. Wicht, 'A Preliminary Account of Rainfall in Jonkershoek', *Transactions of the Royal Society of South Africa*, 28 (1941): 161–73; H. B. Rycroft, 'Random Sampling of Rainfall', *Journal of the South African Forestry Association*, 18 (1949): 71–81; J. C. Meyburgh and C. L. Wicht, 'The Inclination and Bearing of Rainfall Estimated with Vectopluviometers at Jonkershoek', *Forestry in South Africa*, 7 (1966): 71–90, on research employing the Fourcade raingauge; for the statistical regression model, see C. L. Wicht, J. C. Meyburgh, and P. G. Boustead, 'Rainfall at the Jonkershoek Forest Hydrological Research Station', *Annals of the University of Stellenbosch*, 44, Series A, No. 1 (1969): 1–66.

63   This included the innovative Fourcade raingauge.

64   C. L. Wicht, 'Improvements in the Gauging of Rainfall', *Journal of the South African Forestry Association*, 12 (1944): 19–28.

A 10-year record at the base of the Jonkershoek valley, closely correlated with the long record for the Royal Observatory in Cape Town, going back to 1840, told the story of marked year-to-year variation in rainfall at these sites, and the tendency for slower, almost cyclical variation over time. Also available to Wicht was the recently published work by T. E. W. Schumann and W. R. Thompson, showing across the country a secular variation in rainfall—a tendency for periods of several years of comparative drought to alternate with periods of relatively higher rainfall.[65] From his early trials and his work for his 1941 paper,[66] Wicht knew a great deal about the pronounced gradient of rainfall with altitude in the Cape, the year-to-year variation, and the difficulty—if not impossibility—of achieving sufficiently large random samples at local scales (though the work on the latter was published only in 1947).[67] So he believed that rainfall could not be measured accurately enough for water-balance studies, immediately informing his overall research plan (see below). Wicht also thought that trends and fluctuations in rainfall over time (and hence, in other climatic factors such as air temperature) could easily confound findings from experiments that lacked long-term controls.

Accurate measurement of rainfall in mountainous catchments was and is seldom achieved. Measuring rain at any point is bedevilled by errors induced by the aspect of the mountain slope relative to the tracks of rainstorms, and wind-induced turbulence over gauge orifices. Wicht's detailed experiments with different raingauge designs, to combat the effects of wind turbulence, and gauge location and orientation, to take account of storm tracks, provided new knowledge on raingauging and settled many technical questions. This required the choice of gauge design and exposure that would address the effects of wind and slope aspect; some gauges were to be visited only once a week or less often, requiring precaution against evaporation losses from the collection vessel. All these technical problems of measurement were resolved by 1940, after much quantitative experimentation and trial of alternatives.[68]

---

65   T. E. W. Schumann and J. L. Thompson, *A Study of South African Rainfall: Secular Variations and Agricultural Aspects*, University of Pretoria Series No. 1, vol. 28 (Pretoria: Van Schaik, 1934).
66   Wicht, 'A Preliminary Account of Rainfall in Jonkershoek'.
67   Rycroft, 'Random Sampling of Rainfall'.
68   Wicht, 'A Preliminary Account of Rainfall in Jonkershoek'.

Figure 16. A vectopluviometer. This raingauge was designed by Henry Fourcade to measure the inclination (the departure from vertical) of rain and its bearing (the direction from which it arrives), for the full characterisation of rainstorms.

See Fourcade, 'Some Notes on the Effects of the Incidence of Rain on the Distribution of Rainfall over the Surface of Unlevel Ground', and Meyburgh and Wicht, 'The Inclination and Bearing of Rainfall Estimated with Vectopluviometers at Jonkershoek'.

Source: SAFRI Collection, CSIR, Stellenbosch; photograph C. L. Wicht, c. 1940.

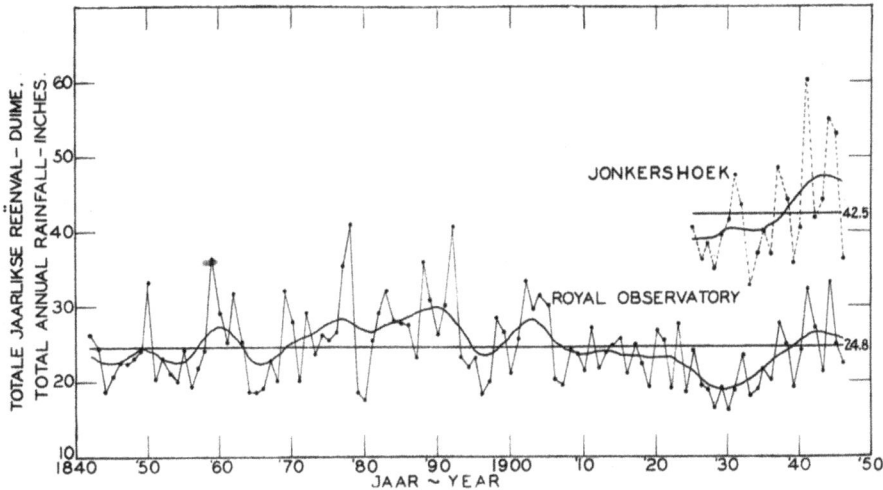

Figure 17. The time series of rainfall at Jonkershoek and Cape Town, showing annual and running mean values and illustrating the climate variability and change that confronted experimental design.

Source: Reproduced from C. L. Wicht, *Forestry and Water Supplies in South Africa* (Pretoria: Department of Forestry, Union of South Africa, 1949).

Measuring the volume of rainfall over the catchment requires adequate sampling across the terrain, with 'irregular mountainsides with variable slopes and aspects', as well as accurate measurement at any one gauging station.[69] Experiments to assess the random sampling of rainfall suggested that the variation was too great to accurately sample the Jonkershoek landscape in this manner.[70] Wicht set out to find a different approach.

Precise measurement of the rainfall received by each experimental catchment was especially elusive, since rainfall varied nearly fourfold (as we now know) from one end of Jonkershoek to the other; this would require accurate measurement at each gauging station as well as adequate, dense sampling across the entire experimental terrain, an almost impossible requirement. For instance, at one experiment in San Dimas in California where researchers sought to track these changes, they maintained a network of 450 gauges for many years to achieve sufficient areal accuracy.[71] But because early measurements showed rainfall to correlate with topography, precipitation as it varied over the landscape could be estimated through a properly distributed network of gauges. The network had 12 gauges initially, finally 19 in all, to give roughly one per 100 ha of experimental terrain, distributed across the landscape; some were visited regularly every rain

---

69   Wicht et al., 'Rainfall at the Jonkershoek Forest Hydrological Research Station', 8.
70   Rycroft, 'Random Sampling of Rainfall'.
71   Dunn et al., *The San Dimas Experimental Forest: 50 Years of Research*, 4.

day, some once a week, the remote sites were reached on foot once a month. Until statistical regression models could be developed for areal estimates, Wicht used the relative measured precipitation as indices for each catchment, rather than accurate volumetric measurements to represent differences in rainfall among catchments and over time.[72] In this way an adequate program would be practically achieved with the limited resources available.

Storms, torrents and small streams

It was six years into the program before some understanding emerged of what determines streamflow in the mountain catchments of Jonkershoek. Streamgauging required a great deal of technical development and fine-tuning, largely to calibrate the chosen compound-notch gauging design properly (chosen to provide accuracy at all anticipated flow levels); although gauging commenced in 1937 (at one station), it was only in 1940 that Wicht's collaboration with engineers finally produced acceptable formulas to convert the recorded water levels in the stilling ponds at the gauging sites to flow rates. This enabled early analyses of the behaviour of the streams, painstaking work that Wicht undertook himself.[73] Careful, very accurate weekly measurements by means of handheld staff gauges provided checks on the continuous records from clockwork chart recorders. Once charts had been validated against the benchmarks, and corrected where necessary, water levels were read off by hand to provide a time series for each station each week. Calculations by entering water-level values into the flow equations, completed by hand-operated calculator, led to tables of hourly flow values. From these, daily, weekly, monthly and annual flows could be summed. Plotting these flow rates as a graph of rate against time provided the hydrograph.

This attention to detail generated many lessons, especially the great diligence required of observers in the field and in the curation and analysis of records to ensure their reliability: 'My experience at Jonkershoek has forced me to conclude that recording apparatus can only run successfully if the greatest possible watchfulness and diligence are exercised'.[74] The norms required to assure this

---

72   This was accomplished in 1969: C. L. Wicht et al., 'Rainfall at the Jonkershoek Forest Hydrological Research Station'. This analysis of the colossal amount of records from 1936 to 1965 has not yet been repeated; Figure I in that report shows the topographic distribution of raingauges; Map III, 34 shows the distribution of raingauges and the isohyetal map generated from the regression model; Map IV, 37 shows seasonal isohyets.
73   Wicht, 'Depletion of Ground-Water Flow in Jonkershoek Streams'; Wicht, 'The Variability of Jonkershoek Streams'.
74   Monthly report, Research Officer: Jonkershoek, May 1938, Jonkershoek Archives, CSIR, Stellenbosch.

quality of observation became the minimum standards for the experimentation, with every participant in the program trained to achieve these, throughout the eventual network of research stations.

A second aspect was the great variation in the flow of Jonkershoek streams. The fact that the Jonkershoek catchments diverged so strongly in physiography and rainfall allowed new understanding of the nature of streamflow from upland watersheds. By 1942, hydrographs were available for three years in the first gauged catchment, and 12 months in three others. From this analysis, Wicht concluded:

> The considerable variation in streamflow caused by differences in the wetness of years and differences in the nature of catchments, complicate the determination of the effects of different systems of catchment management on the flow. Unless the effect of treatment … is very marked, observations over a very long series of years will be required to prove the significance of observed differences …[75]

So the reasons for the variability could be found, but the expression of any experimental effect would be hard to detect among all this variation. Wicht's early work on the hydrographs, against the background of his direct observation of rainfall events and of the rapid rise and fall of extreme stormflows in the Jonkershoek streams, revealed several key facts, not or little known at the time. First, the variation from year to year in any given stream was closely correlated with the rainfall that year, showing that there was little water stored in the catchment and carried over from one year to the next.[76]

Despite the wild torrents that Wicht witnessed—in one event, the flow of the Eerste River ranged more than 200-fold in 24 hours, while its flow varied 1,000-fold over the year—the record told essential information about the processes determining streamflow. The behaviour in catchment streamflow revealed in the relative distribution between stormflow (the portion of the hydrograph attributable to a given rainstorm) and baseflow (the residual once stormflow had subsided) became a central empirical question about forests and streamflow. Wicht, with Hoyt and Troxell, questioned the received scientific proposition that forests tended to decrease peak flows and to increase groundwater flow, i.e. to 'equalise the flow of streams throughout the year'. He argued that the effects of afforestation 'on minimum and peak flows should be investigated separately',[77] anticipating much of the research to follow, and most important, signalling scepticism about the claimed effects of forests on floods.

---

75   Wicht, 'The Variability of Jonkershoek Streams', 17.
76   Ibid., 16.
77   Ibid., 20–1.

Wicht could now build further on earlier findings from Wagon Wheel Gap. At that site, William Hoyt and H. C. Troxell determined that changes in streamflow from deforestation owed to changes in baseflow, not stormflow: forests had no measurable effect on floods. This was disputed immediately among the hydrologist and forestry community, becoming part of a controversy that continued for decades, since the claim that forests diminished floods was a key element in contemporary beliefs about forestry: W. C. Lowdermilk called it 'an attack upon the widely accepted belief that watershed vegetation must be kept intact for the most favorable influence upon streamflow and erosion and flood control, and, that the negative values of the vegetation, because of transpiration losses, are far outweighed by the beneficial effects'.[78]

Another key early finding from Wicht's initial research arose from his curiosity about the behaviour of streams. The work that arose from this anticipated what is now seen as a 'central issue in hydrology today … to establish relationships between hydrological and biological processes in ecosystems'.[79]

Wicht observed at Jonkershoek that during warm rainless periods, stream levels during baseflow as shown on the charts fluctuated daily, a hitherto little-noticed phenomenon anywhere in the world. Sceptical as to the reality of the observed fluctuation, observers then used a special staff gauge to measure water levels in weir ponds to validate the information on the chart recorders, and confirmed that baseflow levels were highest around 8 am, declined to a minimum at about 4 pm, and recovered to about the prior level by about 4 am.[80]

His monthly reports for 1938 and 1939 reveal his intense interest in this phenomenon and his speculations as to its cause. The May 1938 report states: 'at the beginning of the month, when warm weather prevailed, the peculiar 24 [hour] drop in the water level diagram, which although obviously associated with hot days, has not been satisfactorily explained so far, was further investigated'. In catchments in general, different causes may be responsible for these fluctuations: among them, high temperatures, earth-tide effects, atmospheric pressure variations, and daily courses in evapotranspiration.[81]

---

78   W. G. Hoyt and H. C. Troxell, 'Forests and Floods', *Transactions of the American Society for Civil Engineers*, Paper no. 1858 (1932): 1–30; W. C. Lowdermilk, 'Forests and Streamflow: A Discussion of the Hoyt-Troxell Report', *Journal of Forestry*, 31 (1933): 296–307, 296.

79   B. J. Bond, J. A. Jones, G. Moore, N. Phillips, D. Post, and J. J. McDonnell, 'The Zone of Vegetation Influence on Baseflow Revealed by Diel Patterns of Streamflow and Vegetation Water Use in a Headwater Basin', *Hydrological Processes*, 16 (2002), 1671–7.

80   C. L. Wicht, 'Diurnal Fluctuations in Jonkershoek Streams', *Journal of the South African Forestry Association*, 7 (1941): 34–49, 42.

81   Z. Gribovszki, J. Szilágyi, and P. Kalicz, 'Diurnal Fluctuations in Shallow Groundwater Levels and Streamflow Rates and Their Interpretation—a Review', *Journal of Hydrology*, 385 (2010): 371–83. P. C. Inkenbrandt et al., 'Barometric and Earth-Tide Induced Water-Level Changes in the Inglefield Sandstone in Southwestern Indiana', *Proceedings of the Indiana Academy of Science*, 114 (2005): 1–8.

Wicht got his clue from 1930s work in the Santa Ana Valley, where Troxell and others researched similar diurnal fluctuations in groundwater levels and from observation and experiment inferred that these were owing to daily variations in evapotranspiration losses from vegetation drawing moisture from the groundwater.[82] By careful analysis of the hydrographs, comparison between wet and dry years, and comparison between catchments differing in steepness, Wicht concluded that despite differences from the Santa Ana observations, the cause at Jonkershoek was the effect of daily transpiration by plants, the midday maximum effect lagged through delay in transmission of the draft on the groundwater to the stream. This work vindicated the idea behind the policy of leaving riparian zones open in plantations, and set off a line of later studies that fed into standards for plantation forest management throughout South Africa.

Designing a very large experiment

The faults at Jessievale and the experiences gained from attempts to explain complaints of streamflow losses at various sites, at White River and elsewhere, had provided valuable lessons to inform the Jonkershoek project. The first was that the climate, especially rainfall, was so variable and climate change so pronounced that there could be no reliance on short-term observation and common sense to confirm and explain claims about vegetation effects, either way. The lack of modern empirical or theoretical understandings of South Africa's climates was partly lifted by Schumann and Thompson's regional study on rainfall, but local variation in mountainous terrain was largely undocumented, much less the nature of the interaction between climate, vegetation, and streamflow. A method of investigating this complexity and demonstrating effects of afforestation or other treatments, such as veld burning, could not present itself as a cut-and-dried scientific method, but required the lessons of active observation as well as reasoning from the principles of mathematical statistics and an evolution of ideas about catchment hydrology.

Writing several years after his assignment, Wicht reflected on these lessons, arguing that government policy required 'unimpeachable scientific experiments': 'To accept unverified evidence, obtained by quick approximations, and present it as conclusive, will *not* save time or money, but lead to costly failures in catchment conservation, and thus loss of faith in catchment research'.[83] He continued: 'Special requirements needed to be met in stream-flow, or catchment experiments, which were large in scale, necessarily long-term,

82   Wicht, 'Diurnal Fluctuations in Jonkershoek Streams', 34–49.
83   C. L. Wicht, 'Forest Hydrological Research in Africa South of the Sahara', in *Wasserwirtschaft in Afrika* (Köln: Verlag Deutscher Wirtschaftsdienst, 1963).

involving high costs, and should therefore be extremely carefully planned from the outset'.[84] It is obvious that he and his colleagues had learnt from the failures and difficulties encountered in the Jessievale project, and the ambiguities encountered in case studies such as that at White River. They also had in mind the prerequisites set by Smuts and Phillips during the fourth British Empire Forestry Conference.

In 1935, the research questions and hypotheses set for the program were vague and broad. J. J. Kotzé, in his capacity as Chief Forest Research Officer, stated the objective as establishing 'an intensive long-range investigation into the effects of afforestation upon such basic factors as run-off, erosion, soil moisture, humidity, soil and air temperatures, evaporation, transpiration, soil (both physical and chemical), organic matter, plant succession and stream flow'.[85] At the beginning, the treatments to be tested experimentally included not just afforestation with exotic *Pinus radiata*, but also with yellowwoods and other indigenous trees, broadleaved exotics such as oaks and poplars; as well as veld burning in the fynbos, with and without grazing (by goats); and complete protection of the fynbos against fire. 'Observations will determine, quantitatively, the variation in the dispersal of precipitation under different circumstances'; this would be achieved 'by recording rainfall, evaporation, transpiration, run-off, seepage, streamflow and all natural phenomena which might influence the hydrological processes'.[86] Kotzé advocated a form of the so-called 'water balance approach', the approach involving measurement of the relationship between rainfall and all the elements of its dispersal—evaporation, infiltration, transpiration, overland flow, lateral seepage, and exit as streamflow. By balancing the equation for catchments under different types of vegetation cover, it would be possible to examine the role of vegetation in these processes, and from this deduce the aggregate effects of vegetation change.[87]

It was eight years before Wicht and his colleagues were satisfied with the design of the program. This initial, wide brief set in train an intensive process of research design and revision as Wicht confronted the realities on the ground and struggled to find an effective way forward. The final design was the outcome of several iterations. In November 1936, J. J. Kotzé spent four days at Jonkershoek discussing the plan; his successor Ian Craib spent another four days with Wicht in 1938 to discuss a revised program (Craib had also led the

---

84   Wicht, 'Determination of the Effects of Watershed-Management on Mountain Streams', 596.
85   Annual report of the Chief Forest Research Officer (J. J. Kotzé) 1935/36, Jonkershoek Archives, CSIR, Stellenbosch.
86   Ibid.
87   See C. R. Hursh, M. D. Hoover, and P. W. Fletcher, 'Studies in the Balanced Water-Economy of Experimental Drainage-Areas', *Transactions of the American Geophysical Union*, 23 (1942): 509–17, 509.

team searching for the next site at Cathedral Peak, in 1936).[88] The first version of the research program was approved in October 1937; the third revision was 'finally' approved in 1943, having been foreshadowed by Wicht's paper on the experiment in the proceedings of the American Geophysical Union.[89]

A series of papers that began in 1939 (which he began writing in 1937, and submitted in June 1938), and culminated in the paper at the International Symposium on Forest Hydrology at Pennsylvania State University in 1965, on the 'Validity of Conclusions from South African Multiple Watershed Experiments',[90] lays out the steady evolution in Wicht's thinking. His writings reveal a demanding journey through the problems of experimental design. In 1935, Wicht thought he would follow the 'water balance' approach: 'all processes in the water cycle to be studied and measured'.[91] By 1939, the experimental design was to be the 'single basin approach'[92] (see Box 2), in which '[e]ach stream is to be studied independently and compared with itself before and after treatment'.[93] Later he had to exclude this approach, and then the paired-catchment design: his research by this point had 'shown that the variability of streamflow is also normally so great that a single comparison between two watersheds, or between two periods—before and after treatment—will be so burdened with chance-errors that the results must necessarily be uncertain'.[94] By 1943, his preferred design was the multiple catchment experiment. This was a profound evolution in design, motivated by the lessons of earlier trials, the big issues in forest and water policy, and the newly acquired knowledge of the variability in Jonkershoek's streams.

88   Annual Report, Research Officer C. L. Wicht Forest Influences Research Station 1936/37 to Chief Forest Research Officer, Jonkershoek Archives, CSIR, Stellenbosch, digi.nrf.ac.za/dspace/bitstream/handle/10624/378/CLW_Annrep1937p001.pdf?sequence=1 (accessed on 11 August 2013); Annual Report 1937–1938, at digi.nrf.ac.za/dspace/bitstream/handle/10624/380/CLW_AnnRep1938p001.pdf?sequence=1.

89   C. L. Wicht, *Hydrological Research in South African Forestry. British Empire Forestry Conference, 1947, Great Britain* (Pietermaritzburg: Union of South Africa, 1947); Wicht, 'Determination of the Effects of Watershed-Management on Mountain Streams'.

90   C. L. Wicht, 'The Validity of Conclusions from South African Multiple Watershed Experiments', in W. E. Lull and W. H. Sopper (eds), *International Symposium on Forest Hydrology* (Oxford: Pergamon Press, 1967), 749–60.

91   C. L. Wicht, 'Forest Influences Research Technique at Jonkershoek', *Journal of the South African Forestry Association*, 3 (1939): 65–78.

92   See J. Hewlett and L. Pienaar, 'Design and Analysis of the Catchment Experiment', in E.White (ed.), *Symposium on Use of Small Watersheds in Determining Effects of Forest Land Use on Water Quality* (Lexington, KY: University of Kentucky, 1973).

93   Wicht, 'Forest Influences Research Technique at Jonkershoek', 66.

94   Wicht, 'Determination of the Effects of Watershed-Management on Mountain Streams'.

John Hewlett and Leon Pienaar, in a methodological review, identified that catchment experiments may be of four kinds:

The correlation study, in which the hydrological response to presumed causes of the response are correlated from available existing measurements for a group of catchments diverse in terms of their vegetation covers; this differs from the types of catchment experiment summarised below, in that there is no control basin, and observed correlations are taken to show causal connections between streamflow and vegetal cover. [The most well-known of these is the study by Rakhmanov in Russia, still today cited in authoritative works in support of the argument that forests increase total annual river runoff.]

The single watershed experiment (again, one with no control basin) which involves an analysis of an existing set of hydrologic records on a basin that has accidentally sustained some 'treatment', such as afforestation, or alternatively, has been deliberately treated; the response factor of interest is regressed or plotted over one or more 'independent' variables that are changing with time, rainfall, temperature, land use and so on. Both can be classified as time-trend analyses, a search for some sharp break in a plotting of response over time.

The paired watershed experiment: here, conditions in a control basin are kept constant or nearly so, while a second is treated after a period of calibration; the effect of a treatment is evaluated as a difference between expected response, i.e. a value based on correlation between the two catchments before treatment, and the actual response (Y.) measured after treatment on the experimental unit.

The multiple catchment experiment: here, several basins at the same place each receive the same treatment, with treatments in regular sequence over time, and perhaps one basin kept as a control; 'The design might best be described as a sliding replication of paired catchment experiments, with the calibration period telescoped into the treatment period'.

Box 2. Outline of optional designs for catchment experiments.

Source: From J. D. Hewlett and L. Pienaar, 'Design and analysis of the catchment experiment', in E. H. White (ed.), *Symposium on Use of Small Watersheds in Determining Effects of Forest Land Use on Water Quality, May 22 and 23 1973* (Lexington, KY: University of Kentucky, 1973), 90–100.

Table 2. The 1943 multiple-catchment design: treatment of six catchments in the proposed main experiment at Jonkershoek.

Catchment	Gauging commenced	Age of plantation in year					
		1940	1948	1956	1964	1972	1975–80
Bosboukloof	1938	0	8	16	24	32	35–40
Biesievlei	1939	0	0	8	16	24	27–32
Tierkloof	1939	0	0	0	8	16	19–24
Lambrechtsbos B	1943	0	0	0	0	8	11–16
Lambrechtsbos A	'as soon as possible'	0	0	0	0	0	3–8
Langrivier	1940	0	0	0	0	0	0

Source: Table 1 in Wicht, 'Determination of the Effects of Watershed-Management on Mountain Streams', *Transactions of the American Geophysical Union*, 24 (1943): 594–606.

Wicht worked to devise an experimental design that eliminated errors owing to yearly and secular variation in climate, and the physical differences among catchments. He needed to estimate error, analyse empirical information, and adapt to practical challenges and accommodate new information as the research proceeded. By 1942, hydrographs were available for three years from the first gauged catchment, and 12 months later from three others. From these he derived the first quantitative analysis of streamflow variation, and it was immediately clear that the variation within a catchment over time and with 'differences in the wetness of years' meant that treatment effects would be very difficult to detect: 'Unless the effect of treatment ... is very marked, observations over a very long series of years will be required to prove the significance of observed differences'.[95]

Although early work included various studies of processes in the hydrological cycle, Wicht's observations convinced him early on that the only aspect of the water balance that could be reliably measured with accuracy was streamflow (see below). He argued therefore that determining the effects of afforestation, and other treatments, must begin with detection of differences in streamflow attributable to treatment; rainfall, as an index, could be used to extrapolate findings, but neither rainfall nor any of the other terms in the water balance could be measured with adequate accuracy and precision. Further, since the terrain at Jonkershoek meant that the physiography of each catchment differed substantially from that of the next ('no two catchments are even approximately similar'),[96] and thus '[w]atersheds vary so much in their hydrographic character that we can have no idea whether observed differences in flow are due to differences in topography, soil, geology, climate, or some other factor'.[97] Observations were bedevilled by the 'chance errors' of random variation. In addition to all this, his ambitions were necessarily tempered by the constraints of resources and time, especially during the five years of World War II.[98]

In his 1943 paper in the *Transactions of the American Geophysical Union*, Wicht set out a revised and refined, statistically valid experimental design that continued unaltered for the next 40 years (soon after replicated at Cathedral Peak in the summer rainfall, grassland region of the Drakensberg).[99] This came to be known as the multiple-catchment experiment design. The diverse objectives of the earlier version of the program were now gone; now, the research had a single goal: to establish the effects of afforestation with *Pinus radiata* on streamflow,

95 Wicht, 'The Variability of Jonkershoek Streams', 17.
96 Wicht, 'Forest Influences Research Technique at Jonkershoek', 65–77, 66.
97 Wicht, 'Determination of the Effects of Watershed-Management on Mountain Streams', 595.
98 Wicht could not enlist, because of his partial blindness, and soon found himself assigned to do the work of his colleagues who had left for the war.
99 Wicht, 'Determination of the Effects of Watershed-Management on Mountain Streams'.

relative to a catchment under fynbos, protected against fire (or, in the case of Cathedral Peak, *Pinus patula* compared with grassland).[100] The multiple catchment experiment finally supplanted the successive ideas for the water balance study, the single-catchment approach, and the paired-catchment design. Over the following decades Wicht and the members of his larger team would progressively refine this design, especially aspects of the statistical analysis of experimental effects.[101]

The minimum requirement for a sound catchment experiment is a period of calibration, during which the streamflow of one or more experimental catchments is correlated with a control catchment. After an adequate period, the experimental catchments may be treated, and a treatment effect detected as a departure from the prior correlation. Wicht's new design allowed the afforested catchments to be tested against each other, since the first afforested (in 1940) could be calibrated against those yet to be afforested, and so on in sequence (see Table 2); they formed an experiment 'replicated in time'.[102]

The final results would be available in 1980, though valuable findings would emerge in the interim, and 'as the experiments develop these analyses should yield progressively stronger, statistically sounder, evidence of the effects of treatment'.[103] In this way, the calibration period was effectively lengthened to 32 years, in the case of Jonkershoek, thus eliminating the danger of insufficient calibration and neutralising the problem of secular climatic variation. Replication in time accommodated treatments with progressive effects.[104] Supporting studies of other elements of the hydrological cycle, other than streamflow, were merely to help explain the findings of the main experiment: studies of this kind were 'indispensable to the complete understanding of catchment management effects

---

100   Veld burning experiments continued on a small scale, until the major expansion of this aspect to the program in 1965.

101   See, for example, C. L. Wicht and D. E. Schumann, *Experimental Investigation of the Effects of Forests on Stream-Discharge. Paper Presented to the Commonwealth Forestry Conference, Australia and New Zealand, 1957* (Pretoria: Government Printer, Union of South Africa, 1957).

102   H. G. Wilm, 'Notes on Experiments Replicated in Time', *Biometrics Bulletin*, 1 (1945): 16–20.

103   Wicht, 'Determination of the Effects of Watershed-Management on Mountain Streams', 598.

104   Wicht, 'Forest Hydrological Research in Africa South of the Sahara', See corroboration in J. D. Hewlett, H. W. Lull, and K. G. Reinhart, 'In Defense of Experimental Watersheds', *Water Resources Research*, 5 (1969): 306–16. 'This high correlation serves as the best experimental control we can get over climatic variation from season to season and from year to year' (p. 312); Hewlett and Pienaar describe the Jonkershoek design: 'Pine was planted on one basin the first year and for eight years the developing pine stand was matched against five control basins under the more slowly developing fynbos vegetation. In the ninth year, another basin was planted to pine, and in the 17th year still another, and so on. One control basin remains as an index to changing climate and developing fynbos to the end of the experiment. The multiple controls decrease in number while the treatment is replicated through time. One clear advantage is the built-in check upon the quality of control; if one control basin is for any reason a renegade (perhaps a slow subsurface leak is developing) the interrelation among the controls in the absence of treatment will revel it. This advantage, however, is gained at considerable cost and it is difficult to see any other design advantage over a series of paired catchment experiments': Hewlett and Pienaar, 'Design and Analysis of the Catchment Experiment', 99.

as demonstrated by catchment experiments'.[105] Wicht had to defend his approach against repeated challenges. At the outset, Charles Hursh raised 'a point of vital significance … whether or not it is feasible to substitute statistical design and methodology in place of a knowledge of physical processes on experimental drainages'.[106] This question dogged research hydrologists for decades after 1942.

In a paper at the Deutsche Afrika-Gesellschaft conference on '*Wasserwirtschaft in Afrika*', held in Bonn in 1963, Wicht stated emphatically: 'In most catchments it is virtually impossible to determine accurate volumetric values … of all the terms of the hydrological equation'. He criticised catchment experiments then under way in Kenya, on the grounds of inaccurate sampling of rainfall, and the limitations of calculated evapotranspiration in the absence of measurement. Later, generalising, he argued that the sampling required for accurate estimation of rainfall in 'broken country' was impossible.[107]

As well as Charles Hursh, critics of the multiple-catchment approach included Charles Pereira, the originator of the Kenya catchment studies, and Howard Penman of Rothamsted in the UK, at the time the most eminent among scientists of the hydrological cycle. They argued that the physical water-balance approach would give quicker, more meaningful results. Further, they held that the findings from catchment experiments such as those of Wicht had little meaning unless the hydrological cycle was measured and understood.[108] Penman, commenting on Wicht's paper at the International Symposium on Forest Hydrology, had the following, rather stinging comment:

> I like to keep the discussion on a philosophical plane. Leonardo da Vinci states somewhere: "If you know the reason, you have no need of the experiment". Is this long-term experimentation based on a philosophy of despair? That: the botanists, meteorologists, etc., who are working on the fundamental problems in forest hydrology, will, in fact, fail to do their job properly? Somehow behind all this there is a reason for everything that happens. Are we going to put all our energy in just measuring what happens, or shall we put a little more effort in research to try to find out why things happen? When we get that answer we can certainly explain how things happen.[109]

---

105   C. L. Wicht, 'Trends in Forest Hydrological Research', *South African Forestry Journal*, 57 (1966): 17–25.

106   C. M. Hursh in comments on Wicht's paper, *Transactions of the American Geophysical Union*, 24 (1943), 606.

107   Wicht, 'The Validity of Conclusions from South African Multiple Watershed Experiments', 749–60.

108   However, H. C. Pereira did emphasise the need for catchment experiments in East Africa 'so that, in these critical land-use problems, the bright plans for a brave new world in Africa may be based on locally tested fact, rather than on opinion from overseas'; Pereira 1962, in Hewlett et al., 'In Defense of Experimental Watersheds', 314.

109   Lull and Sopper, International Symposium on Forest Hydrology, 760.

In 1973, Hewlett and Pienaar identified the Jonkershoek and Cathedral Peak experiments as the only true multiple-catchment experiments under way in the world.[110] They identified the advantage of the design as lying in the 'built-in check on the control' that comes from having a succession of replicated treatments, but argued that there was no other advantage over the more common paired-catchment design. But over time, Wicht's research strategy has been fully vindicated.[111]

Although research on evaporation processes, soil infiltration, surface runoff and other aspects of the hydrological cycle was part of the program at the start, and accelerated from about the 1970s onward, the catchment experiment formed the core of the investigations, providing the benchmark findings that observation and analysis of the different components of the water cycle—hydrological process research—would serve to explain and extrapolate beyond the experimental site. The most important contribution from hydrological process studies was to enable supplementary research outside the catchments, and to build explanatory knowledge. The multiple catchment experiment also provided the important benefit that researchers with limited resources could focus on trying to produce 'unimpeachable findings', and not have their efforts dissipated by pursuit of too much too soon.

Further, once clear findings emerged from the two major multiple-catchment sites, new smaller-scale paired-catchment trials could be set out elsewhere, and the results securely inferred against the background of prior knowledge from the former. Freed of the impossible demands and the uncertainties of the water balance approach, the program could succeed over time in addressing its central questions across the whole of the South African forestry region, despite stringent constraints in resources. A pragmatic approach to hydrological process studies could then affordably help to understand observed and unambiguous streamflow effects throughout the entire country.

Wicht's early work at Jonkershoek helped to produce a 'model' of hydrological research that could seemingly be transplanted throughout the country. Yet the controversial findings of the program—discussed in the next chapter—could not themselves create forest policies at regional or national scales. The findings at Jonkershoek, secured as they were by Wicht's prestige within South Africa and his experimental design, suggested that forests did use more water than indigenous vegetation such as grassland and fynbos, as Phillips and Smuts had originally argued. The work in experimental research at Jonkershoek became the centrepiece of a wider policy framework that developed from the late 1940s through the 1970s. This policy sought to balance the needs of water users through a national framework.

---

110   Hewlett and Pienaar, 'Design and Analysis of the Catchment Experiment'.
111   Hewlett et al., 'In Defense of Experimental Watersheds'.

Chapter 8

Forest Hydrology in the Policy Domain

Strong dissent surrounded ideas about forests and water in South Africa, in the political, public and intellectual spheres, but the available evidence is that the science in the forest hydrology program in South Africa from 1935 proceeded free of political interference. It did not become the victim of political expediency, as for example did the US study in New Hampshire in 1911–1912. Perhaps that was because of the scientific ethos that prevailed in the community of forest scientists at the time, or the greater 'South Africanisation' drive stimulated by Jan Smuts. It may be that Wicht and his leadership were protected by the intensity of public interest, or perhaps the force of the dissension during the fourth British Empire Forestry Conference stimulated scientists to ask bigger questions than otherwise.

It is also likely that a sense of a discipline of forest hydrology began to take hold, with the simultaneous establishment of rigorous catchment experiments in South Africa and the US, and the self-awareness among members of an emerging discipline with 'a social structure ... whose members are linked together in networks of communication, rivalries, common goals, and agreed-upon norms as to what methods and explanations are legitimate'.[1] We have seen Wicht's participation in creating the discipline of forest hydrology in his early, energetic correspondence with his fellow forest hydrologists around the world. Being engaged in such a fellowship and the consciousness coming from it—together with the collegial diligence over seven years to contrive a research program that would deliver the needed 'unimpeachable' findings—would generate

---

1   Eli Gerson, 1983, paraphrased in Winsor, 'The Practitioner of Science', 231.

the confidence needed to resist interference. On the other hand, there was a continuous interplay between the emerging scientific knowledge about forests and water, and the development of public policy in these fields.

The knowledge that came from Jonkershoek and its satellites had its own scientific value, but its primary purpose was always to bear testimony of the realities of the natural world to policy and practice in the management of South Africa's resources. Over 50 years, during the second half of the twentieth century, a continual interplay marked both the evolution of government policy for the management of forests, catchment and water supplies in South Africa, as well as the scope and course of the research—conceivably, the program would not have survived as long as it did without the pressure for information from policy makers.

Environmental historians of South Africa have recently begun to investigate how these research findings influenced forest science and policy in South Africa during this period.[2] Scholars offer different interpretations regarding the essence and context of research findings and the long-term legacy of Jonkershoek. We suggest that these divergent opinions hew closely to the boundaries of past and present political debates about the impact of exotic trees on hydrological systems; the interest is not exclusively focused on forests, such as studies of fire, and attends more broadly to how research from the program generated and informed a *broader* model of ecosystem management in catchments, as for example outlined by Simon Pooley.[3] This was a model that embraced indigenous and exotic vegetation types; whereas a narrow scholarly focus on forestry, especially exotic trees, emphasises the particular finding from Jonkershoek and its satellite sites, that exotic trees use more water than the indigenous vegetation they replace.

One strain of research situates Jonkershoek within the wider context of catchment management in South Africa. Pooley's research on fire and management highlight the fact that hydrological researchers, especially Wicht, concluded that *all forms of vegetation*, including exotic trees and indigenous fynbos, increased the evaporation from the catchment, relative to the bare soil, and so reduced overall streamflow. Ultimately, findings from Jonkershoek led the 1968 Ministerial Interdepartmental Committee on Afforestation and Water Supplies to conclude, 'that protecting natural vegetation from fire reduced streamflow because [as Wicht had suggested] fires were believed to lower average veld age and reduce evapotranspiration'.[4] This instigated a policy of fire management in

---

2   Showers, 'Prehistory of Southern African Forestry', 295–322; Pooley, 'Recovering the Lost History of Fire in South Africa's Fynbos'.

3   Pooley, 'Recovering the Lost History of Fire in South Africa's Fynbos'.

4   Ibid.

fynbos and grasslands to control vegetation density, streamflow, and invasive plants, within the framework of formal ecosystems management as determined, for example, through the *Mountain Catchment Areas Act* (see below), in turn set within a context of land-use management that encompassed afforestation. This national policy fell into decline in the late 1980s and early 1990s. Pooley's last sentence concludes, 'What remains tantalizing is what the longer term environmental outcomes might have been if the collapse of the apartheid state had not truncated the state conservation forestry research and management program in South Africa in the early 1990s'.[5]

A second strain of research into the history of forest hydrology has focused on the most widely known research findings to come from Jonkershoek—that forests transpire and use more water than South Africa's grasslands and fynbos.[6] Showers points out that the findings from Jonkershoek eventually helped overturn the historical assumption that forests conserved water better than other types of vegetation, such as fynbos and grassland. The article highlights that 'massive tree planting—particularly in the twentieth century—had, indeed, changed South African climates near and in the ground. However, rather than achieving nineteenth-century dreams of moister regimes for plant roots, alien trees were identified as being major contributors to landscape desiccation'.[7] Showers implies that the environmental changes caused by exotic trees, in turn, justified the designation of forestry as a Stream Flow Reduction Activity in the 1998 *National Water Act* and led to the creation of a major exotic tree eradication program, Working for Water.

This chapter positions itself within South African historiographical and environmental and economic policy debates by suggesting that the empirical findings produced by researchers at Jonkershoek and policies drawing from their work cannot be isolated and abstracted from their wider political and social historical contexts and meanings. It argues that the policy recommendations based on findings from Jonkershoek were framed within and contributed to an evolving national water management strategy that sought to account for a variety of forms of land usage, including forestry, agriculture, and indigenous ecosystem conservation. National policy regarding forests and water from the late 1940s onwards, informed by findings from Jonkershoek, sought to direct afforestation to areas with higher rainfall and profitability, and where there was little competition for water use, while encouraging catchment management on public and private lands through various policies and their statutory instruments.

---

5   Ibid., 76.
6   Showers, 'Prehistory of Southern African Forestry', 311–2.
7   Ibid., 312.

This system neither privileged nor discriminated against forestry as a land use at the national level, whereas the current policy framework does discriminate against forestry. Rather, South African policies from the 1960s to the early 1990s were predicated on the assumption that allocation of land use and thus water demand should be determined regionally, according to the comparative economic returns of competing land and water uses, and in relation to geographical patterns of water supply and demand. In this there was a progressive shift, from a policy position in which market forces were allowed to govern afforestation within biophysical potentials, to one where market-based decisions were constrained by regulated geographical planning regimes, and, finally, one marked by slow and burdensome administrative procedures, albeit with the unfulfilled promise of the evolution to market-based policy instruments. This course of development was set initially against the background of the protection of catchment areas, which included the policy that the vast majority of state forest land would remain unforested and the protection of catchments on privately owned land would be governed through the implementation of the *Mountain Catchment Areas Act* of 1970. This policy and legislation framework guided national forest and water policies until new national legislation and regulatory frameworks were created from the late 1980s to the late 1990s, which weakened the ecosystems approach to catchment management and introduced a new, detailed multi-statute bureaucratic regime.

Initially, afforestation was seldom the demonstrable cause of water shortage; the areas afforested were then too small to have had the effects claimed, and many complaints were shown to be groundless—such as cases where there had been no afforestation in the catchments of the streams at issue. J. D. M. Keet, in rebutting I. B. Pole-Evans, analysed the situation:

> ... any indictment of afforestation must have such far-reaching consequences ... [that] it can only be based on the most searching and scientific analysis ... When it is realized how recent has been the policy of rapid extension of afforestation in the Union, the degree to which the application of this policy has coincided with a decade of sub-normal precipitation [and 'a steady diminution in the stream flow throughout South Africa'], the relatively small proportion of areas suitable for afforestation ..., the extreme youth of the majority of afforested areas ..., these antagonists merely express a fear that afforestation will adversely affect water supplies ...[8]

Two factors accounted for the complaints, the rural development drive, with the focus on irrigation farming, and intermittent severe, prolonged drought. But the situation would change progressively, toward real competition, as we shall see.

---

8   J. D. M. Keet, Chief: Division of Forest Management, to Secretary for Agriculture and Forestry, 'The Effects of Forests on Water Supply', C.203 29 May 1935, Wicht Papers, SAFRI Archives, CSIR, Pretoria.

Irrigation settlements in the provinces of Mpumalanga and Limpopo, a product initially of Alfred Milner's rural anglicisation program, relied on water from relatively small streams flowing from the escarpment. The example of the White River Estates (see page 106) illustrates the rapid onset of competition for water, resolved, temporarily, only after the construction of the Longemere dam in 1938–1940. On the Politsi River, another small drainage to the north, the Union government had constructed an earth canal by 1912, the Tzaneen Irrigation District was established in 1918, and by 1926, 99 per cent of normal flow of the Politsi was assigned to the Board, which experienced water shortages during drought years.[9]

These early irrigation settlements had comprehensive ambitions. The Transvaal administration supported the Tzaneen initiative with the establishment of the Tzaneen Government Estate, an estate of about 6,000 acres intended as 'a practical and theoretical training ground for British settlers', with a training syllabus and research program designed to support an agricultural colony of settlers with access to two-thirds of the land on the estate. The syllabus included general farm management, horticulture, forestry and other subjects, but the first enrolment was of just six students, and by 1908, there were none.[10] By 1905, the estate had extensive trials of orchard and field crops under way, including plantation trials of timber trees: 22 species of pine, and *Eucalyptus saligna* (probably *E. grandis*) with several other eucalypt species.[11] The estate continued until 1918, when the land was subdivided into 16 plots and leased off to settlers on the Tzaneen irrigation scheme, the government initiative having been defeated by poor transport links, high production costs, and lack of enterprise.[12]

It was after Union, and especially after World War I, that rural settlement by, preferentially, British returned soldiers accelerated. During the period from 1912 to around 1930, about 400,000 ha of land was put under irrigation through the Irrigation Board's schemes (these were for private farmers), followed by a further 350,000 ha in the period to the 1940s by the government white settlement schemes.[13] The former were located toward the headwaters of smaller catchments, such as the White River (see below), while the latter were located

9   A. R. Turton et al., *A Hydropolitical History of South Africa's International River Basins* (Pretoria: Water Research Commission, 2004), 330–7.
10   Praagh, *The Transvaal and Its Mines*, 181; Menno Klapwijk, *The Story of Tzaneen's Origin* (Unknown Publisher, 1974), 12, 22–31. The fate of the estate is not known, but it does not seem to have lasted long.
11   Klapwijk, *The Story of Tzaneen's Origin*, 26. Klapwijk cites the report of H. S. Altenroxel in the *Transvaal Agricultural Journal*, 4 (1905/1906).
12   Ibid., 31.
13   S. R. Perret, 'Water Policies and Smallholding Irrigation Schemes in South Africa: A History and New Institutional Challenges', *Water Policy*, 4 (2002): 283–300, Table 1.

mainly on major drainages such as the Vaal and Orange rivers. Evidence suggests that earlier irrigation settlements were often located and planned without sufficient knowledge of water supply and the need for reservoir storage.[14]

Despite serious intent, like Tzaneen, the settlements were vulnerable, involving mainly unskilled people settling in an unknown land, having little know-how and little capital, in locations mostly without adequate transport to distant markets. Poor water supply added to their anxieties, and the early round of settlement had little success.[15] The tone of entreaty in the exchange between Captain Palmer, Reitz and Keet about the White River in the early 1930s suggests this vulnerability.

Still, the developments created new farming interest groups in incipient competition with afforestation, a competition that was apparent rather than real initially: it was the series of severe and prolonged droughts from the early 1920s to the early 1930s that were most often cited as cause for the complaints.[16] In the regions of main new afforestation, the eastern escarpment of the Great Plateau in what are now the provinces of Mpumalanga and Limpopo, the running mean annual rainfall declined by 30 per cent or more during in this period (see Figure 10, and other figures in Wicht, 'Afforestation and Water Supplies'). Continued afforestation would in time cause real competition, especially since the plantations were located mainly in higher rainfall regions that were often the upstream sources of locally and regionally important water supply.

By 1949, the area of plantations on state forest land amounted to about 173,000 ha. The total afforested area in South Africa, including private plantations, was 311,000 ha.[17] The area was a small fraction of South Africa's extent, but the plantations were often located within the catchments of important streams. Complaints of loss of supplies to towns arose for the first time, and though these were mostly attributable to rainfall decline, the signs were there of tightening competition for water. And in this respect, Keet was to be proved wrong: the extent of plantings may have been small as a fraction of the country as a whole, but the local and regional effects of afforestation on water resources became the urgent issue for several decades.

---

14   See Chapter 3; F. E. Kanthack in a letter to the Dominions Royal Commission in 1914 reported that 'very complete gaugings exist for some few rivers in the Transvaal, ranging over a period of seven or eight years, in other parts of South Africa information regarding rivers is either entirely wanting, or is now only being acquired', Dominions Royal Commission, *Royal Commission on the Natural Resources*, 52, Appendix V.

15   Fedorowich, 'Anglicisation and the Politicisation of British Immigration to South Africa; Worsfold, *The Reconstruction of the Colonies under Lord Milner*. By 1913 the number of British settlers in Transvaal had fallen to 450 from 550.

16   See C. L. Wicht, *Forestry and Water Supplies in South Africa*, Bulletin No. 33 (Pretoria: Department of Forestry, Union of South Africa, 1949), figures 6–8.

17   W. E. Watt, Director of Forestry, in ibid., Preface.

By the 1960s, the total afforested area approached 1 million ha. Successive analyses showed that, in economic terms, water use upstream in the plantations was more cost-beneficial than for downstream irrigated agriculture, or for alternative land uses in the headwaters, but there was no coherent system guiding the allocation of water, whether through the market or by regulation. Creation of storage by damming the streams allowed both forest development and irrigation, up to a point, and at this point restrictive regulation of afforestation emerged, as the first means of managing the competition for water. The account that follows examines this line of historical development in policy on forests and water.

1949: Forestry and water supplies in South Africa

Although the total area afforested by 1949 was less than 0.3 per cent of South Africa's land area, the Department of Forestry, 'alive to the possible effects which extensive forests of exotic trees may have on water supplies' and aware of the 'considerable concern' among members of the public about this, sought an authoritative statement on the issue. The department, still being far off its goal of around 700,000 ha of sawlog forest, was perhaps concerned that public opinion based on false premises would lead to political action that would curtail the program.[18]

The department wrote 60 letters to 'prominent persons' and received 36 replies with opinions on, or claims of, afforestation effects on water supplies. C. L. Wicht, by this time steeped in the science and informed by his 12 years of experimentation and observation at Jonkershoek, was assigned to develop an authoritative statement through a reconnaissance of catchment areas throughout the country, investigating 'instances where afforestation is alleged to have had a desiccating effect on water supplies'.[19]

Wicht interpreted his brief as being to 'synthesize the meagre and sometimes problematic data available [on forests and water] into a coherent statement, from which practical recommendations can be deduced'.[20] He consolidated early evidence from the analyses of rainfall and streamflow at Jonkershoek with his on-site investigations of 21 claimed cases of desiccation within the forestry regions of South Africa, reported in response to the department's opinion survey, together with a survey of the world's contemporary literature. The outcome was the 1949 report, *Forestry and Water Supplies in South Africa*.

18   Watt in ibid., 58.
19   Watt in ibid., Preface.
20   Wicht, *Forestry and Water Supplies in South Africa*, 1.

The 1934 report by T. E. W. Schumann and W. R. Thompson[21] had provided a scientific account of the geographical patterns of and secular variations in rainfall in South Africa as a whole. Wicht added to this by analysing geographical, secular and seasonal variations in climate, especially rainfall, drought, and temperature within forestry regions, using the growing number of climate records for stations within eight forestry regions delineated for South Africa. With this, Wicht could focus on forestry potential and draw secure inferences about patterns and variations among these forestry regions. At this time, the Jonkershoek afforestation experiment had not yet yielded results, but he could draw on the emerging body of forest hydrological knowledge, including Joseph Kittredge's book, *Forest Influences*, published in 1948,[22] and crucially, on early experimental results from Coweeta in the US, where deforestation treatments yielded quicker results than the converse, afforestation experiments in South Africa. His active program at Jonkershoek had generated important new insights into streamflow responses to rainfall and drought, into the relationship between annual evapotranspiration losses and rainfall and vegetation, and provided knowledge of hydrological processes such as interception and infiltration that allowed critical interpretation of findings from elsewhere.

Of the 21 cases of claimed loss of water supplies owing to afforestation, Wicht found that the greater majority were groundless, either because the area planted at the time was too small to explain the claimed decrease, or because there had been no afforestation in the relevant catchment. In 14 cases he thought the claims were invalid because of lack of evidence or on the grounds that recent drought had caused streamflow decline.[23] For example, he could readily rebut the claim that afforestation had affected the Eerste River (the drainage from Jonkershoek) since only about one-tenth of the rain in the Eerste River catchment fell on slopes to be planted, of which less than half had been planted, and that it was 'quite unlikely that this degree of afforestation could noticeably affect the flow of the river'. He could corroborate this by an early analysis comparing effects of a young plantation at Jonkershoek (where planting had begun in 1941) in a paired-catchment study: he could find no effects. Furthermore, 80 per cent of rainfall in the upper catchments fell on mountain areas never to be afforested, so that the department's program could not materially affect the Eerste River water supply. But Wicht did find cases where dense plantings of eucalyptus grown on a short rotation (around a 10-year cycle) for mining-timber were associated

---

21   Schumann and Thompson, *A Study of South African Rainfall.*
22   Kittredge, *Forest Influences.*
23   See Wicht, *Forestry and Water Supplies in South Africa,* figures 2 and 8.

with streamflow decline, and in these he attributed the loss in water supply to meaningful afforestation effects, despite secular rainfall decline. These were also where rainfall was at the low end for forestry.

Wicht concluded, from the examination of the 'considerable' geographical variation among the forestry regions in rainfall and streamflow together with the analyses of the Jonkershoek experiments, that the amount of evapotranspiration at any place was related to the amount available and to the type of vegetation. Evapotranspiration losses in higher-rainfall regions were much greater than in dry parts, which 'must largely be ascribed to the higher transpiration of the more luxuriant vegetation': at Jonkershoek, in one catchment with several years of record, streamflow varied year to year with rainfall, but evapotranspiration remained quite constant. Where vegetation exists (as it must in regions with enough moisture to sustain streams), transpiration exceeds direct evaporation. Wicht concluded that '[t]he role played by vegetation in water conservation, including its influence on stream discharge has generally been underestimated'.[24] Annual streamflow volumes were principally from baseflow, not stormflow. The early research on diurnal vapour losses, which reduced flows on very sunny days, had shown the clear effects of the drought on baseflow arising from transpiration by vegetation.

From this arose the central concept in the report that '[t]he portion of water returned to the atmosphere is decisive in determining the water cycle or water economy within the catchment'; vegetation cannot exist without transpiration, and the policy question was how transpiration changed with vegetation change, rather than whether the catchment was forested or not.[25] Managing catchments for water supplies involved a trade-off between vegetal cover, or biomass productivity, and water supply, because '[g]round cannot … be preserved unless it is covered by vegetation; its conservation depends, in fact, on maintaining entire ecosystems'. He thus also expressed the idea that catchment management was a matter of ecosystems management: 'A knowledge of ecosystems as wholes is necessary to understand fluctuations in the discharge of streams'.[26]

---

24   Wicht, *Forestry and Water Supplies in South Africa*, 25.
25   Ibid., 22.
26   See also C. L. Wicht, 'Summary of Forests and Evapotranspiration Session', in W. E. Lull and W. H. Sopper (eds), *International Symposium on Forest Hydrology* (Oxford: Pergamon Press, 1967), 493.

Wicht's conclusions from these observations have mostly been borne out by scientific findings since then: under the same conditions, plantations of exotic trees would not use more water than indigenous forests;[27] such plantations would use more water than fynbos or grassland; the quantity of water used by vegetation would depend 'chiefly on the amount of water available in the soil': fast-growing trees would not necessarily use more water than slow-growing ones;[28] and the removal of vegetation from catchments, especially in the riparian zone, would increase streamflow.[29]

From this study arose clear policy lessons for afforestation. Afforestation should be restricted to regions with higher rainfall. Long-rotation timber crops were to be preferred over short-rotation. Sites for afforestation should be chosen and managed with care. Riparian zones should not be afforested where water use downstream was for high-value industries. These findings reinforced existing forestry practice and policies. Since the 1920s, most private and public timber plantations were created in the higher rainfall areas of the Transvaal and Natal and less were created in the arid Cape. South African foresters kept a safe buffer distance—at least 20 metres—between plantations and the riparian zones of streams as a result of the 1932 policy changes.

Despite the lessons and the crucial introduction of the ecosystem concept to catchment management, the 1949 report did not lead to any specific new law or other policy instruments to manage the relation between afforestation and water supplies. Its findings tended to reinforce the policies and practices of the time. The department's annual report noted, 'With the expanding need for increased water supplies, particularly for urban and industrial developments, water supplies from forest reserves had become increasingly important' (noting the examples of Cape Town and Sabie), land continued to be acquired for catchment

---

27   Several authors continue to claim that indigenous forests (as opposed to fynbos or grasses) use less water than exotic trees. But in important forestry climatic zones, when there are similar conditions, water use by plantations and South African indigenous forests does not differ. See M. B. Gush, 'Water-Use, Growth and Water-Use Efficiency of Indigenous Tree Species in a Range of Forest and Woodland Systems in South Africa' (PhD Thesis, University of Cape Town, 2011). For earlier work in East Africa, see H. C. Pereira and P. H. Hosegood, 'Comparative Water-Use of Softwood Plantations and Bamboo Forest', *Journal of Soil Science*, 13 (1962): 299–313.

28   This is true if the measure is water-use efficiency, i.e. the amount of water consumed per unit of growth. In South Africa, water-use efficiency is highest in fast-growing species of *Eucalyptus*: R. M. Wise, P. J. Dye, and M. B. Gush, 'A Comparison of the Biophysical and Economic Water-Use Efficiencies of Indigenous and Introduced Forests in South Africa', *Forest Ecology and Management*, 262 (2011): 906–15.

29   Wicht's early study of daily rise and fall in streams during periods of drought led to several subsequent pieces of research: Wicht, 'Diurnal Fluctuations in Jonkershoek Streams'; H. B. Rycroft, 'The Effect of Riparian Vegetation on Water Loss from a Furrow at Jonkershoek', *Journal of the South African Forestry Association*, 26 (1955): 2–9; C. H. Banks, 'The Hydrological Effects of Riparian and Adjoining Vegetation', *Forestry in South Africa*, 1 (1961): 341–5; D. F. Scott, 'Managing Riparian Zone Vegetation to Sustain Streamflow: Results of Paired Catchment Experiments in South Africa', *Canadian Journal of Forest Research*, 1 (1999): 1149–57. Hursh made a strong point of this in his commentary on Wicht's 1943 paper: C. R. Hursh, in C. L. Wicht, 'Determination of the Effects of Watershed-Management on Mountain Streams', 607.

protection, and afforestation proceeded in both the public and the private sectors, governed it seems by the availability of land in areas with suitable climate, and the finance available to execute the afforestation.[30] From this point on the nature of the evidence required and the terms of the policy arguments about afforestation and water supply in South Africa had been established.

The Ministerial Interdepartmental Committee on Afforestation and Water Supplies, 1968

New innovations in the forestry sector, such as improved species selection and better silvicultural techniques, allowed private and public foresters from the 1940s to 1960s to successfully expand large plantations of pines, eucalyptus and wattle that were located primarily in Natal, Zululand and the Transvaal.[31] The size of afforested areas in South Africa nearly doubled—from about 0.6 million ha in the mid-1940s to 1 million in 1968—and the industry entered a period during which there was an immense shift, when the area of wattle decreased in response to declining market demand and the area under eucalypts and pines grew disproportionately.[32] This rapid afforestation and shift in the composition of plantation resources led to considerable concern amongst water resource managers working for the Department of Water Affairs and among members of the farming community, especially in areas near large-scale government and private plantations in the eastern and northern Transvaal (what are today Mpumalanga and Limpopo provinces) and present-day KwaZulu-Natal. Meanwhile, the first definitive findings on the effects of afforestation on streamflow at Jonkershoek appeared in 1963, 28 years after the program had begun.[33]

Against these concerns, the government appointed the Ministerial Interdepartmental Committee on Afforestation and Water Supplies, which reported in 1968. The committee was partly a response to the publication of the 1961 *Report of the Interdepartmental Committee on the Conservation of Mountain Catchments in South Africa*, the so-called 'Ross Report', based on work that ran from 1952 to 1961. This major report recommended a national plan for managing

30   J. D. M. Keet, *Historical Review of the Development of Forestry in South Africa*, MS available online (Pretoria, c 1970), www2.dwaf.gov.za/webapp/resourcecentre/Documents/Publications_And_Media/Keet_Forestry_History_page_41-66.pdf, 107.
31   Department of Forestry, *Investigation of the Forest and Timber Industry of South Africa: Report on South Africa's Timber Resources, 1960* (Pretoria: Government Printer, 1964), 6.
32   D. W. van der Zel, 'Sustainable Industrial Afforestation in South Africa under Water and Other Environmental Pressures', in *Sustainability of Water Resources under Increasing Uncertainly (Proceedings of the Rabat Symposium S1, April 1997)*, IAHS Publication No. 24 (1997), 217–25, 220. P. J. Dye and D. B. Versfeld, 'Managing the Hydrological Impacts of South African Plantation Forests: An Overview', *Forest Ecology and Management*, 251 (2007): 121–8.
33   C. H. Banks and C. Kromhout, 'The Effect of Afforestation with *Pinus radiata* on Summer Baseflow and Total Annual Discharge from Jonkershoek Catchments', *Forestry in South Africa*, 3 (1963): 43–65.

the catchments on private land, reflecting the fact that most catchment land was private property (80 per cent) with equal percentages in state (10 per cent) and 'Trust' tenure (10 per cent).[34]

The 1968 committee included representatives from five government departments, including Forestry and Water Affairs. Each member was a recognised expert in such fields as water resources management, climatology, agriculture and forestry. The comprehensive terms of reference included investigation of afforestation effects on catchment water yields, compared with alternative vegetation types and land uses, to determine the effects of forestry at different scales, and to make recommendations on how to mitigate or otherwise 'temper' afforestation effects.[35]

The committee canvassed opinion among farmers, water resource managers and others concerned about afforestation effects by analysis of over 100 questionnaires returned by landowners, farmers' associations and other interested parties, and investigated the causes and reason for all the resulting claims of the loss of water supplies. It was from these eastern, summer rainfall forestry regions that the committee received questionnaire returns, and their report focused on this region.

Research in South Africa by this time had progressed to the point that the first analyses of afforestation effects had been published, and the relative effects on floods and baseflow were becoming evident, as was consumption of water by riparian vegetation. The distribution of streamflow in upland catchments between stormflow and baseflow was now well understood. Wicht acted as consultant to the committee and compiled a critical review of knowledge on the subject from work in South Africa as well as relevant evidence from bioclimatically analogous regions elsewhere in the world,[36] to build upon and extend his 1949 report. This critical review seems to have been the first in this field, anticipating by nearly a decade a similar review by the US Forest Service.[37] The committee's

---

34   Wicht, *Forestry and Water Supplies in South Africa*, 27, 39–42. By 'Trust' land is meant land acquired for allocation to the former homelands in terms of *Native Trust and Land Act*, 1936 (Act No. 18 of 1936; subsequently renamed the *Bantu Trust and Land Act*, 1936 and the *Development Trust and Land Act*, 1936).

35   Department of Forestry, *Report of the Interdepartmental Committee of Investigation into Afforestation and Water Supplies in South Africa* (Cape Town: Republic of South Africa, 1968), 2.

36   C. L. Wicht, *Afforestation and Water Supplies: A Review of Literature Prepared for the Interdepartmental Committee on Afforestation and Water Supplies*, Cyclostyled (University of Stellenbosch, 1966).

37   H. W. Anderson, M. D. Hoover, and K. G. Reinhart, *Forests and Water: Effects of Forest Management on Floods, Sedimentation, and Water Supply*, General Technical Report PSW-018 (Berkeley, CA: US Department of Agriculture, Forest Service, Pacific Southwest Forest and Range Experiment Station, 1976), 115.

report thus took account of the best available knowledge on vegetation and water supplies, complemented by the knowledge and perspectives of the committee members on water resource conflicts and outlooks.[38]

The weight of evidence from Wicht's critical review allowed the committee to agree on the hypothesis that 'the hydrological influences of vegetation, all other factors being constant, are correlated with the degree to which it utilizes the site'.[39] The report acknowledged that exotic trees impacted hydrological cycles when planted in catchment areas. But it was not the kind of vegetation, or the species that characterised it, or whether exotic or indigenous, but simply its 'phytomass', the biomass of the vegetation, which determined the effect on the water balance and hence on water supply. In other words, the committee viewed both forests and indigenous vegetation as water users affecting catchments. The issue was not which vegetation type used more water—all vegetation used water—but how the water should best be used. Forestry would fit within a national framework determining the maximum benefit of water rather than seeking to find out merely whether trees used more water than other vegetation or ground covers.

Building on their central hypothesis, the committee accepted a graphical model that represented the relationships between plantation forest cover, forest stand age, stand rotation, and rainfall that could allow estimates of effects of afforestation on streamflow. This model became known as Nänni curves, named after Ugs Nänni, the Secretary to the committee, who constructed the model from available experimental evidence and deductive inferences about the *a priori* physical limits to evapotranspiration under different rainfall regimes in forestry zones.[40]

The report agreed with Wicht's view that riparian vegetation disproportionately influenced streamflow, and the importance of the management of this zone in water conservation. But the committee confirmed that it was essential to manage all types of vegetation in catchments. Yet another of the key findings of the

---

38   Committee members were H. L. Malherbe (Secretary of the Department of Forestry), E. K. Marsh (Chief Forest Research Officer), U. W. Nänni (Forest Research Officer, Cathedral Peak, Secretary), C. E. M. Tidmarsh (Department of Agricultural Technical Services), F. S. Greyvenstein (Department of Water Affairs), J. S. Whitmore (Director of the Hydrological Research Institute, Department of Water Affairs), and J. C. Cox (Department of Water Affairs); Wicht was scientific advisor.

39   Department of Forestry, *Report of the Interdepartmental Committee of Investigation into Afforestation and Water Supplies in South Africa*, 29–31.

40   U. W. Nänni, 'Trees, Water and Perspective', *South African Forestry Journal*, 75 (1970): 9–17.

committee was that fire should be used to control the biomass of indigenous vegetation in catchment areas.[41] The committee could also agree that in South Africa, forests did not affect rainfall.[42]

Additionally, the committee compared the economics of water use across different sectors (e.g. forestry, agriculture, etc.) and found that forest enterprises were, with few exceptions, more beneficial than alternative upland land uses in terms of financial returns to water consumption.[43] By using the Nänni curves to predict streamflow alongside an economic analysis of water consumption, the committee offered an evidence-based model to determine what types of land use were best suited to South Africa's different geographies. The Nänni curves would become embedded in policy frameworks that flowed out of the committee's report. The model, subsequently refined by Diek van der Zel,[44] was the key to the afforestation permit system, legislated for the purposes of regulating new afforestation (see below).

The committee concluded with a series of detailed policy recommendations, which had far-ranging consequences, not only with respect to policy and practice on forests and water, but also to the science program.

The long-run consequences of the findings of the Committee on Afforestation and Water Supplies: Afforestation permits and catchment planning

The principal conclusions of the 1968 committee informed South African forestry policies for the next 30 years. Most importantly, the committee recognised the need to create ways for resolving water-use conflicts among competing interest groups. The report served to advance thinking about water resources toward the concept that is now called 'integrated water resources management', building on Wicht's argument that catchment management should be aligned with broader water supply objectives.[45] Though disagreements over water uses continued to happen after the 1968 report, there emerged clear legislative methods for determining water allocation in South Africa based on these findings.

---

41   Pooley, 'Recovering the Lost History of Fire in South Africa's Fynbos', 67.

42   Department of Forestry, *Report of the Interdepartmental Committee of Investigation into Afforestation and Water Supplies in South Africa*, 30–1, 37–40.

43   See Nänni, 'Trees, Water and Perspective'.

44   D. W. van der Zel, 'Accomplishments and Dynamics of the South African Afforestation Permit System', *South African Forestry Journal*, 172 (1995): 49–58; van der Zel, 'Sustainable Industrial Afforestation in South Africa under Water and Other Environmental Pressures'.

45   C. L. Wicht, 'The Effects of Timber Plantations on Water Supplies in South Africa', in *Proceedings of the Symposium of Hannoversch-Münden, 8–14 September 1959* (Hannoversch-Münden: International Association of Scientific Hydrology, n.d.), 238–44.

A related government decision in 1966 appointed the Commission of Enquiry into Water Matters. Given South Africa's situation as a semi-arid country, and the anticipated growth in water use, the commission investigated and advised on 'all aspects of water provision and utilisation within the Republic', with a specific requirement to determine the areas that should be allocated for afforestation and timber production.[46] The findings of the committee on Afforestation and Water Supplies fed into the work of the commission. Among other things the commission's findings outlined a strategy for managing the use of water resources in the country.

Subsequent legislation enshrined aspects of the committee and commission's findings into the statutory instruments of forests and water policy. The 1968 report recommended that the state should further extend catchment management in mountain catchment areas. Since not all necessary catchment land could be acquired by the state, this recommendation led to the first national legislation for catchment management on land in private hands. The promulgation of the *Mountain Catchment Areas Act* 63 of 1970 provided for the demarcation of catchment areas, including private land, and placing such areas under joint management plans involving the Forestry Department and private landowners (and which, with the *Abolition of Racially Based Land Measures Act* 108 of 1991, now applies to all land in South Africa). The Act gave effect to Wicht's 1949 concept of catchment management being ecosystems management. The plans would deal with the conservation of land for the purposes of catchment protection, for example, through the management of fire in vegetation, and by an amendment in 1981, specifically the prevention of soil erosion and the control of 'intruding vegetation', i.e. alien invasive plant species.[47] Foresters—under this and concurrent legislation—were responsible for managing catchments for downstream users, determining which catchment areas could be afforested for economic benefits, and maintaining the sustainability of indigenous ecosystems.[48]

An amendment in 1972 to the *Forest Act* created an afforestation permit system that required landowners to apply for permits prior to afforesting. Diek van der Zel refined the Nänni curves to improve predictions of the effects of proposed

---

46   Republic of South Africa, *Report of the Commission of Enquiry into Water Matters* (Pretoria: Department of Water Affairs, 1970), 1.

47   *Mountain Catchment Areas Act* (No. 63) of 1970 as amended Section 3.

48   By 1981, approximately 4,000 km² of private land in fynbos ecosystems had been proclaimed Mountain Catchment Area; ultimately, nearly 20,000 km² of private and state land was to be managed for water conservation in terms of this Act: F. J. Kruger, 'Use and Management of Mediterranean-Type Ecosystems in South Africa: Current Problems', in C. E. Conrad and W. C. Oechel (eds), *Proceedings of the Symposium on Dynamics and Management of Mediterranean-Type Ecosystems; June 22–26, 1981; San Diego, CA* (Berkeley, CA: Pacific Southwest Forest and Range Experiment Station, Forest Service, US Department of Agriculture, 1982), 42–8.

afforestation.[49] The amendment also created a consultative framework that engaged water, forestry, agriculture and environmental sectors in decision-making processes of where new plantations would be located. The system was administered by an interdepartmental Central Afforestation Permit Committee, which considered applications that had been examined by equivalent provincial committees.

To determine what areas would be suitable for afforestation, an Interdepartmental Committee for the Indication of Priority Areas for Afforestation (the Afforestation Priorities Committee) published a land classification system in 1975, employing van der Zel's refinement of the Nänni curves, vegetation surveys, and land capability and water resources assessments.[50] They assessed the extent and location of land suitable and potentially available for afforestation (excluding land suited to agriculture) and stipulated three classes of catchment: Category I, in which no new afforestation would be allowed; Category II, where further afforestation would be permitted, to the degree that 5 per cent of streamflow could be reduced; and Category III, in which further afforestation to the point of a 10 per cent reduction in streamflow would be permitted. From 1972 to 1994, the permit system prevented afforestation in catchments where competition for water was severe and directed afforestation to other areas where water was more freely available.

Guide planning for expansion of the plantation forest resource followed soon after to support the institution of the Afforestation Permit System. The Afforestation Priorities Committee reported in December 1975 on the area of land that would be required to satisfy South Africa's timber needs, as projected to the year 2000. From their projected estimates of timber demand (which were optimistic, suggesting 32 million cubic metres consumption by 2000, whereas the actual in that year was around 18 million), they estimated that the area of plantation forests would need to increase 2.3-fold over, from about 1 million ha in 1972 to 2.34 million ha. From reconnaissance surveys of each forest region, the committee reckoned that a little more than 1 million ha was suited to and potentially available for additional afforestation.[51]

---

49   Van der Zel, 'Accomplishments and Dynamics of the South African Afforestation Permit System', 49–58; van der Zel, 'Sustainable Industrial Afforestation in South Africa under Water and Other Environmental Pressures'.

50   Departement van Bosbou, *Verslag van Die Interdepartementele Komitee vir die Aanduiding van Prioriteitsgebiede vir Bebossing* (Pretoria: Republic of South Africa, 31 December 1975).

51   Ibid., Tables 1, 2, paragraph 10 5–6, paragraph 37, 39; the estimates of area for new afforestation included land in the then 'homelands', excluded land in Category I catchments, and took account of the constraints in the other categories of catchments.

The case of afforestation in the Eastern Transvaal is an interesting illustration of this guide planning approach. The Water Planning Committee for the Eastern Transvaal (appointed by the Minister of Water Affairs, and including two senior representatives of the Department of Forestry) compiled a plan for the distribution of water supplies among competing sectors based upon estimates of the water-resource balances for the eastward-flowing rivers of the then Transvaal, including the White and Sabie rivers.[52] At the time, the extent of plantations in the area the committee investigated was 341,000 ha, 2.6 per cent of the whole area, but concentrated in the narrow zone with average annual rainfall exceeding about 800 mm per year, along the escarpment from Swaziland to the Soutpansberg. The Afforestation Priorities Committee had found that in this region perhaps as much as a further 430,000 ha was suitable and potentially available for afforestation, though only about 300,000 ha had good potential, mostly located in the catchments of the Crocodile and Komati rivers.

This committee held public hearings during 1973 during which participants made representations on the advantages and disadvantages of forestry (including spokespersons for the 40 irrigation districts that existed within the area). Expert analyses to inform the committee included an analysis by the new Institute of Hydrology, which found that the 57 per cent afforestation of the White River catchment (all state plantations, which had by then matured) had caused a 37 per cent reduction in annual streamflow. They based the planned distribution of water use among sectors on the water supply and demand situation projected to the year 2000, at which time they anticipated that there would still be an overall surplus of water supply, but expected shortages in certain catchments. Their water resource balance sheet provided for environmental flows to the Kruger National Park, as part of consumption in primary 'rural' use, as well as for towns, industries, and power generation, and finally, as lowest priority, irrigated agriculture and plantations.

With the estimated actual water use in 1970 (1.67 million cubic metres per year) as baseline, the committee projected a requirement of 3.27 million cubic metres per year in 2000, almost a doubling. Consumptive use of 432,000 cubic metres was attributable to plantations in 1970, while the projection provided for 592,000 cubic metres, i.e. from 26 per cent to 18 per cent equivalent of use by other sectors—equivalent in the sense that the consumption was the estimated reduction in streamflow attributable to plantations, whereas other uses were proportions of assured supply, two variables that are not commensurate. The plan provided for resources required in Mozambique and Swaziland, and allowed for a threefold increase in the provisions overall for primary rural use,

---

52   Excluding the southernmost catchment in the region, the Usuthu; Waterbeplanningskomitee vir Oos-Transvaal, Waterbeplanning vir Oos-Transvaal, *Verslag Waterbeplanning Vir Oos-Transvaal* (Pretoria: Republic of South Africa, April 1980).

towns, industry and power generation, 1.5-fold for irrigation, and 1.4-fold for forestry.[53] To provide this, the capacity of water infrastructure (such as dams and reticulation systems) would need to be doubled.

The Eastern Transvaal Committee used the work of the Afforestation Priorities Committee to plan for the regulation of Eastern Transvaal afforestation 'in the national interest' to supply the country's timber needs without putting other sectors at a disadvantage. Calculations from the Nänni curves and local rainfall data, the assessed water use by currently afforested areas, the national afforestation potential assessment, the catchment classification of the Central Afforestation Permit Committee, and their assessments of current and future water supply and demand allowed appraisals for each of the 17 catchments within their planning region. They estimated that the afforested area could increase from the current (1972) area of about 342,000 ha, to 439,000 in 1985 and 500,200 in the year 2000. Of the 17 catchments, two had no plantations and should not be afforested at all, and in a further six, no further afforestation could be allowed, despite their potential. The Letaba catchment, where Wicht in 1949 had found mining-timber plantations of *Eucalyptus grandis* to be the culprits of water reduction, was one of these, and another was the Sabie, where most government sawlog plantations had been established from the 1930s onward; and the White River, a sub-catchment of the Crocodile, was also 'closed'. Just four of the 17 were eligible for substantial increases in plantation area, and here afforestation would be allowed to account for nearly 140,000 of the envisaged 160,000 ha of new afforestation.[54] As the development of forestry unfolded, the area actually afforested by 2008 was 400,000 ha, 100,000 short of what the Eastern Transvaal Committee allowed for.[55]

Up to the time of the work of the Eastern Transvaal Committee, in areas such as the Sabie and White River catchments, most afforestation had been by the state. A period of rapid investment by the private sector followed. Sappi, which had completed the pulp and paper mill at Ngodwana on a tributary of the Crocodile River in 1963 and had begun buying farmland for afforestation in 1961, enlarged the mill in 1985 having expanded its plantations in parallel.[56] Hunt Leuchars

53   Ibid., 2–3, 57–63.

54   Waterbeplanningskomitee vir Oos-Transvaal, *Waterbeplanning vir Oos-Transvaal*, 57–63.

55   Department of Agriculture and Forestry, *State of the Forests Report 2011*, www2.dwaf.gov.za/webapp/resourcecentre/Documents/Reports/Stateoftheforestsreport_web.pdf (accessed 14 October 2013), 9. The 1980 committee report did not encompass the Usuthu catchment in the south of Mpumalanga, which in 2002 had 160,000 ha plantation, which we subtracted from the total area of 560,000 ha stated for Limpopo and Mpumalanga in the *State of the Forests Report 2011*—see section 7.3 in Department of Water Affairs and Forestry, *Overview of the Water Resources of the Usutu-Mhlathuze Water Management Area* (Pretoria: Republic of South Africa, 2002), www.dwaf.gov.za/sfra/SEA/usutu-mhlathuze%20wma/Hydro-Economic%20Component/Overview%20of%20 water%20resources%20of%20the%20U-M%20WMA.pdf (accessed 14 October 2013).

56   Sappi, 'Company History', www.sappi.com/regions/sa/group/Pages/Company-history.aspx (accessed 16 October 2013).

and Hepburn, an old Natal timber and forestry firm, began to buy up farms in the White River region in the mid-1970s, when orchard farming there was undermined by disease, to plant eucalypts for mining-timber.[57] In the Crocodile catchment, where these developments unfolded, plantations amounted to about 144,000 ha by 1972, and the Eastern Transvaal Committee envisaged a further 48,000 (excluding the catchment of the White River). Under the administration of the Afforestation Permit System, the plantation area grew to its present 177,500 ha,[58] around 15,000 ha less than the committee suggested.

The case of the Eastern Transvaal, and the Crocodile catchment within it, illustrates the general course of forestry development that followed on the adoption of the Afforestation Permit System. Diek van der Zel reported in 1995 that between 1972 and 1994, the Central Committee received about 4,300 applications for permits, for a total proposed area of afforestation of 1.1 million ha. Of these, the committee approved nearly 3,900 applications, for an area of about 940,000 ha, and of this 430,000 ha was planted. Clearly, as van der Zel concludes and as the Eastern Transvaal history illustrates, the permit system as administered did not hinder investment in new plantations, but rather directed these plantations toward sites of higher forest productivity, where competition for water was not yet a constraint, and away from catchments where the water was no longer available.[59]

Within these planning guidelines, the afforestation of eucalypts and pines continued consistently in the Transvaal (currently Mpumalanga and Limpopo provinces) and Natal (currently KwaZulu-Natal) until the early 1990s, when the afforested area in South Africa culminated at nearly 1.5 million ha.[60] Forest products and industrial production of them was one of the fastest growing sectors of the South African economy from the 1960s to the early 1990s.[61] By 1993, forestry and forest products contributed 2.01 per cent of South Africa's GDP.[62]

---

57   Witt, '''Clothing the Once Bare Brown Hills of Natal''', 107; Interview, Stoney Steenkamp, Bedrock Mining Support, White River, 9 October 2013. Bedrock Mining Support is a private firm that acquired the farms originally afforested by Hunt, Leuchars and Hepburn, as well as mining-timber plantations from Sappi, and with 30,000 ha under *Eucalyptus grandis* supplies 68 per cent of South Africa's mining-timber requirement of about 800,000 tonnes per year.

58   Department of Water Affairs and Forestry, *Internal Strategic Perspectives: Inkomati Water Management Area – Version 1 (March 2004)* (Pretoria: Department of Water Affairs and Forestry, 2004), www.dwaf.gov.za/Documents/Other/WMA/5/optimised/INKOMATI%20REPORT.pdf, 41.

59   Van der Zel, 'Accomplishments and Dynamics of the South African Afforestation Permit System'.

60   Roger Godsmark, Forestry SA, email with statistical tables, 23 September 2013.

61   Louw, 'General History of the South African Forest Industry: 1975 to 1990'.

62   Institute for Natural Resources, *Pilot State of the Forest Report: A Pilot Report to Test The National Criteria and Indicators, March 2005* (Pretoria: Department of Water Affairs and Forestry, 2005), 7.

Chapter 9

1965 to 1995: Fluctuating Fortunes and Final Dividends

Expanding the experimental network throughout the forestry regions

While Jonkershoek was the source of method and technique, the real need for hydrological knowledge lay in the summer-rainfall forestry regions. Establishing Cathedral Peak in 1948[1] was a major step to fill this gap, but it was only the first, and the progressive extension of the forest hydrology program led eventually to a network of eight sites representing the upland forestry regions most important for water supplies, from Jonkershoek in the south-west to Westfalia in the far north-east (Table 3). Jonkershoek and Cathedral Peak were multiple catchment experiments, but with these as the benchmarks to assure 'unimpeachable' findings, subsequent sites took the form of the more economical paired-catchment design.

The experimental treatments included not just afforestation, but also prescribed fire in natural fynbos or grassland. The new experiments with prescribed fire treatments built on the early catchment trials at Jonkershoek, and these as well as afforestation experiments accelerated from 1965 once interdepartmental rivalries over the domains of hydrology and ecology were reconciled, the 1968 Committee on Afforestation and Water Supplies had emphasised the need, and resources became available. As new observational technology became available, so the science was deepened, but the catchment experiment remained as the foundation. The program was sustained, long enough to yield unexpected and valuable findings that would otherwise have remained in ignorance. From this long-term success came the answers to some of the intriguing questions that motivated the program back near the start of the twentieth century.

---

1   Initial work by Mike de Villiers in 1939 had included the building of an access road through precipitous terrain, work only completed after World War II; Söhnge, 'Bosboupionier', 12–3.

Table 3. Summary of catchment experiments in South Africa.

Experiment set	Location and climate	Date begun	Treatments	Present status	Outcomes and illustrative examples of publications
Jonkershoek	Western Cape Province, near Stellenbosch: 33°57'S 18°15'E, 274–1530 m amsl; Mediterranean, MAP 1200–2260 mm; fynbos	1935 (rain gauging) 1937 (stream gauging)	Nine gauged catchments in a multiple-catchment design with time-serial replicated afforestation treatments (Pinus radiata). Rain chemistry, erosion and nutrient deposition and exports.	Gauging of four catchments terminated 1991 and 1992, others maintained but treatment protocol suspended, network now being restored by SAEON.	Findings on effects of afforestation, riparian zone treatment, veldfire all tested and reported. C.L. Wicht, 'The Validity of Conclusions from South African Multiple Watershed Experiments'; D.F. Scott, D.B. Versfeld and W. Lesch, 'Erosion and Sediment Yield in Relation to Afforestation and Fire in the Mountains of the Western Cape Province, South Africa', South African Geographical Journal, 80 (1998): 52–9; D.F. Scott and F.W. Prinsloo, 'Longer-Term Effects of Pine and Eucalypt Plantations on Streamflow', Water Resources Research, 44 (2008); D.W. van der Zel and F.J. Kruger, 1975. 'Results of the Multiple Catchment Experiments at the Jonkershoek Research Station, South Africa. Influence of Protection of Fynbos on Stream Discharge in Langrivier', Forestry in South Africa, 16 (1975): 13–8.
Cathedral Peak	KwaZulu-Natal Province near Bergville: 23°04'S 30°04'E, 1829–2439 m amsl; temperate climate with dry winter (mountain), MAP 1400 mm; grassland	1948–1956, 1963, 1975–1976 (rain and stream gauging)	15 gauged catchments, 13–190 ha, multiple-catchment design with time-serial replicated afforestation treatments (Pinus patula), replicated veld burning treatments, unburnt grassland control. Rain chemistry, erosion and nutrient deposition and exports.	Progressive closure of gauging sations from 1992, all observations terminated in year 1997, network now being restored by SAEON	Findings on effects of afforestation, riparian zone treatment, veldfire, catastrophic forest fire. J.M Bosch, 'Treatment Effects on Annual and Dry Period Streamflow at Cathedral Peak', South African Forestry Journal, 108 (1979):, 29–38; D.B. van Wyk, 'The Influence of Catchment Management on Sediment and Nutrient Exports in the Natal Drakensberg', in R.E. Schulze, Proceedings of the Second National Hydrological Symposium, Pietermaritzburg, University of Natal (Pietermaritzburg: University of Natal, 1986), 266–75; C.S. Everson, 'The Water Balance of a First Order Catchment in the Montane Grasslands of Natal', Journal of Hydrology, 241 (2000): 110–3.

Experiment set	Location and climate	Date begun	Treatments	Present status	Outcomes and illustrative examples of publications
Mokobulaan	Mpumalanga Province, near Mashishing: 25°17'S 30°34'E, 1292–1494 m amsl; transitional sub-tropical summer rainfall climate, MAP 1167 mm; transitional savanna and grassland	1956 (rain and stream gauging)	Paired-catchment, 3 gauged catchments (26–36 ha), 2 treatments (*Eucalyptus grandis*, Pinus *patula*) with grassland control.	Terminated 2004	Findings on effects of afforestation, deforestation and reforestation. W.S. van Lill, F.J. Kruger and D.B. van Wyk, 'The Effect of Afforestation with *Eucalyptus grandis* Hill Ex Maiden and *Pinus patula* Schlecht. Et. Cham. on Streamflow from Experimental Catchments at Mokobulaan, Transvaal', *Journal of Hydrology*, 48 (1980): 107–18; D.F. Scott and W. Lesch, 'Streamflow Responses to Afforestation with *Eucalyptus grandis* and *Pinus patula* and to Felling in the Mokobulaan Experimental Catchments, Mpumalanga Province, South Africa', *Journal of Hydrology*, 199 (1997): 360–77.
Witklip	Mpumalanga Province, near White River: 25°14'S 30°53'E, 1000–1470 m amsl; sub-tropical summer rainfall, MAP 1475 mm; grassland.	1974–1975 (rain and stream gauging)	Eight gauged catchments, 47–197 ha, serial paired design, treatments of riparian clearing and deforestation (*Pinus patula*). Rain chemistry, erosion and nutrient deposition and exports.	Terminated 1991	Findings on effects of deforestation, riparian clearing and the mineral balance of the catchments under plantations. R.E. Smith, 'Effect of Clearfelling Pines on Water Yield in a Small Eastern Transvaal Catchment, South Africa', *Water SA*, 17 (1991): 217–24.
Westfalia	Limpopo Province, near Tzaneen: 23°44'S 30°04'E, 1050–1420 m amsl; tropical Wet and Dry, MAP 1253 mm; transitional savanna and forest	1972 (rain gauging) 1974 stream gauging)	Three gauged catchments, 16–40 ha, serial paired design, treatments of riparian clearing and afforestation (*Eucalyptus grandis*), indigenous forest as control.	Terminated 2008	Findings on effects of riparian clearing, of indigenous forest cover, and the short- and long-term effects of afforestation with *Eucalyptus grandis*. D.F. Scott and W. Lesch, 'The Effects of Riparian Clearing and Clearfelling of an Indigenous Forest on Streamflow, Stormflow, and Water Quality', *South African Forestry Journal*, 175 (1996): 1–14; Scott and Prinsloo, 'Longer-Term Effects of Pine and Eucalypt Plantations on Streamflow'.

Experiment set	Location and climate	Date begun	Treatments	Present status	Outcomes and illustrative examples of publications
Jakkalsrivier	Western Cape Province, near Grabouw: 34°09'S 19°09'E, 656–1190 m amsl; Mediterranean-type climate, MAP 961 mm; fynbos	1968 - 1971	Nine gauged catchments, 2.8 –24 ha, multiple-catchment design, replicated fire treatments, protected fynbos as control. Rain chemistry, erosion and nutrient deposition and exports.	Terminated 1992	Findings on effects of burning on streamflow, erosion, mineral balance. Scott, 'The Hydrological Effects of Fire in South African Mountain Catchments'; Scott et al., 'Erosion and Sediment Yield in Relation to Afforestation and Fire in the Mountains of the Western Cape Province, South Africa'.
Zachariashoek	Western Cape Province, near Franschhoek: 33°50'S 19°03'E, 240–850 m amsl; Mediterranean, MAP 1400 mm; fynbos	1965	Three adjacent gauged catchments, two with tandem gauges, paired-catchment design, replicated fire treatments, protected fynbos as control. Rain chemistry, erosion and nutrient deposition and exports.	Terminated 1995	Findings on effects of burning on streamflow, erosion, mineral balance. Scott, 'The Hydrological Effects of Fire in South African Mountain Catchments'; Scott et al., 'Erosion and Sediment Yield in Relation to Afforestation and Fire in the Mountains of the Western Cape Province, South Africa'.
Moordkuil	Western Cape Province, near Mosselbaai: 33°52'S 22°04'E, 442–979 m amsl; Transitional Mediterranean-type climate, MAP 1200 mm; fynbos	1984	Three gauged catchments, fire treatments, protected fynbos control.	Terminated 1991	No definite findings.

Only experiments with controls included.[2] For an omnibus analysis of the experiments, see D. F. Scott and others, *A Re-Analysis of the South African Catchment Afforestation Experimental Data*, 2000, www.wrc.org.za/Knowledge%20Hub%20Documents/Research%20Reports/810-1-00.pdf (accessed 11 February 2015). Examples of syntheses include J. M Bosch and J. D. Hewlett, 'A Review of Catchment Experiments to Determine the Effect of Vegetation Changes on Water Yield and Evapotranspiration', *Journal of Hydrology*, 55 (1982): 3–23; J. D. Hewlett and J. M Bosch, 'The Dependence of Stormflow on Rain Intensity and Vegetal Cover in South Africa', *Journal of Hydrology*, 75 (1984): 365–81; D. F. Scott, 'The Hydrological Effects of Fire in South African Mountain Catchments', *Journal of Hydrology*, 150 (1993): 409–32; D. F. Scott and R. E. Smith, 'Preliminary Empirical Models to Predict Reductions in Annual and Low Flows Resulting from Afforestation', *Water SA*, 23 (1997): 135–40; Scott, 'Managing Riparian Zone Vegetation to Sustain Streamflow'; M. B. Gush and others, 'A New Approach to Modelling Streamflow Reductions Resulting from Commercial Afforestation in South Africa', *Southern African Forestry Journal*, 196 (2002): 27–36; (MAP: mean annual precipitation; amsl: above mean sea level; SAEON: South African Environmental Observation Network).

A program at hazard

Contrary to the impression of a smooth path of development, crisis in South Africa's forest hydrology was more the rule than the exception. We have seen how in the early stages J. J. Kotzé, Ian Craib and J. D. M. Keet, and no doubt others, intervened at critical moments to give support and keep up the momentum of the work at Jonkershoek. Although funding was always a problem, the most severe constraint was availability of competent scientists and technicians. At the beginning of the program, Wicht worked with only untrained assistants. In March 1938, Mike de Villiers, a forestry graduate assigned to Jonkershoek to train for the project at Cathedral Peak, assisted in the field and the laboratory, but left for Cathedral Peak in 1939.[3]

At the outbreak of World War II, Wicht filled in for his colleagues who had entered military service, inspecting silvicultural research experiments in the Cape program up to 600 km away (while also continuing with lectures in the forestry program at the University of Stellenbosch).[4] After the war, university-trained assistance became available intermittently, but still by 1963, the Forestry Research Institute would state plainly 'The staff employed in hydrological research is inadequate to carry out the approved research program'.[5] In 1965,

---

2   From R. E. Smith, 'Effect of Clearfelling Pines on Water Yield in a Small Eastern Transvaal Catchment, South Africa', *Water SA*, 17 (1991): 217–24; D. F. Scott et al., *A Re-Analysis of the South African Catchment Afforestation Experimental Data*, 2000; email, G. Forsyth, CSIR, Stellenbosch, 9 September 2013; email, G. Forsyth, CSIR, Stellenbosch, 5 September 2013; email, N. Allsopp, SAEON, 9 October 2013; email, Sue van Rensburg, SAEON, 8 November 2013.

3   Around 1936, an investigation team led by Ian Craib explored on horseback and foot the foothills and slopes of the Drakensberg range, and finally recommended Cathedral Peak as the second forest influences research site.

4   For example Annual Report, District Forest Officer, Jonkershoek, 1937/38, digi.nrf.ac.za/dspace/bitstream/handle/10624/380/CLW_AnnRep1938p001.pdf?sequence=1 (accessed 11 August 2013); Wicht had volunteered for military service, but was rejected because of his eyesight (email, Susan Wicht Clark, 7 September 2010).

5   Annual Report, Hydrological Research, 1962–63, file R.7090, SAFRI Archives, CSIR, Pretoria.

defending the program against criticism from H. L. Penman, Wicht said: 'We are trying to do what we can ... There have been very serious manpower problems. The original programs of research have not always been carried out ... because of interruptions due to the war and the impossibility of obtaining trained staff afterwards. I regret that fundamental problems have not been gone into more, but we are hoping to be able to do so'.[6] Again, in 1967, Ugs Nänni warned 'Staff difficulties have again hampered progress in forest hydrology ... At Cathedral Peak the situation was serious ... For the greater part of the year there was no professional officer on the station ... at Jonkershoek work was seriously handicapped by the lack of funds'.[7]

At the same time, ongoing efforts to establish a more secure institutional arrangement for hydrological research overall in South Africa had little consequence. Wicht as Chief Forest Research Officer had in 1948 made detailed proposals to Meiring Naudé, President of South Africa's CSIR, for a new Hydrological Research Institute, met with Minister of Transport Paul Sauer (responsible also at the time for meteorology and forestry) in May 1952 about the issue and wrote in 1952 (now as Professor of Forestry) to Sauer with the same proposals, and offering his services to direct the new organisation. This led to a civil service investigation, and the upshot was that an Interdepartmental Coordinating Committee for Hydrological Research was appointed in 1962 to optimise the national effort (later to be supplanted by the Water Research Commission), while the forest hydrology program remained as a forestry research function.[8]

One key outcome of the formation of the Interdepartmental Committee was agreement among participating government departments that Forestry should expand catchment research to include the options for managing natural vegetation—fynbos and grassland—in properly designed experiments, with the collaboration of experts from different departments. In May 1963, Ugs Nänni led a team that identified Lebanon and Zachariashoek as new catchment research sites, the network at Cathedral Peak was expanded, and from this began the new impetus that would lead to the highly productive programs in fire ecology and invasion biology described by Simon Pooley and Brian van

---

6   C. L. Wicht, 'The Validity of Conclusions from South African Multiple Watershed Experiments', Discussion, 760.

7   Annual Report 1996–97: Forest Hydrology, V. 2000, SAFRI Archives, CSIR, Pretoria.

8   C. L. Wicht to Minister of Transport, 30 June 1952, Wicht Papers, SAFRI Archives, CSIR Pretoria; Wicht to C. Meiring Naudé, President, CSIR, 7 February 1956 and reply 12 March 1956; CSIR memorandum *'Oorsig van besprekinge en aanbevelings oor uitbreiding en koördinering van hidrologiese navorsing in Suid-Afrika'*, at digi. nrf.ac.za/dspace/bitstream/handle/10624/409/CLW_Oorbes1955001.pdf?sequence=1 (accessed 12 November 2013).

Wilgen,[9] as well as, later, the passage of the *Mountain Catchment Areas Act*. (Though the issue of the right organisational home for the program did not go away—in 1977 the transfer of the forest hydrology program together with its growing ecological research activity to the agriculture department was strongly mooted by the latter department, a proposal that was rejected, but not before creating uncertainty among researchers.)[10]

By the time of the work of the 1968 Committee on Afforestation and Water Supplies the difficulties had become acute, since now several years of experimental data were available, and the network had been expanded to include Cathedral Peak and Mokobulaan. Yet just two papers had been published presenting results from the catchment experiments, one in 1963 and one in 1965, and both dealt only with Jonkershoek.[11] Because the committee found 'a dearth of reliable data on catchment management, its effects on water supplies, and the economic implications of water conservation measures', they had substantial recommendations with respect to research. The most immediate need was to analyse and interpret the data yielded by existing investigations in order 'to produce the reliable, quantitative, hydrological data needed to prescribe practical catchment management with assurance'. They recommended increasing staff and other resources to work quickly on 'the copious data which have been accumulated from current investigations at research stations'.[12] The committee continued, recommending studies of fundamental hydrological processes, to supplement the data yielded by catchment experiments and planned surveys, to explain overall effects on water yields of treatments applied in catchments, and to provide criteria for extrapolation of experimental results. This should include investigations of the water relations of plant communities, and research on the methods of vegetation control should be expanded.

The committee recommended multidisciplinary teams, and linkages with universities. Because knowledge of vegetation management as an essential part of catchment management was inadequate, the research program required ecologists and physiologists. Forestry and agricultural practices in important catchments should be specially investigated to ensure well-maintained soils and water resources. But 'nothing will be achieved if the urgent need for research is conceded, but steps are not taken simultaneously to provide the facilities, funds

9   Annual Report, Hydrology to the Inter-Departmental Coordinating Committee for Hydrological Research, 1962–63, R.7090, SAFRI Archives, CSIR, Pretoria; Pooley, 'Recovering the Lost History of Fire in South Africa's Fynbos'; B.W. van Wilgen, 'The Evolution of Fire and Invasive Alien Plant Management Practices in Fynbos', *South African Journal of Science*, 105 (2009): 335–41.

10   Director Forestry Research W. H. van der Merwe to C. L. Wicht, 13 December 1977, at digi.nrf.ac.za/ dspace/bitstream/handle/10624/403/CLW_Briwic1978001.pdf?sequence=1 (accessed 7 October 2013).

11   Banks and Kromhout, 'The Effect of Afforestation with *Pinus radiata*'; Wicht, 'The Validity of Conclusions from South African Multiple Watershed Experiments'.

12   Department of Forestry, *Report of the Interdepartmental Committee of Investigation into Afforestation and Water Supplies in South Africa* (Cape Town: Republic of South Africa, 1968), 101, 103.

and staff to do the work'.[13] Below, we shall see what effect these recommendations were to have. With a clear mandate, and questions of its institutional place and role settled, the then Forest Research Institute, collaborating with Wicht at the University of Stellenbosch and others, could now proceed with the successful resourcing of the program, thorough analysis of catchment experimental findings, and research on an ecosystems approach to catchment management.

In 1972, after the recommendations in the 1970 report of the Commission of Enquiry into Water Matters, government formed what became the Hydrological Research Institute in the Department of Water Affairs, but with a brief that excluded forest hydrology.[14] Government also established the Water Research Commission in 1971 'to promote and expedite the country's water research purposefully'. The commission, financed by levies raised on irrigation and the supply of water from government schemes, brought fresh resources to hydrological research, but not to fund or supplant research conducted as part of the 'normal functions' in extant programs, such as forest hydrology. Following an extensive investigation, the commission developed the 'National Master Research Plan and National Priority Research Program' which flagged management of catchments as a factor influencing river flow, and nominated as research priorities such elements as the management and conservation of catchments, effects of afforestation, natural vegetation and veld burning, as well as flood control[15] The commission identified and funded new centres of excellence at universities to fill gaps, one of which became the Agricultural Catchments Research Unit (ACRU) at the then University of Natal, which complemented the forest hydrology program and continues this work today.

---

13   Ibid.
14   Memorandum by CSIR Liaison Division Paper M.9 of 2 September 1948 summarising Wicht's proposal and official responses to it; Wicht to Minister of Transport, 30 June 1952; Wicht to S. Meiring Naudé, President, CSIR, 7 February 1956; Naudé to Wicht, 212 March 1956; Wicht Papers, SAFRI Archives, CSIR, Pretoria; see also Wicht's reports on diverse committee meetings and their recommendations at digi.nrf.ac.za/dspace/bitstream/handle/10624/409/CLW_Oorbes1955001.pdf?sequence=1 (accessed 7 October 2013); J.S. Whitmore, 'The founding of the Hydrological Research Institute: Recollections and Reflections, 21 October 1993', www.dwa.gov.za/iwqs/iwqshistory/hristory.asp (accessed 5 November 2013); *Water Research Commission Annual Report 1 September 1971 to 31 March 1972*, Water Research Commission, Pretoria, 29, at www.wrc.org.za/Knowledge%20Hub%20Documents/Annual%20Reports/Annual%20Report%201972.pdf (accessed 8 November 2013), 6–8, 17, 20; in 1994 this institute became the Institute for Water Quality Studies and, in 2003, the Resource Quality Services entity, a directorate in the Water Resource Information Management.
15   *Water Research Commission Annual Report 1 April 1973 to 31 March 1974*, Water Research Commission, Pretoria, 30, at www.wrc.org.za/Knowledge%20Hub%20Documents/Annual%20Reports/Annual%20Report%201973.pdf (accessed 8 November 2013), 7.

The International Symposium on Forest Hydrology and international collaboration

The pressures arising from intensified competition for water resources in South Africa during the 1960s coincided with global efforts to advance the science of hydrology. The global community launched the International Hydrological Decade on 1 January 1965, under the leadership of UNESCO. The initiative continued for the following 10 years, as a 'concerted international effort' which had as one of its prime purposes the bringing together of the fragmented and 'laggard science' of hydrology.[16] A key event in this initiative and in the development of forest hydrology was the International Symposium on Forest Hydrology, convened at the Pennsylvania State University, Pennsylvania, USA during August – September 1965. Its purpose was for forest hydrologists to establish the current state of knowledge, to define research needs and trends, and to speculate about the future direction of forest hydrology research.[17] That the meeting was timely was clear from John Hewlett's later assessment: 'As valuable as this proceedings has been, its chapters nevertheless reveal disjointed purposes, terminology and methods among the workers in several countries, each with its own background and attitudes toward hydrologic problems'.[18]

The assembly included 87 scientists from 22 countries. From South Africa, C. L. Wicht and Ugs Nänni attended. Wicht delivered two papers, and compiled the report on the session on forests and evapotranspiration.[19] For the first time the forest hydrology researchers from around the world (at that time, the USA, South Africa, Japan and Kenya had noteworthy catchment research programs under way) were able to expose their ideas to a global assembly. It also engaged the leading figures in water balance studies, such as Howard Penman, author of the seminal 1963 book, *Vegetation and Hydrology*; Charles Pereira, who then led the program in East Africa; and Albert Baumgartner, from Germany, a leader in evapotranspiration studies. The result was a forum for penetrating review and healthy debate. Pereira and others welcomed the meeting as being one in which, for the first time, concepts, methods and terminology in forest hydrology could

16   W. L. Nace, 'Hydrology Comes of Age: Impact of the International Hydrological Decade', *Eos, Transactions American Geophysical Union*, 61 (1980), 1241; W. L. Nace, 'Water Resources: A Global Problem with Local Roots', *Environmental Science and Technology*, 1 (1967): 550–60, 551.

17   W.T. Swank, 'Models in Forest Hydrology: An Overview', in *Proceedings IUFRO Workshop on Water and Nutrient Simulation Models, Swiss Federal Institute of Forestry Research, Birmensdorf, 1981* (1981), 13–20, 13.

18   Hewlett, 'The Relation of Forests and Forestry to Water Resources', 6.

19   C. L. Wicht, 'Forest Hydrology Research in the South African Republic', in W. E. Lull and W. H. Sopper (eds), *International Symposium on Forest Hydrology* (Oxford: Pergamon Press, 1967), 75–84; Wicht, 'The Validity of Conclusions from South African Multiple Watershed Experiments'; Wicht, 'Summary of Forests and Evapotranspiration Session', 493.

be discussed in a representative international forum. This forum served as an important opportunity for critical review of South Africa's work by eminent hydrologists.

The sessions on technique—instrumentation and analytical problems and their solutions—solved problems and achieved standards that assisted progress in South Africa. An example is the key step forward in hydrograph analysis offered by John Hewlett and Alden Hibbert (p. 281 of the proceedings), of the construct of 'constant separation slope' of 0.033 cubic metres per minute per square kilometre per hour (the number is the rate of change in the rate of flow) as the means to separate baseflow from stormflow on the storm hydrograph, thus obviating fruitless and time-consuming effort to find for each catchment the 'baseflow separation curve'. This device allowed more rapid and efficient digital analysis of hydrograph charts, and, adopted immediately in South Africa, sped up analysis of experimental effects. (Ugs Nänni's master's research, in which he could find no baseflow recession curve for the Cathedral Peak catchments, had already prepared the ground.)[20]

The South African findings had a good reception. For example, the concept of the catchment as an ecosystem had important influence: some time after, Wayne Swank, one of the United States' leading forest hydrologists, commended Wicht's position as 'perhaps one of the most perceptive observations' at the meeting, when he noted that 'a complete, integrated whole—the ecosystem of the forest' is the appropriate level for understanding evapotranspiration.[21]

But it also had to withstand criticism. As we have seen earlier, Penman's criticisms were especially sharp, and he caused a heated debate on whether transpiration was a purely physical process (Penman: 'I firmly believe, and I am going on with my job on the assumption, that the water use by plants is not a vital process') or whether it was biologically mediated ('vital' – Wicht). Wicht stood his ground; A. J. Rutter sought to mediate: 'Could we sort of reconcile this by saying that I must agree with Penman that the plant can do nothing to evaporate water and I must agree with Wicht that the resistances in the system are under living or physiological control'.[22] Later in the meeting Penman criticised the multiple-catchment design, emphasising the water balance approach as being necessary (see earlier; Wicht by this time seems to have become impatient: 'We are trying to do what we can with the means at our disposal').

---

20   U. W. Nänni, 'Base-Flow in Cathedral Peak Streams' (MSc (Forestry) Thesis, University of Stellenbosch, 1957). Nänni's analyses refuted J. E. Horton's theory of baseflow recession, and rejected Horton's mathematical model for baseflow recession curves, as well as Wicht's earlier findings (see Wicht, 'Depletion of Ground-Water Flow in Jonkershoek Streams').

21   Swank, 'Models in Forest Hydrology: An Overview', 13.

22   Wicht, 'Summary of Forests and Evapotranspiration Session', 494.

Although Wicht maintained an exchange over time with Penman's colleagues Perreira and others,[23] such as Albert Baumgartner, on the question of hydrological methodology and fundamental studies, it was with US forest hydrologists that the most important links emerged. Wicht had corresponded with US counterparts from the beginning, and cited Charles Hursh from Coweeta as long ago as 1943, but it was at this meeting that the hydrologists from South Africa would meet their counterparts from the US for the first time. Wicht and Nänni visited Coweeta during a study tour after the conference. They came away affirmed in their belief in the design of the South African program. 'The basic designs and observation techniques applied in the South African experimental investigations is sound and often better than those seen in the United States', though the staff was too small and facilities inadequate, while there was an urgent need for advanced modern computational aids. There was a sense of urgency about getting the results out from South Africa's 'sound foundation for watershed management research', and an emphasis on the value of cooperative research that is 'free and by mutual agreement' which later paid dividends.[24]

From this conference and tour, Wicht began to realise that place played a key role in design and methodology. The focus on fundamental hydrological studies promoted by Penman originated not only in what was called the 'philosophical' approach, but 'also by circumstances in the region where hydrological research is undertaken'. In the US and South Africa, extensive regions were [then] available 'where many experimental watersheds could be selected and experimentally investigated for long periods', but not so in Britain and European countries, where in consequence 'disjunctive, fundamental research must be relied upon'. Though the results of investigations of individual hydrological processes may then be integrated to estimate the aggregate effects of vegetation on water supplies, '*What* happens when vegetal cover is modified, replaced or removed is therefore *not* observed',[25] while the concern about the impossibility of accurate measurement at catchment scale of any water balance term other than streamflow remained.

23   For example H. C. Perreira to Wicht, 11 January 1962, re Wicht's concerns about the East African hydrological research methodology; Wicht Papers, SAFRI Archives, CSIR, Pretoria.
24   C. L. Wicht 1965. A report on a study tour of forestry research and educational institutions in the Eastern United States of America. R7090 SAFRI Archives, CSIR, Pretoria.
25   Ibid.

Forests, water yield and floods: The cooperation with John Hewlett

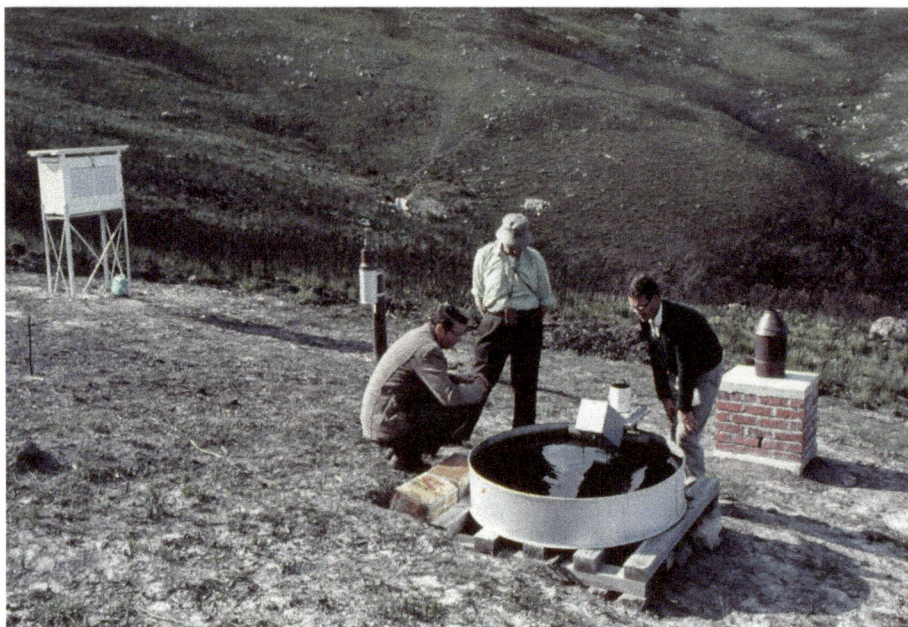

Figure 18. John Hewlett, Christiaan Wicht and Diek van der Zel at the high-altitude weather station in the Jakkalsrivier experimental catchment, May 1970, at the time of Hewlett's visit to South Africa.

Source: Jonkershoek archives, CSIR, Stellenbosch. Photographer: Fred Kruger.

Soon after 1965, Leon V. Pienaar, who had completed his Master's thesis on rainfall interception in *Pinus radiata* under Wicht's supervision, took up a position at the University of Georgia, where he collaborated with John Hewlett—together, they later produced a definitive overview of the design of small catchment experiments.[26] Hewlett toured South Africa in 1970 for five weeks as a guest of the Water Research Commission, visiting hydrological research centres as well as participants involved in the land, forests and water contest.[27] In 1977, one of us (FJK) visited Hewlett and Pienaar at the University of Georgia, as well as Coweeta, to explore opportunities for collaboration. The outcome was that in 1979 the SA Forestry Research Institute sent Jan Bosch to Athens to work with Hewlett for a year. The work that followed generated fundamental contributions to global knowledge of forest hydrology.

---

26 Hewlett and Pienaar, 'Design and Analysis of the Catchment Experiment'.
27 See Hewlett, 'The Relation of Forests and Forestry to Water Resources. Address'.

Jan Bosch arrived in Athens with a very broad brief—he was to receive mentorship; with no fixed agenda, he and Hewlett initially simply had daily discussions on the philosophy of forest hydrology.[28] Eventually, the idea emerged of addressing two salient problems in the field, yielding two highly influential papers. The first, 'A Review of Catchment Experiments to Determine the Effect of Vegetation Changes on Water Yield and Evapotranspiration' (1982), built on earlier, tentative findings of Alden Hibbert[29] and is considered the classic review in the field, while the second, 'The Dependence of Stormflow on Rain Intensity and Vegetal Cover in South Africa' (1984) was a landmark study of the question of vegetation and floods.[30]

Their meta-analysis of research on the effect of vegetation change assembled and analysed the findings from 94 catchment experiments around the world (in 1965, Hibbert could find results for 39 experiments). They found that increases in vegetation cover consistently resulted in decreases in water yield, i.e. streamflow, and that with the data available, it was now possible to estimate the approximate magnitude of the effects: pine and eucalypt forest types caused on average 40-mm change in water yield per 10 per cent change in cover, deciduous hardwood and scrub about 25 and 10 mm, respectively.

The findings supported both Hibbert's earlier findings, and Wicht's hypothesis about vegetation density and streamflow change. At the time of the 50th anniversary of the *Journal of Hydrology* this paper was the third most-cited article in the journal's history. In 2001, the findings were corroborated, when Lu Zhang and co-authors analysed the findings from 250 catchment experiments worldwide, and again by Alice Brown and co-authors in 2005.[31] Bosch and Hewlett's work was crucial in the line of evidence that addressed the long-standing question of the role of vegetation in determining water supplies; the refined answers to broad question of the relationship between forests, or vegetation in general, now included quantitative assessments for different vegetation types in different environments.

---

28   Jan Bosch, email to F. J. Kruger, 17 December 2010.
29   A. R. Hibbert, 'Forest Treatment Effects on Water Yield', in W. E. Lull and W. H. Sopper (eds), *International Symposium on Forest Hydrology* (Oxford: Pergamon Press, 1967): 527–43.
30   J. M. Bosch and J. D. Hewlett, 'A Review of Catchment Experiments to Determine the Effect of Vegetation Changes on Water Yield and Evapotranspiration', *Journal of Hydrology*, 55 (1982): 3–23; J. D. Hewlett and J. M. Bosch, 'The Dependence of Stormflow on Rain Intensity and Vegetal Cover in South Africa', *Journal of Hydrology*, 75 (1984): 365–81. A third paper had less impact: J. M Bosch and J. D. Hewlett, 'Sediment Control in South African Forests and Catchments', *South African Forestry Journal*, 115 (1980): 50–5.
31   L. Zhang, W. R. Dawes, and G. R. Walker, 'Response of Mean Annual Evapotranspiration to Vegetation Changes at Catchment Scale', *Water Resources Research*, 37 (2001): 701–8; A. E. Brown et al., 'A Review of Paired Catchment Studies for Determining Changes in Water Yield Resulting from Alterations in Vegetation', *Journal of Hydrology*, 310 (2005): 28–61.

Bosch and Hewlett's second major paper centred on another great question in forest hydrology, the relationship between forests and flooding, and their work was amongst the first to do this.[32] Wicht had an inkling of the answer when he wrote that 'During catastrophic storms it has however been observed that the influence of vegetation becomes reduced to insignificance',[33] but the experimental data were lacking or controversial until the 1980s: forests were still believed to mitigate floods, and this turned on the belief in Hortonian overland flow as the mechanism that generates stormflow from the catchment (many, however, still believe this, though without reason).[34] They argued that if stormflow generation depended on overland flow (and the inverse, that forests as opposed to other land cover promoted infiltration and minimised overland flow), then stormflow would respond directly to rainfall intensity, and that forests would attenuate that relationship. Hewlett and Bosch used South African experimental catchment data in an empirical test of the Hortonian overland flow hypothesis and compared catchments to assess the role, or absence of a role, of forests or other vegetation in mitigating floods. This was the second test of its kind, after Hewlett and co-authors in 1977,[35] and validated the work argued to be 'one of the most important papers published in the field of forest hydrology'.[36] But whereas Hewlett's prior study involved 545 stormflow hydrographs from a single catchment, this one encompassed hydrograph analyses for 1,546 stormflows from eight catchments in South Africa, with accompanying rainstorm data (i.e. of rainstorms larger than 20 mm, large enough to generate stormflow). The forms of vegetation in this set of catchments included pine forest, sclerophyllous shrubland (fynbos), grassland burnt regularly, and grassland where fire had been excluded. Great effort on the part of analysts back in South Africa was needed to capture this huge data set in the time available, for a comprehensive and unprecedented analysis, which occupied Jan Bosch for a year.

---

32   FAO and Cifor, *Forests and Floods: Drowning in Fiction or Thriving on Facts?* RAP Publication 2005/03 Forest Perspectives 2 (Bogor: Food and Agriculture Organization of the United Nations, Bangkok, and Center for International Forestry Research, 2005), 5.

33   Wicht, *Forestry and Water Supplies in South Africa*, 36.

34   See in I. R. Calder, 'Forest and Floods: Moving to an Evidence-Based Approach to Watershed and Integrated Flood Management', *IWRA, Water International*, 31 (2006): 1–3.

35   J. D. Hewlett, J. C. Fortson, and G. B. Cunningham, 'The Effect of Rainfall Intensity on Storm Flow and Peak Discharge from Forest Land', *Water Resources Research*, 13 (1977): 259–66; J. D. Hewlett and A. R. Hibbert, 'Factors Affecting the Response of Small Watersheds to Precipitation in Humid Areas', in W. E. Lull and W. H. Sopper (eds), *International Symposium on Forest Hydrology* (Oxford: Pergamon Press, 1967): 279–90.

36   J. J. McDonnell, 'Classics in Physical Geography Revisited: Hewlett, J. D. and Hibbert, A. R. 1967: Factors Affecting the Response of Small Watersheds to Precipitation in Humid Areas. In Sopper, W. E. and Lull, H. W., Editors, *Forest Hydrology*, New York: Pergamon Press, 275–90', *Progress in Physical Geography*, 33 (2009), 288.

The key finding was that very little of the rain in any storm emerged as stormflow, just 5 per cent of the average 38-mm rainstorm:[37] thus, these basins (which under natural conditions yield up to 50 per cent and more of precipitation as streamflow) are deep and permeable hydrologically, seldom yielding appreciable stormflows. And since the values for stormflows are small, 'there is little variation in storm flow to account for, and that variation is due primarily to the gross amount of storm rainfall'. Hourly rain intensity, though statistically significant, 'played a minor quantitative role'. Rather than the Hortonian model, the variable source-area concept of stormflow generation provided the explanation for the stormflow behaviour they found:

> When a basin's average response to rainfall is small, below ~ 5%, the channel and its nearby saturated zones supply virtually all of the storm flow. Channel zones occupying only a few percent of the total basin area quickly convey intense bursts of rain to the mouth of the basin. These tiny storm flows are thereby subject to hourly variations in rainfall intensity … In forested and other well-vegetated land, most of the flow path is sub-surface.

Neither afforestation nor the biennial burning of grass veld had an important influence on stormflow.[38]

Although many important findings followed, the collaboration with John Hewlett and with others more remotely in the program at Coweeta, as well as Wicht's confidence—after comparing the South African work with that in the US—that his overall strategy was correct, helped to invigorate the program during the period from 1965 onward. Jonkershoek, with its satellite sites, and Coweeta, with its set of experimental catchments, together formed the places from which key concepts in forest hydrology emerged, but they also formed the foci for a network of relationships through which the ideas flowed and formed.[39]

Surprises in the long term

The commitment to long-term hydrological and ecological research is always regarded as a good idea, but seldom easy to justify once the cost of the commitment becomes apparent, and mounts up. Once there is commitment (and we have

---

37   Langrivier at Jonkershoek, with its upper 30 per cent consisting of high cliffs, was the exception: there, 39 per cent of the average 57-mm rainstorm was released as stormflow, exemplifying the flash flooding reported by Wicht for the Jonkershoek Valley.

38   David Scott reported increases of 62 per cent in stormflow volumes (which he termed quickflow) after clearfelling of Bosboukloof at Jonkershoek, but since stormflows were a tiny 0.9 mm of the average 50-mm rainstorm, the increase in flow was trivial in terms of flooding. D. F. Scott, 'The Contrasting Effects of Wildfire and Clearfelling on the Hydrology of a Small Catchment', *Hydrological Processes*, 11 (1997), 551.

39   L. Cameron and D. Matless, 'Translocal Ecologies: The Norfolk Broads, the "Natural", and the International Phytogeographical Excursion, 1911', *Journal of the History of Biology*, 44 (2011): 15–41, 17 on 'place' as a network of relations.

seen how difficult this proved to sustain in South Africa), it seems obvious that experiments should be maintained for as long as it takes to achieve the original objectives, but the question as to how much longer always arises, and is not easily answered by appeal to a vague argument about ignorance of the future. In some measure, the long-term observations in the South African catchment studies provide an answer. The social gains from a long-term commitment are immense, two features perhaps being paramount: first, the tradition that emerges of sustained, high-quality observation under diverse ordeals and contingencies. And second, the intergenerational, virtual school of inquisitive scientists, with its network of relationships and institutional memory, that is built and which allows for continual, progressive accumulation of scientific knowledge within a coherent sense of purpose. But still, it is the discovery about the ecosystem over the long term that must justify a costly commitment. Two instances of such discovery illustrate the value of the long-term research in the South African program: forests and erosion, and the hydrological function of old even-aged forests.

The question of forests and erosion was an important founding motive for the program. At the outset, Wicht set up plot trials to examine hydrological processes that may have informed the erosion question, but these experiments did not yield results.[40] Dirk Versfeld next completed plot studies that in 1981 confirmed that Hortonian overland flow at Jonkershoek was negligible, and so also the potential for erosion.[41] But it was only after a technical advance in the late 1970s that real progress became possible. This was when Danie van Wyk developed and proved an ingenious apparatus, driven hydraulically by the rise and fall in water levels at the streamgauging weir, enabling the automatic, routine and accurate sampling of water during stormflows.[42] This allowed accurate estimates of net outflows from catchments of sediment and dissolved minerals, which are borne mainly during stormflows. These apparatuses were installed in several

---

40   Wicht, 'Forest Influences Research Technique at Jonkershoek', 73.

41   D. B. Versfeld, 'Overland Flow on Small Plots at the Jonkershoek Forestry Research Station', *South African Forestry Journal*, 119 (1981): 35–40. Versfeld concluded 'Overland flow was found to be negligible on small plots ... Treatments such as burning or hoeing of fynbos and thinning of a plantation had no significant effect ... Runoff never exceeded 0.05% of rainfall, even with large storms (exceeding 125 mm) immediately after treatment'. He showed that at Jonkershoek in the Western Cape of South Africa, land cover has little effect on the generation of overland flow and soil erosion.

42   D. B. van Wyk, 'Apparatus for Sampling of Streams for Chemical Quality and Sediment', *Water SA*, 9 (1983): 88–92.

catchments across the entire network, and the observations showed how in undisturbed catchments (those with plantations included), net erosion and export of minerals was trivial.[43]

High-intensity accidental wildfires gave the opportunity to investigate the effects of ecological catastrophe. Wildfires passed through one of the afforested catchments at Cathedral Peak in 1981, when the plantation was 26 years old, through Bosboukloof at Jonkershoek in 1986, and a catchment monitored by Agricultural Catchment Research Unit (ACRU) in the Drakensberg foothills near Cathedral Peak 1987. In comparison with fires in catchments under fynbos and grassland, the plantation fires transformed erosion and the hydrological regime absolutely. The catchment conditions found in the stormflow study by Hewlett and Bosch—permeable catchments with high infiltration capacity, minimal overland flow, and small stormflows—were gone, substituted by Hortonian conditions. Stormflow volumes doubled, but most important, erosion, measured as sediment delivered from the catchment per hectare per year, multiplied 20- to 100-fold and more.

David Scott's careful investigation in the field soon after the fires uncovered the reason for the change. The very intense fires, arising from the fact that conditions favoured the combustion of coarse fuels on the ground, had consumed soil organic matter and volatilised organic compounds, which then condensed in layers of the soil at some depth below the surface. These layers of condensate caused a water repellent condition, effectively neutralising the infiltration capacity of the soil; the resulting overland flow caused the relative increase in stormflow, and with the loss of soil structure, carried the surface soil to the stream. The nearby catchments, under fynbos or grassland and also burnt, did not respond this way, responding to rainstorms very much as they had before fire.[44] True, the catchments recovered comparatively quickly—in Bosboukloof, stormflow volume had declined by two-thirds by the second year after the

---

43   See D. F. Scott and D. B. van Wyk, 'The Effects of Wildfire on Soil Wettability and Hydrological Behaviour of an Afforested Catchment', *Journal of Hydrology*, 121 (1990): 239–56, 239. Scott, 'The Contrasting Effects of Wildfire and Clearfelling on the Hydrology of a Small Catchment', 553; D. B. van Wyk, 'The Influence of Prescribed Burning as a Management Tool on the Nutrient Budgets of Mountain Fynbos Catchments in the South-Western Cape, Republic of South Africa', in C. E. Conrad and W. C. Oechel (eds), *Proceedings of the Symposium on Dynamics and Management of Mediterranean-Type Ecosystems; June 22–26, 1981; San Diego, CA* (Berkeley, CA: Pacific Southwest Forest and Range Experiment Station, Forest Service, US Department of Agriculture, 1982), 390–6; D. B. van Wyk, 'The Influence of Catchment Management on Sediment and Nutrient Exports in the Natal Drakensberg', in R. E. Schulze, *Proceedings of the Second National Hydrological Symposium, Pietermaritzburg, University of Natal*, ACRU Report No. 22 (Pietermaritzburg: University of Natal, 1986), 266–75; D. F. Scott and D. B. van Wyk, 'The Effects of Fire on Soil Water Repellency, Catchment Sediment Yields and Streamflow', in B.W. van Wilgen, D. M. Richardson, F. J. Kruger, and H. J. van Hensbergen (eds), *Fire in South African Mountain Fynbos*, Ecological Studies 93 (Berlin, Heidelberg: Springer Verlag, 1992), 216–39.
44   Scott and van Wyk, 'The Effects of Wildfire on Soil Wettability and Hydrological Behaviour of an Afforested Catchment'; D. F. Scott, 'The Hydrological Effects of Fire in South African Mountain Catchments', *Journal of Hydrology*, 150 (1993): 409–32.

fire—but the pronounced effects illustrate the serious environmental risks of wildfire in plantations. Taken with the work by John Hewlett and Jan Bosch on stormflows and Danie van Wyk on erosion in normal, stable catchments, the chance events of intense wildfire revealed valuable insights into the nature of ecosystems when intensely disturbed, as well as underscoring the counterfactual state against which the stable catchment conditions may be understood.

A second demonstration of the value of long-term observation arose when answers began to emerge on one of the old forestry ideas, that of the 'true forest'. The notion had two aspects. First, the development of a plantation forest ecosystem resembled the process of ecological succession that generated the natural forest. In the case of the eucalyptus plantation, if the stand is thinned and grown on a long rotation, 'then true forest conditions are approached', indigenous forest species invade the plantation as the eucalypts are 'not making excessive demands on water' (Wicht here revives arguments expressed during the 1935 British Empire Forestry Conference).[45] Wicht argued that in 'true forestry sites' mature eucalyptus plantations would be unlikely to use more water than indigenous forests. Other observers went further, arguing that 'true forest' conditions would favour water supplies; for example, Keet in his comments on Jessievale wanted analysis to wait until 'the plantations have reached an older stage and forest-like conditions are attained'.[46] The White River Valley Farmers' Association wanted temporary relief from the perceived transient loss of water supply, until the plantations had formed the 'requisite water conserving mulch'.[47]

The second, related aspect was the benefit expected from afforestation of degraded sites; the landscapes around Sabie and the forestry sites to the north, with vulnerable granite soils, had been badly eroded and incised by *dongas* from the effects of gold mining, the exploitation of forests and open woodlands for fuel and mine supports, as well as seasonal stock grazing tied with veld burning, and the Sabie River ran red with sediment during the rains: E. B. Glaeser describes how in 1908 when approaching Graskop to take up his post as state forester, he was confronted by the 'appalling' sight of *dongas*, a 'huge waste' 'exposing the

---

45   Wicht, *Forestry and Water Supplies in South Africa*, 47. Wicht illustrates the invasion by forest species for two cases.
46   J. D. M. Keet, Conservator of Forests Transvaal and OFS 8 December 1923 to Chief Conservator of Forests C. E. Legat on R7649/711 SAFRI Archives, CSIR, Pretoria; Keet wrote of attainment of 'proper forest conditions' in plantations, when plantations 'cannot affect water supplies other than beneficially': J. D. M. Keet memorandum to Chief Forest Research Officer *'Invloed van Bos op Stand van Water'*, 21 October 1933, Wicht Papers, SAFRI Archives, CSIR, Pretoria.
47   White River Valley Farmers' Association to Deneys Reitz, Minister for Agriculture, 18 May 1934, Wicht Papers, SAFRI Archives, CSIR, Pretoria.

bare red subsoil on which nothing grew'.[48] The government justified the return of such land to the state and the Forestry Department in the 1930s, claiming that afforestation would rehabilitate the catchments:

> the acquisition of all farms on the eastern escarpment from Mariepskop to Witwater for the protection of indigenous forests and conservation of their water resources in the interests of properties in the Lowveld and the game reserve (Kruger National Park): conservation measures were to include … re-afforestation of destroyed forests, afforestation of the open ground and the prevention of grazing on the forest reserves; reclamation of dongas and eroded land; and prevention of further damage to watercourse land. With one exception the properties were acquired by purchase and the work was begun.[49]

Fifteen years later the department expressed in its annual report its satisfaction with progress: *donga* reclamation in the Sabie area and rehabilitation of the indigenous forests had effected the 'remarkable recovery of catchment kloofs' 'with streams running all year'.[50] These ideas are close to George Perkins Marsh's picture of forest hydrology: 'The vegetable mould, resulting from the decomposition of leaves and wood, carpets the ground with a spongy covering which obstructs the evaporation from the mineral earth below, drinks up the rains … that would otherwise flow rapidly over the surface … and then slowly gives out, by evaporation, infiltration, and percolation, the moisture thus imbibed'.[51]

These were intuitions without scientific justification. Recently, afforestation experiments that had been maintained beyond the normal cycle of industrial forestry afforded the opportunity to test for long-term hydrological effects: once the trees reached maturity, water use declined, even to the point where original flows were recovered. David Scott and Eric Prinsloo used paired-catchment analyses of Bosboukloof at Jonkershoek, involving afforestation with *Pinus radiata* that had been maintained for 43 years, and another with *Eucalyptus grandis* at Westfalia, maintained for 21 years. In both cases, the stands had had successive thinnings to a density suited for sawlog production, and both were on sites that Wicht classified as 'true forest' sites. They found that the sharp decreases in streamflow that result from rapid establishment and growth of plantation trees reached a maximum and then reversed, to levels that prevailed before afforestation, or nearly so.

---

48   E. B. Glaeser in Olivier, *There Is Honey in the Forest*, 159–60. Olivier excerpts the interview recorded with Glaeser by Winkler of the then Sabie Forestry Museum (text in the documents of the Sabie Forestry Museum at Lydenburg Museum, Mashishing).

49   Keet, 'Historical Review of the Development of Forestry in South Africa', 83

50   Keet, 'Historical Review of the Development of Forestry in South Africa', 110.

51   G. P. Marsh, *Man and Nature*, David Lowenthal (ed.) (Cambridge, MA: The Belknap Press of Harvard University, [1864] 1965), 145.

The *Pinus radiata* plantation took six years to significantly reduce streamflows and maximum reductions in flows were evident between the tenth and twentieth years after planting. However, at around 20 years streamflow reductions diminished, and by 45 years the streamflows approached the initial condition. Maximum water use in the faster-growing *Eucalyptus grandis* plantation was realised between 6 and 14 years, but flows recovered to approach initial flows over the period of 15–21 years after planting. The succession is not unique: a similar finding comes from catchments studies in Mountain Ash forests in Australia, where initial declines in streamflow after tree stands had been killed in fires were later reversed to levels that prevailed under the mature forests.[52] And the results are consistent with new knowledge on the decline in transpiration as the tree, or the whole forest, ages.[53] Scott and Prinsloo concluded that 'trees may have a useful role in catchment restoration provided they are managed on long rotations'.[54] The research provides new insight into what ecologists are now calling novel ecosystems, 'those types of ecosystems containing new combinations of species that arise through human action, environmental change, and the impacts of the deliberate and inadvertent introduction of species from other parts of the world';[55] the forester's intuitions, inferred principally from their observations of the colonisation of the site beneath older plantation stands in 'true forest' environments, begin to carry weight. Such insights may also add new dimensions to what Scott Carroll has called 'conciliation biology', the evolutionary management of such novel ecosystems.[56]

The late bonus of hydrological process studies

The question of the water balance approach was set aside after the early trials at Jonkershoek showed that the terms in the water balance could not each be measured simultaneously with sufficient accuracy. Even 40 years after the start of Jonkershoek, leading policy makers preferred that the program continue to prioritise empirical experiments, with the explanatory fundamental studies as

---

52   K. J. Langford, 'Change in Yield of Water Following a Bushfire in a Forest of *Eucalyptus regnans*', *Journal of Hydrology*, 29 (1976): 87–114; L. Bren, P. Lane, and G. Hepworth, 'Longer-Term Water Use of Native Eucalyptus Forest after Logging and Regeneration: The Coranderrk Experiment', *Journal of Hydrology*, 384 (2010): 52–64, 62.

53   See M. G. Ryan and B. J. Yoder, 'Hydraulic Limitations to Tree Growth', *BioScience*, 47 (1997): 235–42, 240.

54   D. F. Scott and F. W. Prinsloo, 'Longer-Term Effects of Pine and Eucalypt Plantations on Streamflow', *Water Resources Research*, 44 (2008), 7.

55   R. J. Hobbs et al., 'Novel Ecosystems: Theoretical and Management Aspects of the New Ecological World Order', *Global Ecology and Biogeography*, 15 (2006): 1–7.

56   S. P. Carroll, 'Conciliation Biology: The Eco-Evolutionary Management of Permanently Invaded Biotic Systems', *Evolutionary Applications*, 4 (2011): 184–99.

complementary, rather than having priority.[57] But the difficulty of measuring water use by different types of vegetation within the forestry regions of the country, beyond the set included in the eight sites of catchment experiments—grassland, fynbos, and plantation forests—remained, and required some form of direct measurement beyond the experimental catchment.

The old question of water use by indigenous forests illustrates the difficulty: there were no sites where water balance of plantation catchments could be compared with adjoining catchments under indigenous forests. Technical limitations on early attempts to overcome this through physiological studies, by measuring water loss from cut branches, did not deliver acceptable results. John Phillips hinted at experiments in the 1920s to examine the water physiology of exotic and indigenous trees,[58] but it was in the 1940s that Margaret Henrici first reported on a systematic analysis of indigenous tree and shrub water physiology using the cut-shoot method.[59] This work received severe criticism from Wicht and others,[60] and no reasonable answer emerged for the next 30–40 years, awaiting new ideas and new techniques.

The approach that led to answers emerged around 1980. A way of measuring water use in the field was needed, which would allow estimates for the whole plant or the entire vegetation cover. This became possible only with the development of sophisticated electronic data systems and environmental sensors from the mid-1970s onward. The first study of transpiration and photosynthesis in the field, conducted in the late 1970s, was on species of *Protea* in the fynbos.[61] Conducted through a team of physiologists at Stanford University, it introduced the new technology of porometry to South Africa, and with this, the demanding technical skills to execute the work. Soon afterward, this technique was in use to study *Eucalyptus grandis*. From a tall tower built to give access to the crowns of the trees, researchers measured transpiration and photosynthesis on a sample of leaves attached to the trees, and then projected these values to estimate the water losses from the whole canopy. This then opened the way to validating and improving techniques of measuring transpiration as the flow of water up the stem of the tree, using the co-called sap-flow velocity technique. In turn, this

---

57   Secretary for Water Affairs J. P. Kriel to C. L. Wicht, 12 July 1974, Wicht Papers, SAFRI Archives, CSIR, Pretoria and digi.nrf.ac.za/dspace/bitstream/handle/10624/400/CLW_Brikri1974001.pdf?sequence=1 (accessed 9 August 2014).

58   *Proceedings of the Fourth British Empire Forestry Conference*, 126.

59   M. Henrici, *The Transpiration of South African Plant Associations – Part III: Indigenous and Exotic Trees in the Drakensberg Area* (Pretoria: Union of South Africa Department of Agriculture, Government Printer, 1945).

60   Wicht, 'Review "Transpiration of South African Plant Associations by M Henrici"', *Journal of South African Botany*, 1947; Wicht, *Forestry and Water Supplies in South Africa*, 31–4.

61   H. A. Mooney et al., 'Photosynthetic Characteristics of South African Sclerophylls', *Oecologia*, 58 (1983): 398–401.

allowed in-field estimates of tree growth and water use, and direct estimates of water-use efficiency,[62] with the results appearing in a major series of scientific reports.[63]

The availability of techniques to measure transpiration in the plant and to scale up to the canopy of the vegetation augmented measurements in experimental catchments and together these allowed the validation of techniques to measure evapotranspiration across the whole landscape. This created the capability to measure evapotranspiration directly without having to set up new catchment experiments, using the available gauged catchments as essential benchmarks. Introduction of methods of measuring evapotranspiration above the vegetation ran in parallel to techniques of measurement at the level of the leaf and the whole tree. This project began through linkages with scientists at CSIRO in Australia, and through the trial and adoption of apparatus to measure the Bowen Ratio above grasslands, following a period one of us (FJK) spent on sabbatical in Australia in 1974–1975. Later, Peter Dye and Colin Everson collaborating with university colleagues, introduced the eddy-correlation and the scintillometry[64] technologies, so completing a set of techniques required for the direct measurement of evapotranspiration over extensive areas of vegetation of different types—after trial and calibration on the gauged catchments, which included the gaining of important new technical and scientific skills.[65]

---

62   Water-use efficiency here is defined as the unit mass of wood produced per unit mass of water consumed.

63   P. J. Dye and B. W. Olbrich, 'Estimating Transpiration from 6-Year-Old *Eucalyptus grandis* Trees: Development of a Canopy Conductance Model and Comparison with Independent Sap Flux Measurements', *Plant Cell Environment*, 16 (1993): 45–53; P. J. Dye, 'Response of *Eucalyptus grandis* Trees to Soil Water Deficits', *Tree Physiology*, 16 (1996): 233–8; P. J. Dye, S. Soko, and A.G. Poulter, 'Evaluation of the Heat Pulse Velocity Method for Measuring Sap Flow in *Pinus patula*', *Journal of Experimental Botany*, 47 (1996): 975–81; P. J. Dye, 'Water-Use Efficiency in South African Eucalyptus Plantations: A Review', *South African Forestry Journal*, 189 (2000): 17–26; M. B. Gush and P. J. Dye, 'Water-Use Efficiency within a Selection of Indigenous and Exotic Tree Species in South Africa as Determined Using Sap Flow and Biomass Measurements', *Acta Horticulturae*, 846 (2009): 323–30; B. W. Olbrich et al., 'Variation in Water Use Efficiency and $\partial13C$ Levels in *Eucalyptus grandis* Clones', *Journal of Hydrology*, 150 (1993): 615–33; J. M. Campion, P. J. Dye, and M. C. Scholes, 'Modelling Maximum Canopy Conductance and Transpiration in *Eucalyptus grandis* Stands Not Subjected to Soil Water Deficit', *Southern African Forestry Journal*, 202 (2004): 3–11.

64   The Bowen Ratio technique measures the micrometeorological variables required to solve the standard Penman-Monteith equation for the estimation of evapotranspiration over a surface of land; the eddy correlation technique measures by ultrasonic anemometers and an infrared gas analyser the vertical turbulent fluxes of water vapour through the boundary layer of the atmosphere over vegetation, and scintillometry is the measurement of gas exchange, including water vapour, over transects of several kilometres by the observation of scintillation in the atmosphere, that is, the fast small-scale variation in the refection of light by air as its temperature and vapour content varies; all require sophisticated electronic and high-speed data capture systems.

65   M. J. Savage, C. S. Everson, and B. R. Metelerkamp, 'Bowen Ratio Evaporation Measurement in a Remote Montane Grassland: Data Integrity and Fluxes', *Journal of Hydrology*, 376 (2009): 249–60. M. J. Savage et al., 'Measurement of Grassland Evaporation Using a Surface-Layer Scintillometer', *Water SA*, 36 (2010): 1–8.

These developments soon led to direct comparisons of water use by plantation forests as contrasted with natural vegetation, such as grassland, and other crops, such as sugarcane. However, it was only in 2004 that researchers measured evapotranspiration from an indigenous forest. Working with a scintillometer in the Groenkop Forest near George, together with necessary supplementary weather observations, researchers were able in three campaigns to gather direct observations for 18 days across 3.2 km of forest surface, sufficient for the calibration of the evapotranspiration formula and so the calculation of annual water use. Mark Gush concluded from this work that 'the hypothesis that the water-use of the indigenous forest in this study is less than that of an introduced plantation growing under similar conditions can not be conclusively supported': the calculated evapotranspiration amounted to about 1,000 mm per year, while rainfall averaged about the same, so the indigenous forest consumed nearly all the rainfall equivalent.[66]

Mark Gush's findings provided at least an initial answer to the question that troubled researchers for decades, that of whether or not exotic plantation forests use more water than indigenous forests. Of course, the answer depends on how one frames the question: is the evapotranspiration from the plantation greater than that of the indigenous forest? Is the rate of transpiration from an exotic tree greater than that from an indigenous tree? Is the amount of water consumed per unit of growth different between the two types?

Thus far, we have no evidence that, under like conditions, our indigenous forests use less water than plantations of exotic trees, but it is true that the forests as well as the individual trees of the native species, slow growing as they are, use water much less efficiently than do the exotics. Mark Gush found that in the indigenous species he studied, the water-use efficiency averaged 1.1 g wood produced for every litre of water consumed, whereas in the exotic plantations species, the efficiency was 2.6 g per litre, while Russell Wise working with Peter Dye and Mark Gush report generally much higher water-use efficiencies in plantations in South Africa than indigenous forests.[67]

Concluding, the new environmental observation techniques that became available from the 1970s onward permitted scientists to complement the catchment studies and help to explain the findings from those experiments, as well as to extend assessments of vegetation effects beyond the confines of the experimental catchments. These techniques became available not only because of technical innovation, but also because the program entered a period when

---

66   Gush, 'Water-Use, Growth and Water-Use Efficiency of Indigenous Tree Species in a Range of Forest and Woodland Systems in South Africa'.
67   R. M. Wise, P. J. Dye, and M. B. Gush, 'A Comparison of the Biophysical and Economic Water Use Efficiencies of Indigenous and Introduced Forests in South Africa', *Forest Ecology and Management*, 262 (2011): 906–15.

it was relatively free from uncertainties about its organisational home and the availability of financial and professional capacity, from about 1965 to 1995. It was also in this period, the catchments and their treatments having been maintained as planned for up to nearly 50 years, that comprehensive analyses could yield novel insights. The flow of ideas and expertise through a larger network of relationships generated an emergent body of knowledge about the catchment ecosystem. Hydrological behaviour in diverse physiographies and under diverse vegetations was well quantified and explained, exemplified perhaps by the joint work of Jan Bosch and John Hewlett. The long-term ecosystem response to afforestation has become apparent in David Scott's work with his several colleagues, as has the response of the catchment ecosystem to the severe disturbance of intense forest fires. These are findings that would neither have come by inference from 'fundamental' hydrological process studies, nor without a commitment to the long term.

Chapter 10
Devolution, Drift and
New Directions, 1990–2014

The period from 1986 to the present day marked a significant change in South African funding priorities, in national legislation and policy, and in institutions relating to forests and water. The excitement of the 1980s—which saw the creation of a coordinated national policy on catchments, plantations, invasive species, and biodiversity preservation—dissipated remarkably quickly as a result of several factors: shifting priorities in catchment management, for example, to recreation management, and the emerging litigious ethos arising from the changing fire risk profile as new commercial ventures penetrated the mountain landscapes,[1] but especially owing to changes in government from the late 1980s until today. Changing government funding priorities for environmental research and new legislation dramatically changed forest research and the forest industry in South Africa. This affected every aspect of forest policy, from water conservation, plantation policy, indigenous vegetation management, and control of invasive species, and accelerated the demise of the integrated forestry paradigm established 50 years before.

Though fragmentation and drift has been the norm, there is a hopeful renewed commitment to aspects of the long-term research program at Jonkershoek and at Cathedral Peak, and to seeking to create an integrated framework for managing catchments holistically. This arises from the creation of South African Environmental Observation Network (SAEON), which made a small start in 2002 but has quickly grown its program, and which is resuming data collection and analysis at Jonkershoek and Cathedral Peak, as well as a vigorous resumption of lines of research that build on the earlier work. A second development is

---

1   Pooley, *Burning Table Mountain*, 111–2.

the initiative to manage a catchment in KwaZulu-Natal under the *National Environmental Management: Biodiversity Act*, facilitated by the South African National Biodiversity Institute (SANBI) and premised on the idea of maintaining the ecological infrastructure that supports catchment services.

South African policies for forests, water, and biodiversity are at a crossroads. While there is still the institutional and individual capacity to once again create a coordinated framework for forest management, this would require collaboration across disciplines that have become institutionally and intellectually fragmented. Yet this coordinated framework is exactly what is needed. A continuation of the status quo could put South Africa further away from its economic and ecological goals, endangering the prosperity and sustainability of the nation in the future.

1986–2014

The administrations and policies that governed South African forestry from the late 1960s to the late 1980s began to shift in response to changing financial and political pressures and imperatives brought on by the decline of the apartheid regime. The institutions that governed catchment management and afforestation permit approvals changed jurisdictions in the late 1980s and early 1990s. The President's Minute 1109 in November 1986 transferred the management of mountain catchments from the national government to provincial governments in April 1987.[2] This transfer was a response to the erosion of state revenue caused by South Africa's recession, induced in part by international sanctions against the National Party's apartheid regime. Provincial governments were unable to maintain the Forestry Department's highly successful program of fire management, a policy that underpinned water conservation and invasive species control efforts.

Jonkershoek was a bellwether for these wider changes. The 1990s saw the people and the program at Jonkershoek scattered to the wind, owing to government restructuring within CSIR and the Forestry Department. In 1992, SAFRI merged with the Timber Research Institute at CSIR to form Forestek, which effectively continued the work of SAFRI until 1995, when the contract with the government ended. After this Jonkershoek and its satellite sites were funded through competitive government grants and a small funding from CSIR, thanks to former researchers who continued in other roles. SAFRI–Forestek was at the time the largest employer of forestry researchers—80 out of 188 nationally—and it had the largest remit, including the economic, environmental, and technical

---

2   Pooley, 'Recovering the Lost History of Fire in South Africa's Fynbos', 74.

aspects of trees.[3] This reflected declining investment in research, both at a national level and within the Department of Water Affairs and Forestry. The Department of Water Affairs and Forestry provided 13 million rands for research and development in the 1989–1990 budgetary year, but only 8 million in 1994–1995 and 5 million in 1995–1996. Several researchers moved into the academy to become experts in invasive species, fire, botany, tree breeding, biodiversity, and other fields. Others remained within CSIR to lead national research on similar subjects.

After devolving power to provinces in 1987, South Africa had no coordinated program for managing invasive species in catchments during the first half of the 1990s, until the creation of Working for Water in 1995. Working for Water, now one of the environmental programs of the Department of Environmental Affairs, was established to serve the dual purpose of creating employment and controlling invasive species. The political mandate makes the program popular with the ruling African National Congress, but it dilutes the effectiveness of control efforts. Whereas before, trained forestry professionals instituted catchment management which embodied coordinated invasives control and prescribed burning, Working for Water employs unskilled workers on short-term contracts, who do mainly manual labour in controlling invasive species, moving on from one campaign to the next. About 40 per cent of Working for Water's budget goes to labour, which after overheads and operating costs leaves only 2.6 per cent for biological control,[4] which is the increasingly effective means of controlling certain invasive trees. Working for Water, which was predicated on aspects of Jonkershoek's research, chose not to contribute to the continuation of long-term research at Jonkershoek because they felt that the work until then had already yielded sufficient evidence to justify their program.[5]

Administration of afforestation permits was transferred in 1991 from the national forest authority to the national water authority, and then in 1998 afforestation was regulated through provisions in the *National Water Act*.[6] The *National Water Act* directly reflects the Commission of Enquiry into Water Matters' recommendation for provisions for catchment management agencies to regulate water usage in strategic areas. Yet, benefits of this law have been slow in coming, with some areas of implementation 'stalled completely'; for example, as of 2012

---

3   F. J. Kruger, 'Research in Times of Austerity: Experiences in South Africa', in *Proceedings of IUFRO XX World Congress, 6–12 August 1995, Tampere, Finland 'Caring for the Forest: Research in a Changing World'*, www.metla.fi/iufro/iufro95abs/rsp16.htm (accessed 9 August 2014).

4   B. W. van Wilgen and W. J. de Lange, 'The Costs and Benefits of Biological Control of Invasive Alien Plants in South Africa', *African Entomology*, 19 (2011), 504–14, 10.

5   Email, Pat Manders, Acting Executive Director, CSIR, April 2015.

6   Kruger et al., 'The Regulation of Water-Use Impacts of Forestry in South Africa', 5–7.

only two catchment management agencies have been created for South Africa's 19 water management areas gazetted at the time.[7] Afforestation rates fell sharply as a result of new government policies towards plantations.[8]

A central set of provisions in the Act stipulates diverse categories of water use, each requiring a licence. These kinds of use range from abstraction of water from a stream, to altering the channel of a stream, to the discharge of wastewater. One category is afforestation, stipulated as a streamflow reduction activity, which as the law stands now, is the only form of upstream land use determined to be a streamflow reduction activity.[9] Although most, if not all, plantation forests received water-use licences under the provisions for existing lawful use,[10] any application for new afforestation has since 1998 been subject to onerous and costly procedures, in which not only the provisions of the *National Water Act* apply, but also environmental and other legislation.[11] Once licensed, the plantation forest is subject to water-use charges, the only catchment land use subject to this tariff. Since 1999, the area of plantation forests has declined, by about 150,000 hectares, from 1.4 million hectares, and net afforestation has been nil or nearly so; the balance of trade in forest products has deteriorated in parallel.[12] This has been attributed to several factors, but the regulatory regime is one of them, and most frequently cited by both government and forest sector representatives.[13]

Through these provisions the *National Water Act* supplanted the old *Forest Act*, and its successor, the *National Forests Act*. The decentralisation of water-use allocation was to occur through the catchment management agencies, according to their catchment management strategies,[14] and this would substitute for the catchment prioritisation that followed on the 1968 report on afforestation and water supplies; but these agencies (with the exception of two) have yet to manifest. Meanwhile, afforestation licensing is guided by provincial committees, which 'integrate the requirements of interlocking legislation into their

---

7   M. Muller, 'Lessons from South Africa on the Management and Development of Water Resources for Inclusive and Sustainable Growth', 22; Bourblanc and Blanchon, 'The Challenges of Rescaling South African Water Resources Management'.
8   Department of Agriculture and Forestry, *State of the Forests Report 2011*, 9–12; Department of Agriculture, Forestry and Fisheries, *State of the Forests Report 2007–2009*, 9–12.
9   *National Water Act*, section 21; there are suggestions that this provision should also apply to dryland sugar plantations.
10   *National Water Act*, Part 3.
11   Kruger et al., 'The Regulation of Water-Use Impacts of Forestry in South Africa', 5–7.
12   Department of Agriculture and Forestry, *State of the Forests Report 2011*, 26–7.
13   Ibid., 9–12; this report notes that some of the apparent decline in area may have been owing to reporting errors.
14   *National Water Act*, Chapter 7.

procedures',[15] but the work has become mired in bureaucracy, environmental resistance, and uncertainty, despite national forest-sector development aspirations.[16]

The effects of this legislation have helped to distort land-use and development, and effectively to stop the planting of exotic trees for commercial plantations.[17] The government has initiated a National Afforestation Program in response to the need to expand the timber resource to minimise timber imports and to optimise enterprise development opportunities. Recent projections of timber supply and demand indicated that about 785,000 ha of new afforestation is needed to make up the estimated timber shortfall from domestic sources.[18] However, there has been no further expansion of exotic plantations over the past 15 years,[19] despite the government's commitment to empowerment of emergent players through the Forest Sector Code, which stipulates that the government would expedite afforestation licensing procedures to achieve new afforestation of 100,000 ha by black participants at 10,000 ha per annum.[20]

There are several reasons for the lack of investment. The regulation required to afforest an area under the *National Water Act* has become enmeshed in often complex and difficult bureaucracy, described by Mike Muller and his colleagues as 'administrative challenges'.[21] Existing consultative processes for catchment management admit a large number of stakeholders, and can allow the loudest voices to win. There is a popular idea that the plantation forests have developed 'down an environment-destroying path' using 'water-hungry alien species'[22]

15   Department of Water Affairs and Forestry, *Water-Use Licensing: The Policy and Procedure for Licensing Stream Flow Reduction Activities* (Pretoria: Department of Water Affairs and Forestry, 1999), www.dwaf.gov.za/SFRA/Licensing/pdf/Policy%20&%20Procedure%20on%20Water-use%20Licensing.pdf; I.J. van der Walt, A. Struwig, and J.R.J. van Rensburg, 'Forestry as a Streamflow Reduction Activity in South Africa: Discussion and Evaluation of the Proposed Procedure for the Assessment of Afforestation Permit Applications in Terms of Water Sustainability', *GeoJournal*, 61 (2004): 178–9.

16   Dye and Versfeld, 'Managing the Hydrological Impacts of South African Plantation Forests: An Overview', 127; Kruger et al., 'The Regulation of Water-Use Impacts of Forestry in South Africa', 41; Department of Agriculture, Forestry and Fisheries, *Forestry 2030 Roadmap (Forestry Strategy 2009–2030)*,16–7, 19.

17   See, for example, Dye and Versfeld, 'Managing the Hydrological Impacts of South African Plantation Forests', 127; Jacobson, 'Wood vs. Water', 31–5.

18   Department of Water Affairs and Forestry, *Annual Report* 2007/2008, www.dwaf.gov.za/documents/AnnualReports/ANNUALREPORT2007-2008.pdf (accessed 9 November 2013), 77.

19   Department of Agriculture and Forestry, *State of the Forests Report 2011*, 11.

20   The Code is law under South Africa's *Broad-Based Black Economic Empowerment Act* (no. 53) of 2003, at www.thedti.gov.za/economic_empowerment/docs/BEE-SECTOR_CHARTERS/FORESTRY/forestry_1.pdf (accessed 2 October 2013), 20.

21   Muller et al., 'Water Security in South Africa', 33–4.

22   D. T. Tewari, 'Is Commercial Forestry Sustainable in South Africa? The Changing Institutional and Policy Needs', *Forest Policy and Economics*, 2 (2001): 333–53, 335, 337.

or 'greedy' water users, which shapes local and national opinion.[23] This overall shift in perceptions has led to a 'neutral to negative' view of popular perceptions of forestry nationally in South Africa.[24]

Because of this haltingly implemented new regime, Keet's integrated forestry paradigm, the historic policy system that sought to harmonise land use and water conservation by balancing afforestation with the needs of other sectors, biodiversity and water catchment conservation, has fallen away. In effect, afforestation has been stopped for water conservation purposes in spite of the fact that afforestation effects account for just 2.7 per cent of the country's mean annual run-off[25] and the country has a negative forest products trade balance.[26] The concept of catchment management, as defined in the *National Water Act*, focuses on 'the protection, use, development, conservation, management and control of water resources within its water management area', with 'water resource' defined as including 'a watercourse, surface water, estuary, or aquifer'.[27] While the government admits to the confusion in policy arising from 'use of different terms regarding water resources management contained in the various policies, acts and guidelines', it recognises that though 'strictly speaking the [catchment management agency] is in the business of [integrated water resources management], this cannot be achieved without consideration for land- and water-based activities that impact on the resource base'.[28] But notwithstanding this, examination of the catchment management strategies, for example in the *Water Resources Assessment for the Usutu-Mhlathuze Catchment*, reveals at best only passing attention to 'land- and water-based activities that impact on the resource base'.[29]

23   See, for example, the statement that tree plantations 'squander our precious water'; www.geasphere. co.za/southafrica.htm#AreAllTreesGreenThespotlightonforestry (accessed 10 March 2013).

24   Department of Water Affairs and Forestry, *Annual Report, 1 April 2008 to 31 March 2009* (R.P. 163/2009), 39. Available at www.dwaf.gov.za/documents/AnnualReports/ANNUALREPORT2008-2009.pdf (accessed 13 March 2013); see also Department of Agriculture, Forestry and Fisheries, *Forestry 2030 Roadmap*, 16.

25   Department of Agriculture, Forestry and Fisheries, *Forestry 2030 Roadmap*, 6.

26   Ibid., 16–7.

27   *National Water Act*, Part 2; section 1(1), xxvii.

28   Department of Water Affairs and Forestry, *Introduction and Orientation Guidelines for the development of Catchment Management Strategies in South Africa*, www.dwaf.gov.za/Documents/Other/CMA/CMSFeb07/ CMSFeb07Ed1Ch1.pdf (accessed 16 October 2013), 4.

29   Department of Water Affairs and Forestry, *Overview of the Water Resources of the Usutu-Mhlathuze Water Management Area* (Pretoria: Republic of South Africa, 2002), www.dwaf.gov.za/sfra/SEA/usutu-mhlathuze%20wma/Hydro-Economic%20Component/Overview%20of%20water%20resources%20of%20 the%20U-M%20WMA.pdf.

Future prospects?

There are hopeful signs that policy makers are coming to grips with issues similar to those faced by Wicht and researchers at Jonkershoek. Jonkershoek's recording weirs were kept alive through CSIR funding through the efforts of former SAFRI researchers who transferred to CSIR in 1992 and continued to work there, and this function has now been assumed by South African Environmental Observation Network (SAEON). This means that there is still continual data recording at the site from the mid-1930s.

The issue of long-term research came to the forefront of South African scientific discussion in the late 1990s as a result of renewed global interest in measuring anthropogenic climate change. In 1999, South African participants met at Skukuza in the Kruger National Park to discuss prompts by the International Long-Term Ecological Research Network to establish a research network in South Africa.[30] The network's focus was broad—environmental observation—but it had particular emphasis on climate change and ecology.[31] In 2002, the then Department of Arts, Culture, Science and Technology, through the National Research Council, provided pilot funding of 21.8 million rand for the SAEON, a network of institutions, agencies, and researchers across South Africa.[32]

In 2007, CSIR approached the new SAEON for assistance in maintaining Jonkershoek. SAEON provided help on a limited scale because SAEON at that time was still in the process of being established. In 2009 SAEON become custodians of the data and site, but it was only in 2011 when the stations were handed over to SAEON, who rehabilitated the network where necessary and upgraded the apparatus, and in 2012 SAEON proceeded with the comprehensive curation and digitisation of the old streamflow and weather records; for the field program, SAEON has support from the provincial nature conservation agency as well as the forestry company that now operates the plantations. SAEON intends resuming some of the gauging stations at Jakkalsrivier. Meanwhile, the Department of Water Affairs refurbished the weirs at Zachariashoek and has resumed streamflow and rainfall monitoring there.[33]

---

30   H. C. Biggs, G. I. H. Kerley, and T. Tshiguvho, 'A South African Long-Term Ecological Research Network: A First for Africa?', *South African Journal of Science*, 95 (1999): 244–6.

31   G. F. Midgley, S. L. Chown, and B. S. Kgope, 'Monitoring Effects of Anthropogenic Climate Change on Ecosystems: A Role for Systematic Ecological Observation?', *South African Journal of Science*, 103 (2007): 282–6.

32   A. S. van Jaarsveld, et al., 'South African Environmental Observation Network: Vision, Design and Status', *South African Journal of Science*, 103 (2007): 289–94, 289.

33   Email, Nicky Allsopp, SAEON, 9 October 2013; see also www.saeon.ac.za/enewsletter/archives/2010/june2010/doc07/ (accessed 8 November 2013).

SAEON is also in the process of regenerating part of the original Cathedral Peak site. It is extending its upper catchment streamflow and weather monitoring program to incorporate four of the original streamflow stations, with perhaps another two to be added, while the weather station and the original raingauging network are being re-instrumented. This has involved the major task of repairing the access road through Mike's Pass, named after Mike de Villiers whose first task it was to align and build this route in 1939. Funding has come from the National Research Foundation's Strategic Research Infrastructure Grant and a Water Research Commission contract research grant, and the initiative has the support of the provincial nature conservation agency and Working for Water.[34]

SAEON's goals at these sites are broad, stated as 'to develop unique, well instrumented, integrated monitoring platforms, yielding critical data relevant to global change science that address key policy and development challenges, through collaboration'. There is a strong emphasis on a program of networked collaboration, across a wide range of disciplines. Continuity with the forest hydrology program has been promoted by engaging with researchers from that program where possible, and the digitisation and proper curation of prior records of all kinds. Although experimental interventions, such as afforestation, are not within the scope of the new program, everything indicates that a potentially effective successor to Wicht's original concept is being created.

There is now an early initiative to recover an ecosystems approach to catchment management, not through the *National Water Act*, but through the *National Environmental Management: Biodiversity Act,* adopting an inclusive scope—extending beyond the mountain catchments, to the entire system down to the sea. The initiative centres on the catchment of the uMngeni River, the principal supply to the metropolitan municipality of Ethekwini, which includes the city of Durban, a conurbation housing 3.4 million people. The uMngeni catchment has been the subject of a long series of systematic assessments of water supply and demand, intended to support catchment management, and including the needs for plantation forests.[35]

The intention of the proposal is to institute the 'management and restoration of ecological infrastructure' in the catchment and by 'ecological infrastructure' is meant the ecological conditions in the catchment that ensure its proper function. This would serve to 'enhance the efficiency of water service delivery through improving water quality, reducing sediment loads, reducing flood risk,

34   www.saeon.ac.za/enewsletter/archives/2013/august2013/doc01 (accessed 8 November 2013).

35   Nänni, 'Trees, Water and Perspective'; D. W. van der Zel, 'Umgeni River Systems Analysis', *Water SA*, 1 (1975): 70–5; Department of Water Affairs and Forestry, *Internal Strategic Perspective: Mvoti to Mzimkulu Water Management Area Version 1: June 2004,* 2004, www.dwaf.gov.za/Documents/Other/WMA/11/ optimised/MVOTI%20TO%20MZIMKULU%20REPORT.pdf (accessed 8 July 2014).

and increasing yield through increased winter baseflows'.[36] Measures intended include the removal of 30,000 ha of alien vegetation, improved rangeland practices (about 74 per cent of the catchment remains under natural vegetation), and rehabilitation of wetlands, as well as other measures aimed at mitigating the deteriorated water quality. The intervention is to be instituted through a partnership including relevant national, provincial, and local government entities and the South African National Biodiversity Institute, as well as Working for Water and Working for Wetlands, the appropriate entities in the natural resources component of the Extended Public Works Program. This initiative may prove to open the way forward toward a recovery of the ecosystems management approach to catchments.

There remains the question, however, of what the right institutions are for catchment management. Previously, mountain catchments were recognised as physiographically distinct zones, as in the Ross Report. This distinction reflects the characteristic hydrology of mountain catchments (Chapter 7). Further, we have seen that the mountain catchments, hydrologically deep and porous, constitute systems that, under either natural vegetation or plantations, assure a streamflow regime in which spateflows are a very small fraction of total annual streamflows—the catchments yield well-regulated flows—of clean water, with negligible rates of erosion. But the catchment research has also shown that this characteristic hydrology, as stable as it may be otherwise, may be easily lost under extreme disturbance, such as the intense, soil-disrupting wildfires reported by David Scott and his colleagues (Chapter 8). Finally, the mountain catchment zones are the locus of low-revenue land uses—conservation, plantation forestry, or pasturage—and a management regime distinguished by strict cost control is essential. These considerations suggest that mountain catchment management should be institutionally distinct from downstream management, such as in the uMngeni; that present-day conservation agencies should have a stronger remit for catchment management in the uplands of their provinces; and that the implementation of the *Mountain Catchment Areas Act* should be revitalised.

This in turn may require a review and restructuring of the organisations and institutions needed for the effective management of the upper catchments, as part of the 'rescaling' of resource (not just water) governance to match organisations and institutions more closely to catchment tenure, physiography, and ecology. An innovative resource-management dispensation would also need to embrace a holistic approach to the legacy effects of plantation forest development, where costs to catchment protection have intensified over time as plantation species

---

36   Email, Kristal Maze, Chief Executive Biodiversity Planning, South African National Biodiversity Institute, Pretoria, 17 October 2013.

have invaded the neighbouring landscapes,[37] and finally yield a proper response to C. L. Wicht's 1949 argument for catchment management being ecosystems management, and resolve the concerns about invasive species he originally discerned as long ago as 1945, in his Royal Society 'Report of the Committee on the Preservation of the South Western Cape'.

As to the approach to research in this field, again, this history may inform the question of the most effective institutional model. South African forest hydrology and catchment ecosystems research emerged from the hard lessons of decades of inquiry, a vigorous dissent from which emerged a strong common purpose, and the thinking and work of a cohort of highly educated, argumentative scientists who pursued this common purpose. Currently, the preferred institutional model in South Africa is the network, a model that in most existing forms (e.g. SAEON) is only loosely linked with natural resource economic sectors, such as forestry. Whether such a loosely structured approach will encourage the required degree of alignment and cross-disciplinary engagement to solve complex environmental and economic problems remains a serious question. The paradigm established from 1935 to the early 1990s offers one example of a model that allowed for the advancement of basic research without losing sight of the broader policy and economic imperatives of scientific research. Such integrated perspectives will be required to develop South Africa's economy in the twenty-first century without causing undue harm to its landscapes and ecosystems.

---

37   For example T. Kraaij, R. M. Cowling, and B. W. van Wilgen, 'Past Approaches and Future Challenges to the Management of Fire and Invasive Alien Plants in the New Garden Route National Park', *South African Journal of Science*, 107 (2011): 15–25.

Selected Bibliography

Andréassian, V., 'Review: Waters and Forests: From Historical Controversy to Scientific Debate', *Journal of Hydrology*, 291 (2004), 1–27.

Anker, P., *Imperial Ecology Environmental Order in the British Empire, 1895–1945* (Cambridge, MA: Harvard University Press, 2002).

Austin, G., 'Resources, Techniques, and Strategies South of the Sahara: Revising the Factor Endowments Perspective on African Economic Development, 1500–2000', *The Economic History Review*, 61 (2008), 587–624, dx.doi.org/10.2307/40057603.

Barton, G., *Empire Forestry and the Origins of Environmentalism* (Cambridge: Cambridge University Press, 2002).

Barton, G. A., and B. M. Bennett, 'Edward Harold Fulcher Swain's Vision of Forest Modernity', *Intellectual History Review*, 21 (2011), 135–50.

Beattie, J., *Empire and Environmental Anxiety: Health, Science, Art and Conservation in South Asia and Australasia, 1800–1920* (Basingstoke: Palgrave Macmillan, 2011).

Beinart, W., 'Conservation and Ideas about Development: A Southern African Exploration, 1900–1960', *Journal of Southern African Studies*, 11 (1984), 52–83.

———, *The Rise of Conservation in South Africa: Settlers, Livestock, and the Environment 1770–1950* (Oxford: Oxford University Press, 2008).

Beinart, W., K. Brown, and D. Gilfoyle, 'Experts and Expertise in Colonial Africa Reconsidered: Science and the Interpenetration of Knowledge', *African Affairs*, 108 (2009), 413–33, dx.doi.org/10.1093/afraf/adp037.

Beinart, W., and K. Middleton, 'Plant Transfers in Historical Perspective: A Review Article', *Environment and History*, 10 (2004), 3–29.

Beinart, W., and L. Wotshela, *Prickly Pear: The Social History of a Plant in the Eastern Cape* (Johannesburg: Wits University Press, 2011).

Bennett, B. M., 'The El Dorado of Forestry: The Eucalyptus in India, South Africa, and Thailand, 1850–2000', *International Review of Social History*, 55 (2010), 27–50.

———, 'A Global History of Australian Trees', *Journal of the History of Biology*, 44 (2011), 125–45.

———, 'An Imperial, National and State Debate: The Rise and near Fall of the Australian Forestry School, 1927–1945', *Environment and History*, 15 (2009), 217–44.

———, 'Model Invasions and the Development of National Concerns Over Invasive Introduced Trees: Insights from South African History', *Biological Invasions*, 16 (2104), 499–512.

———, 'A Networked Approach to the Origins of Forestry Education in India, 1855–1885', in B. M. Bennett and J. M. Hodge (eds), *Science and Empire: Knowledge and Networks of Science Across the British Empire 1800–1970* (Basingstoke: Palgrave Macmillan, 2011), 81–84.

———, *Plantations and Protected Areas: A Global History of Forest Management* (Cambridge, MA: MIT Press, 2015).

Bennett, B. M., and J. M. Hodge, *Science and Empire: Knowledge and Networks of Science across the British Empire, 1800–1970* (Basingstoke: Palgrave Macmillan, 2011).

Bennett, B. M., and F. J. Kruger, 'Ecology, Forestry and the Debate over Exotic Trees in South Africa', *Journal of Historical Geography*, 42 (2013), 100–109.

Biggs, H. C., G. I. H. Kerley, and T. Tshiguvho, 'A South African Long-Term Ecological Research Network: A First for Africa?', *South African Journal of Science*, 95 (1999), 244–46.

Bosch, J. M, and J. D. Hewlett, 'A Review of Catchment Experiments to Determine the Effect of Vegetation Changes on Water Yield and Evapotranspiration', *Journal of Hydrology*, 55 (1982), 3–23.

———, 'Sediment Control in South African Forests and Catchments', *South African Forestry Journal*, 115 (1980), 50–55.

Cajander, A. K., 'Über Walden Typen', *Acta Forestalia Fennica*, 1 (1909).

Cameron, L., and D. Matless, 'Translocal Ecologies: The Norfolk Broads, the "Natural," and the International Phytogeographical Excursion, 1911', *Journal of the History of Biology*, 44 (2011), 15–41.

Carroll, S. P., 'Conciliation Biology: The Eco-Evolutionary Management of Permanently Invaded Biotic Systems', *Evolutionary Applications*, 4 (2011), 184–99.

Carruthers, J., and L. Robin, 'Taxonomic Imperialism in the Battles for Acacia: Identity and Science in South Africa and Australia', *Transactions of the Royal Society of South Africa*, 65 (2010), 48–64.

Clements, F. E., and V. E. Shelford, *Bio-Ecology* (New York: J. Wiley & Sons, Inc., 1939).

Comaroff, J., and J. L. Comaroff, 'Naturing the Nation: Aliens, Apocalypse and the Postcolonial State', *Journal of Southern African Studies*, 27 (2001), 627–51.

Craib, I. J., 'The Place of Thinning in Wattle Silviculture and Its Bearing on the Management of Exotic Conifers', *Zeitschrift für Weltforstwirtshaft*, 1 (1931), 77–108.

Cushman, G. T., 'Humboldtian Science, Creole Meteorology, and the Discovery of Human-Caused Climate Change in South America', *OSIRIS*, 26 (2011), 19–44.

Dargavel, J., *The Zealous Conservator: A Life of Charles Lane Poole* (Perth: University of Western Australia Press, 2008).

Darrow, W. K., *David Ernest Hutchins: A Pioneer in South African Forestry* (Pretoria: Department of Forestry, 1977).

Delius, P., *This Land Belongs to Us: The Pedi Polity, the Boers and the British in the Nineteenth Century Transvaal* (London: Heinemann, 1984).

Denoon, D., *A Grand Illusion: The Failure of Imperial Policy in the Transvaal Colony during the Period of Reconstruction 1900–1905* (London: Longman, 1973).

Dickie, I. A., B. M. Bennett, L. E. Burrows, M. A. Nuñez, D. A. Peltzer, A. Porté, and others, 'Conflicting Values: Ecosystem Services and Invasive Tree Management', *Biological Invasions*, 16 (2014), 705–19, dx.doi.org/10.1007/s10530-013-0609-6.

Dodson, B., 'A Soil Conservation Safari: Hugh Bennett's 1944 Visit to South Africa', *Environment and History*, 11 (2005), 35–53.

Donaldson, J. E., C. Hui, D. M. Richardson, M. P. Robertson, B. L. Webber, and J. R. U. Wilson, 'Invasion Trajectory of Alien Trees: The Role of Introduction Pathway and Planting History', *Global Change Biology*, 20 (2014), 1527–37, dx.doi.org/10.1111/gcb.12486.

Dubow, S., 'Colonial Nationalism, The Milner Kindergarten and the Rise of "South Africanism" 1902–10', *History Workshop Journal*, 43 (1997), 53–86.

———, *A Commonwealth of Knowledge: Science, Sensibility, and White South Africa 1820–2000* (Oxford: Oxford University Press, 2006).

———, 'A Commonwealth of Science: The British Association for the Advancement of Science, 1905–1929', in S. Dubow (ed.), *Science and Society in Southern Africa* (Manchester: Manchester University Press, 2000), pp. 66–99.

———, *Scientific Racism in Modern South Africa* (Cambridge: Cambridge University Press, 1995).

Dunn, P. H., S. C. Barro, W. G. II Wells, M. A. Poth, P. M. Wohlgemuth, and C. G. Colver, *The San Dimas Experimental Forest: 50 Years of Research* (Berkeley, CA: US Department of Agriculture, Pacific Southwest Forest and Range Experiment Station, 1988).

Dye, P. J., and D. B. Versfeld, 'Managing the Hydrological Impacts of South African Plantation Forests: An Overview', *Forest Ecology and Management*, 251 (2007), 121–28.

FAO and Cifor, *Forests and Floods: Drowning in Fiction or Thriving on Facts?*, RAP Publication 2005/03 Forest Perspectives 2 (Bogor: Food and Agriculture Organization of the United Nations, Bangkok, and Center for International Forestry Research, 2005).

Fedorowich, K., 'Anglicisation and the Politicisation of British Immigration to South Africa, 1899–1929', *Journal of Imperial and Commonwealth History*, 19 (1991), 22–46.

Foster, J., *Washed with Sun: Landscape and the Making of White South Africa* (Pittsburgh: University of Pittsburgh Press, 2008).

Freund, Bill, 'South Africa: The Union Years, 1910–1948 – Political and Economic Foundations', in R. Ross, A. K. Mager, and B. Nasson (eds), *The Cambridge History of South Africa Volume II: 1885–1994* (Cambridge: Cambridge University Press, 2011), pp. 211–54.

Giliomee, H., *The Afrikaners: Biography of a People* (Cape Town, South Africa and Charlottesville, VA: Tafelberg Publishers and the University of Virginia Press, 2003).

Giliomee, H., and B. Mbenga, *A New History of South Africa* (Cape Town: Tafelberg, 2007).

Grove, R., 'Early Themes in African Conservation: The Cape in the Nineteenth Century', in R. Grove and D. Anderson (eds), *Conservation in Africa: People, Policies and Practice* (Cambridge: Cambridge University Press, 1987), 21–39.

———, *Green Imperialism: Colonial Expansion, Tropical Island Edens and the Origins of Environmentalism, 1600–1860* (Melbourne: Cambridge University Press, 1996).

———, 'Scotland in South Africa: John Croumbie Brown and the Roots of Settler Environmentalism', in T. Griffiths and L. Robin (eds), *Ecology and Empire: Environmental History of Settler Societies* (Seattle: University of Washington Press, 1997), 139–53.

Grundlingh, A. M., '"God Het Ons Arm Mense Die Houtjies Gegee": Towards a History of the "Poor White" Woodcutters in Southern Cape Forest Area, c. 1900–1939', in R. Morrell (ed.), *White but Poor: Essays on the History of Poor Whites in Southern Africa, 1880–1940* (Pretoria: Unisa Press, 1992), 40–56.

Grut, M., *Forestry and Forest Industry in South Africa* (Cape Town: A. A. Balkema, 1965).

Gush, M. B., 'Water-Use, Growth and Water-Use Efficiency of Indigenous Tree Species in a Range of Forest and Woodland Systems in South Africa' (unpublished PhD thesis, University of Cape Town, 2011).

Hancock, W. K., *Smuts: The Fields of Force, 1919–1950* (Cambridge: Cambridge University Press, 1962).

———, *Smuts: The Sanguine Years 1870–1919. Vol 1* (Cambridge: Cambridge University Press, 1962).

Hewlett, J. D., *The Principles of Forest Hydrology* (Athens, OH: The University of Georgia Press, 1982).

———, 'The Relation of Forests and Forestry to Water Resources. Address to the Annual General Meeting of the South African Institute of Forestry, Stellenbosch, 15 May 1970', *South African Journal of Forestry*, 75 (1970), 4–8.

Hewlett, J. D., and J. M. Bosch, 'The Dependence of Stormflow on Rain Intensity and Vegetal Cover in South Africa', *Journal of Hydrology*, 75 (1984), 365–81

Hewlett, J. D., J. C. Fortson, and G. B. Cunningham, 'The Effect of Rainfall Intensity on Storm Flow and Peak Discharge from Forest Land', *Water Resources Research*, 13 (1977), 259–66.

Hewlett, J. D., and A. R. Hibbert, 'Factors Affecting the Response of Small Watersheds to Precipitation in Humid Areas', in W. E. Lull, and W. H. Sopper (eds), *International Symposium on Forest Hydrology* (Oxford: Pergamon Press, 1967), 279–90.

Hewlett, J. D., H. W. Lull, and K. G. Reinhart, 'In Defense of Experimental Watersheds', *Water Resources Research*, 5 (1969), 306–16.

Hewlett, J. D., and L. Pienaar, 'Design and Analysis of the Catchment Experiment', in E. H. White (ed.), *Symposium on Use of Small Watersheds in Determining Effects of Forest Land Use on Water Quality, May 22 and 23, 1973* (Lexington, KY: University of Kentucky, 1973), 90–100.

Hobbs, R. J., S. Arico, J. Aronson, J. S. Baron, P. Bridgewater, V. A. Cramer, and others, 'Novel Ecosystems: Theoretical and Management Aspects of the New Ecological World Order', *Global Ecology and Biogeography*, 15 (2006), 1–7, dx.doi.org/10.1111/j.1466-822X.2006.00212.x.

Hodge, J. M., *Triumph of the Expert: Agrarian Doctrines of Development and the Legacies of British Colonialism* (Athens, OH: Ohio University Press, 2007).

Hoyt, W. G., and W. B. Langbein, *Floods* (Princeton: Princeton University Press, 1955).

Hoyt, W. G., and H. C. Troxell, 'Forests and Floods', *Transactions of the American Society for Civil Engineers*, 99 (1934), 1–30.

Hursh, C. R., M. D. Hoover, and P. W. Fletcher, 'Studies in the Balanced Water-Economy of Experimental Drainage-Areas', *Transactions of the American Geophysical Union*, 23 (1942), 509–17.

Hyam, R., and P. Henshaw, *The Lion and the Springbok: Britain and South Africa Since the Boer War* (Cambridge: Cambridge University Press, 2003).

Immelman, W. F. E., C. L. Wicht, and D. P. Ackerman (eds), *Our Green Heritage: A Book About Indigenous and Exotic Trees in South Africa, About Trees and Timber in Our Cultural History and About Our Extensive Silvicultural, Forestry and Timber Industries* (Cape Town: Tafelberg, 1973).

Keet, J. D. M., 'Historical Review of the Development of Forestry in South Africa' (Pretoria, 1970) www2.dwaf.gov.za/webapp/resourcecentre/Documents/Publications_And_Media/Keet_Forestry_History_page_41-66.pdf.

————, 'Rainfall and Streamflow at the Cape', *Journal of the South African Forestry Association*, 4 (1940), 15–20.

Kohler, R. E., 'Practice and Place in Twentieth-Century Field Biology: A Comment', *Journal of the History of Biology*, 45 (2012), 579–86.

Kraaij, T., R. M. Cowling, and B. W. van Wilgen, 'Past Approaches and Future Challenges to the Management of Fire and Invasive Alien Plants in the New Garden Route National Park', *South African Journal of Science*, 107 (2011), 15–25, dx.doi.org/10.4102/sajs. v107i9/10.633.

Krige, A. V., 'An Examination of the Tertiary and Quaternary Changes of Sea-Level in South Africa, with Special Stress on the Evidence in Favour of a Recent World-Wide Sinking of Ocean-Level', *Annals of the University of Stellenbosch*, 5 Series A (1927), 81.

Krikler, J., *The Rand Revolt: The 1922 Insurrection and Racial Killing in South Africa* (Johannesburg: Jonathan Ball, 2005).

————, *White Rising: The 1922 Insurrection and Racial Killing in South Africa* (Manchester and New York: Manchester University Press, 2005)

Kruger, F. J., 'Research in Times of Austerity: Experiences in South Africa', in *Proceedings of IUFRO XX World Congress, 6–12 August 1995, Tampere, Finland 'Caring for the Forest: Research in a Changing World'*, 1995, www. metla.fi/iufro/iufro95abs/rsp16.htm.

————, 'Use and Management of Mediterranean-Type Ecosystems in South Africa: Current Problems', in C. E. Conrad, and W. C. Oechel (eds), *Proceedings of the Symposium on Dynamics and Management of Mediterranean-Type Ecosystems; June 22–26, 1981; San Diego, CA* (Berkeley, CA: Pacific Southwest Forest and Range Experiment Station, Forest Service, US Department of Agriculture, 1982), 42–48.

Kruger, F. J., and B. M. Bennett, 'Wood and Water: An Historical Assessment of South Africa's Past and Present Forestry Policies as They Relate to Water Conservation', *Transactions of the Royal Society of South Africa*, 68 (2013), 163–74, dx.doi.org/10.1080/0035919X.2013.833144.

Kruger, F. J., J. Crafford, and A. Ginsburg, 'The Regulation of Water-Use Impacts of Forestry in South Africa: Appraisal of the Development of Policy and Governance', in *Workshop on Forest Governance & Decentralization in Africa 8–11 April 2008, Durban, South Africa* (Bogor, Indonesia: Cifor, 2008), www. cifor.org/publications/pdf_files/events/documentations/durban/papers/ Paper25Krugeretal.pdf.

Kull, C., and H. Rangan, 'Acacia Exchanges: Wattles, Thorn Trees, and the Study of Plant Movements', *Geoforum*, 39 (2008), 1258–72.

Lawes, M., H. A. C. Ealey, C. M. Shackleton, and B. G. S. Geach (eds), *Indigenous Forests and Woodlands in Southern Africa: Policy, People, and Practice* (Pietermaritzburg: University of KwaZulu-Natal Press, 2004).

Lester, A., and D. Lambert (eds), *Colonial Lives across the British Empire: Imperial Careering in the Long Nineteenth Century* (Cambridge: Cambridge University Press, 2006).

Louw, W. J. A., 'General History of the South African Forest Industry: 1975 to 1990', *Southern African Forestry Journal*, 200 (2004), 77–86.

——, 'General History of the South African Forest Industry: 1991 to 2002', *Southern African Forestry Journal*, 201 (2004), 65–76

Von Maltitz, G., L. Mucina, C. J. Geldenhuys, M. Lawes, H. Eeley, H. Adie, and others, *Classification System for South African Indigenous Forests: An Objective Classification for the Department of Water Affairs and Forestry*, Environmentek Report (Pretoria: CSIR, 2003).

Marks, S., 'War and Union, 1899–1910', in R. Ross, A. K. Mager, and B. Nasson (eds), *The Cambridge History of South Africa Volume II: 1885–1994* (Cambridge: Cambridge University Press, 2011), 157–210.

Marks, S., and S. Trapido, 'Lord Milner and the State Reconsidered', in M. Twaddle (ed.), *Imperialism, the State and the Third World* (London: British Academic Press, 1992), 80–94.

Maylam, P., 'The Rand Revolt', *Journal of African History*, 47 (2006), 341–43.

McCracken, D. P., 'Dependence, Destruction and Development: A History of Indigenous Timber Use in South Africa', in M. Lawes, H. A. C. Ealey, C. M. Shackleton, and, B. G. S. Geach (eds), *Indigenous Forests and Woodlands in Southern Africa: Policy, People, and Practice* (Pietermaritzburg: University of KwaZulu-Natal Press, 2004), 277–308.

——, *Gardens of Empire: Botanical Institutions of the Victorian British Empire* (London and Washington DC: Leicester University Press, 1997).

——, 'The Indigenous Forests of Colonial Natal and Zululand', *Natalia*, 16 (1986), 19–38.

McDonnell, J. J., 'Classics in Physical Geography Revisited: Hewlett, J. D. and Hibbert, A. R. 1967: Factors Affecting the Response of Small Watersheds to Precipitation in Humid Areas. In Sopper, W. E. and Lull, H. W., Editors, Forest Hydrology, New York: Pergamon Press, 275–90', *Progress in Physical Geography*, 33 (2009), 288–93.

McGlynn, B. L., and J. J. McDonnell, 'Quantifying the Relative Contributions of Riparian and Hillslope Zones to Catchment Runoff', *Water Resources Research*, 39 (2003), dx.doi.org/10.1029/2003WR002091.

McGuire, K. J., and G. E. Likens, 'Historical Roots of Forest Hydrology and Biogeochemistry', in D. F. Levia, D. Carlyle-Moses, and T. Tanaki (eds), *Forest Hydrology and Biogeochemistry: Synthesis of Past Research and Future Directions* (Berlin: Springer-Verlag, 2011), 3–26.

Meredith, M., *Diamonds, Gold, and War: The British, the Boers, and the Making of South Africa* (London: PublicAffairs, 2007).

Midgley, G. F., S. L. Chown, and B. S. Kgope, 'Monitoring Effects of Anthropogenic Climate Change on Ecosystems: A Role for Systematic Ecological Observation?', *South African Journal of Science*, 103 (2007), 282–86.

Mucina, L., and M. C. Rutherford, *The Vegetation of South Africa, Lesotho, and Swaziland*, Strelitzia, 19 (Pretoria: South African National Biodiversity Institute, 2006).

Muller, M., 'Lessons from South Africa on the Management and Development of Water Resources for Inclusive and Sustainable Growth', in *European Report on Development 2011/2012: Confronting Scarcity: Managing Water, Energy and Land for Inclusive and Sustainable Growth*, 2012, 43, www.die-gdi.de/fileadmin/user_upload/pdfs/dauerthemen_spezial/European-Commission_European-Report-on-Development-2011-2012.pdf.

Muller, M., B. Schreiner, L. Smith, B. van Koppen, H. Sally, and M. Aliber, *Water Security in South Africa*, Working Paper Series No. 12 (Midrand: Development Planning Division Development Bank of Southern Africa, 2009).

Niehaus, I., *Witchcraft, Power and Politics: Exploring the Occult in the South African Lowveld* (Cape Town, London, and Sterling, VA: David Philip and Pluto Press, 2001).

Olivier, W., *There Is Honey in the Forest: The History of South African Forestry*, 2nd edn (Pretoria: Southern African Institute of Forestry, 2010).

Oosthoek, J., *Conquering the Highlands: A History of the Afforestation of the Scottish Uplands* (Canberra: ANU E Press, 2013).

Packard, R., '"Malaria Blocks Development" Revisited: The Role of Disease in the History of Agricultural Development in the Eastern and Northern Transvaal Lowveld, 1890–1960', *Journal of Southern African Studies*, 27 (2001), 591–612.

Pakenham, T., *The Boer War* (Weidenfeld and Nicolson, 1997).

Palmer, R., H. Timmermans, and D. Fay, *From Conflict to Negotiation: Nature-Based Development on South Africa's Wild Coast* (Pretoria: Human Sciences Research Council, 2002).

Phoofolo, P., 'Face to Face with Famine: The BaSotho and the Rinderpest, 1897–1899', *Journal of Southern African Studies*, 29 (2003), 503–27.

Pooley, S., *Burning Table Mountain: An Environmental History of Fire on the Cape Peninsula* (Basingstoke: Palgrave Macmillan, 2014).

———, 'Jan van Riebeeck as Pioneering Explorer and Conservator of Natural Resources at the Cape of Good Hope (1652–62)', *Environment and History*, 15 (2009), 3–8.

———, 'Pressed Flowers: Notions of Indigenous and Alien Vegetation in South Africa's Western Cape, c. 1902–1945', *Journal of Southern African Studies*, 36 (2010), 599–618.

———, 'Recovering the Lost History of Fire in South Africa's Fynbos, c.1910–90', *Environmental History*, 17 (2012), 55–83, dx.doi.org/10.1080/03057070.2010.507565.

Porter, A. N., *The Origins of the South African War: Joseph Chamberlain and the Diplomacy of Imperialism, 1895–99* (New York: St Martin's, 1980).

Powell, J. M., *An Historical Geography of Modern Australia* (Melbourne: Cambridge University Press, 1988).

Radkau, J., *Nature and Power: A Global History of the Environment* (New York: Cambridge University Press, 2008).

Reekie, W. D., 'The Wood from the Trees: Ex Libri Ad Historiam Pertinentes Cognoscere', *South African Journal of Economic History*, 19 (2004), 67–99.

Richardson, D. M., 'Forestry Trees as Invasive Aliens', *Conservation Biology*, 12 (1998), 18–26.

Richardson, D. M., P. Pyšek, M. Rejmánek3, M. G. Barbour, F. D. Panetta, and C. J. West, 'Naturalization and Invasion of Alien Plants: Concepts and Definitions', *Diversity and Distributions*, 6 (2000), 93–107.

Righter, R. W., *The Battle over Hetch Hetchy: America's Most Controversial Dam and the Birth of Modern Environmentalism* (New York: Oxford University Press, 2005).

Roach, T. R., 'The White Labour Forest Settlement Programme in South Africa 1917–1938' (unpublished MA thesis, Witwatersrand, 1989).

Roche, M., 'Colonial Forestry at Its Limits: The Latter Day Career of Sir David Hutchins in New Zealand 1915–1920', *Environment and History*, 16 (2010), 431–54.

———, 'Forestry as Imperial Careering New Zealand as the End and Edge of Empire in the 1920s–40s', *New Zealand Geographer*, 68 (2012), 201–10.

———, 'The New Zealand Timber Economy 1840 to 1935', *Journal of Historical Geography*, 16 (1990), 295–313.

———, 'The State as Conservationist, 1920–60: "Wise Use" of Forests, Lands and Water', in T. Brooking and E. Pawson (eds), *Environmental Histories of New Zealand* (Melbourne: Oxford University Press, 2002), 183–99.

Roche, M., and J. Dargavel, 'Imperial Ethos, Dominions Reality: Forestry Education in New Zealand and Australia, 1910–1965', *Environment and History*, 14 (2008), 529–31.

Schirmer, S., 'Enterprise and Exploitation in the 20th Century', in P. Delius (ed.), *Mpumalanga History and Heritage* (Pietermaritzburg: University of KwaZulu-Natal Press, 2007), 522.

Scott, D. F., 'The Contrasting Effects of Wildfire and Clearfelling on the Hydrology of a Small Catchment', *Hydrological Processes*, 11 (1997), 543–55.

———, 'The Hydrological Effects of Fire in South African Mountain Catchments', *Journal of Hydrology*, 150 (1993), 409–32, dx.doi.org/10.1016/0022-1694(93)90119-T.

———, 'Managing Riparian Zone Vegetation to Sustain Streamflow: Results of Paired Catchment Experiments in South Africa', *Canadian Journal of Forest Research*, 1 (1999), 1149–57.

Scott, D. F., and F. W. Prinsloo, 'Longer-Term Effects of Pine and Eucalypt Plantations on Streamflow', *Water Resources Research*, 44 (2008), dx.doi.org/10.1029/2007WR006781.

Scott, D. F., F. W. Prinsloo, G. Moses, M. Mehlamakulu, and A. D. A. Simmers, *A Re-Analysis of the South African Catchment Afforestation Experimental Data* (Pretoria: Water Research Commission, 2001), www.wrc.org.za/Knowledge%20Hub%20Documents/Research%20Reports/810-1-00.pdf.

Scott, D. F., and R. E. Smith, 'Preliminary Empirical Models to Predict Reductions in Annual and Low Flows Resulting from Afforestation', *Water SA*, 23 (1997), 135–40.

Scott, D. F., and D. B. van Wyk, 'The Effects of Fire on Soil Water Repellency, Catchment Sediment Yields and Streamflow', in B. W. van Wilgen, D. M. Richardson, F. J. Kruger, and H. J. van Hensbergen (eds), *Fire in South African Mountain Fynbos*, Ecological Studies 93 (Berlin and Heidelberg: Springer, 1992), 216–39, dx.doi.org/10.1007/978-3-642-76174-4_12.

————, 'The Effects of Wildfire on Soil Wettability and Hydrological Behaviour of an Afforested Catchment', *Journal of Hydrology*, 121 (1990), 239–56.

Shaughnessy, G. L., 'Historical Ecology of Alien Woody Plants in the Vicinity of Cape Town, South Africa' (unpublished PhD thesis, Cape Town, 1980).

Showers, K., 'From Forestry to Soil Conservation: British Tree Management in Lesotho's Grassland Ecosystem', *Conservation and Society*, 4 (2006), 1–35.

————, *Imperial Gullies: Soil Erosion and Conservation in Lesotho* (Athens, OH: Ohio University Press, 2005).

————, 'Prehistory of Southern African Forestry: From Vegetable Garden to Tree Plantation', *Environment and History*, 16 (2010), 295–322.

Sundnes, F., 'Scrubs and Squatters: The Coming of the Dukuduku Forest, an Indigenous Forest in KwaZulu-Natal, South Africa', *Environmental History*, 18 (2013), 277–308.

Tewari, D. T., 'Is Commercial Forestry Sustainable in South Africa? The Changing Institutional and Policy Needs', *Forest Policy and Economics*, 2 (2001), 333–53.

Thompson, J. L., *Forgotten Patriot: A Life of Alfred, Viscount Milner of St. James's and Cape Town* (Madison, NJ: Fairleigh Dickinson University Press, 2007).

Tomlinson, B. R., 'Empire of the Dandelion: Ecological Imperialism and Economic Expansion 1860–1914', *Journal of Imperial and Commonwealth History*, 26 (1988), 84–99.

Torrance, D. E., *The Strange Death of the Liberal Empire: Lord Selborne in South Africa* (Montreal: McGill-Queen's University Press, 1996).

Tropp, J., 'Displaced People, Replaced Narratives: Forest Conflicts and Historical Perspectives in the Tsolo District, Transkei*', *Journal of Southern African Studies*, 29 (2003), 207–33, dx.doi.org/10.1080/0305707032000060458.

———, *Natures of Colonial Change in the Making of the Transkei* (Athens, OH: Ohio University Press, 2006).

Turton, A. R., R. Meissner, P. M. Mampane, and O. Seremo, *A Hydropolitical History of South Africa's International River Basins* (Pretoria: Water Research Commission, 2004).

Van der Walt, I. J., A. Struwig, and J. R. J. van Rensburg, 'Forestry as a Streamflow Reduction Activity in South Africa: Discussion and Evaluation of the Proposed Procedure for the Assessment of Afforestation Permit Applications in Terms of Water Sustainability', *GeoJournal*, 61 (2004), 178–79.

Van der Zel, D. W., 'Accomplishments and Dynamics of the South African Afforestation Permit System', *South African Forestry Journal*, 172 (1995), 49–58.

———, 'Sustainable Industrial Afforestation in South Africa under Water and Other Environmental Pressures', in *Sustainability of Water Resources under Increasing Uncertainly (Proceedings of the Rabat Symposium S1, April 1997)*, 1997, 217–25.

———, 'Umgeni River Systems Analysis', *Water SA*, 1 (1975), 70–75.

Van Jaarsveld, A. S., J. C. Pauw, S. Mundree, S. Mecenero, B. W. T. Coetzee, and G. F. Alard, 'South African Environmental Observation Network: Vision, Design and Status', *South African Journal of Science*, 103 (2007), 289–94.

Van Koppen, B., B. Schreiner, and S. Fakir, 'The Political Social and Economic Context of Changing Water Policy in South Africa Post-1994', in B. Schreiner, and R. Hassan (eds), *Integrated Water Resource Management in South Africa* (Resources for the Future Publications, 2010).

Van Sittert, L., 'From Mere Weeds and Bosjes to a Cape Floral Kingdom: The Re-Imagining of Indigenous Flora at the Cape, c.1890–1939', *Kronos*, 28 (2002), 102–26.

———, 'Making the Cape Floral Kingdom: The Discovery and Defense of Indigenous Flora at the Cape, c.1890–1939', *Landscape Research*, 28 (2003), 113–29.

————, 'Nation Building Knowledge: Dutch Indigenous Knowledge and the Invention of White South Africanism, 1890–1901', in D. Gordon and S. Krech (eds), *Indigenous Knowledge and the Environment in Africa and North America* (Athens, Ohio: Ohio University Press, 2012), 94–109.

————, '"Our Irrepressible Fellow-Colonist": The Biological Invasion of Prickly Pear (*Opuntia ficus-indica*) in the Eastern Cape c.1890–c.1910', *Journal of Historical Geography*, 28 (2002), 397–419.

————, '"The Seed Blows About in Every Breeze": Noxious Weed Eradication in the Cape Colony, 1860–1909', *Journal of Southern African Studies*, 26 (2000), 655–74.

Van Wilgen, B. W., 'The Economic Consequences of Alien Plant Invasions: Examples of Impacts and Approaches to Sustainable Management in South Africa', *Environment, Development and Sustainability*, 3 (2001), 145–68.

————, 'The Evolution of Fire and Invasive Alien Plant Management Practices in Fynbos', *South African Journal of Science*, 105 (2009), 335–41.

————, 'Introduction to John F. V. Phlllips' Article', *Fire Ecology*, 8 (2012), 1–2

————, 'Plantation Forestry and Invasive Pines in the Cape Floristic Region: Towards Conflict Resolution', *South African Journal of Science*, 111 (2015), 1–2.

Van Wilgen, B. W., and W. J. de Lange, 'The Costs and Benefits of Biological Control of Invasive Alien Plants in South Africa', *African Entomology*, 19 (2011), 504–14.

Van Wyk, D. B., 'Apparatus for Sampling of Streams for Chemical Quality and Sediment', *Water SA*, 9 (1983), 88–92.

————, 'The Influence of Catchment Management on Sediment and Nutrient Exports in the Natal Drakensberg', in R. E. Schulze (ed.), *Proceedings of the Second National Hydrological Symposium, Pietermaritzburg, University of Natal* (Pietermaritzburg: University of Natal, 1986), 266–75.

————, 'The Influence of Prescribed Burning as a Management Tool on the Nutrient Budgets of Mountain Fynbos Catchments in the South-Western Cape, Republic of South Africa', in C. E. Conrad, and W. C. Oechel (eds), *Proceedings of the Symposium on Dynamics and Management of Mediterranean-Type Ecosystems; June 22–26, 1981; San Diego, CA* (Berkeley, CA: Pacific Southwest Forest and Range Experiment Station, Forest Service, US Department of Agriculture, 1982), 390–96.

Vetter, J. (ed.), *Knowing Global Environments: New Historical Perspectives on the Field Sciences* (Piscataway: Rutgers University Press, 2011).

Visser, N., 'A Space for Conflict: The Scab Acts of the Cape Colony, circa 1874–1911' (unpublished PhD Thesis, University of Cape Town, 2011).

Waag, I., 'Rural Struggles and the Politics of a Colonial Command: The Southern Mounted Rifles of the Transvaal Volunteers, 1905–1912', in S. Miller (ed.), *Soldiers and Settlers in Africa: 1850–1918* (Leiden: Brill Press, 2009), 251–86.

Wicht, C. L., 'Afrikaanse Bosboomname', *Journal of the South African Forestry Association*, 5 (1940), 41–61.

———, 'Depletion of Ground-Water Flow in Jonkershoek Streams', *Journal of the South African Forestry Association*, 8 (1942), 50–63.

———, 'Determination of the Effects of Watershed-Management on Mountain Streams', *Transactions of the American Geophysical Union*, 24 (1943), 594–606.

———, 'Diurnal Fluctuations in Jonkershoek Streams', *Journal of the South African Forestry Association*, 7 (1941), 34–49.

———, 'The Effects of Timber Plantations on Water Supplies in South Africa', in *Proceedings of the Symposium of Hannoversch-Münden, 8–14 September 1959* (Hannoversch-Münden: International Association of Scientific Hydrology, 1959), 238–44.

———, 'Forest Hydrological Research in Africa South of the Sahara', in *Wasserwirtschaft in Afrika* (Köln: Verlag Deutscher Wirtschaftsdienst, 1963).

———, 'Forest Hydrology Research in the South African Republic', in W. E. Lull, and W. H. Sopper (eds), *International Symposium on Forest Hydrology* (Oxford: Pergamon Press, 1967), 75–84.

———, 'Forest Influences Research Technique at Jonkershoek', *Journal of the South African Forestry Association*, 3 (1939), 65–78.

———, *Forestry and Water Supplies in South Africa* (Pretoria: Department of Forestry, Union of South Africa, 1949).

———, 'Forestry Research – Elixir of the Industry', *South African Forestry Journal*, 53 (1965).

———, *Hydrological Research in South African Forestry. British Empire Forestry Conference, 1947, Great Britain* (Department of Forestry, Union of South Africa, 1947).

————, 'Improvements in the Gauging of Rainfall', *Journal of the South African Forestry Association*, 12 (1944), 19–28.

————, *Land En Lewe: Die Behoud van Grond, Water, Plante, En Diere*, Excelsiorboekies, VI No. 7 (Afrikaanse Pers-Boekhandel, 1952).

————, 'A Preliminary Account of Rainfall in Jonkershoek', *Transactions of the Royal Society of South Africa*, 28 (1941), 161–73.

————, 'Preservation of the Vegetation of the South Western Cape', *Transactions of the Royal Society of South Africa*, 30 (1943), 7–18, dx.doi. org/10.1080/00359194309519824.

————, 'A Statistically Designed Experiment to Test the Effects of Burning on a Sclerophyll Scrub Community. I. Preliminary Account', *Transactions of the Royal Society of South Africa*, 31 (1948), 479–501.

————, 'Summary of Forests and Evapotranspiration Session', in W. E. Lull, and W. H. Sopper (eds), *International Symposium on Forest Hydrology* (Oxford: Pergamon Press, 1967), 493.

————, 'The Validity of Conclusions from South African Multiple Watershed Experiments', in W. E. Lull, and W. H. Sopper (eds), *International Symposium on Forest Hydrology* (Oxford: Pergamon Press, 1967), 749–60.

————, 'The Variability of Jonkershoek Streams', *Journal of the South African Forestry Association*, 10 (1943), 13–22.

————, 'Trends in Forest Hydrological Research', *South African Forestry Journal*, 30 (1966), 17–25.

————, *Zur Methodik Des Durchforstungsversuchs. Mitteilung Aus Der Sächsischen Forstlichen Versuchsanstalt, Abteilung Für Ertragskunde* (Dresden: Bufra, Buchdruckerei Otto Franke, 1934).

Wicht, C. L., J. C. Meyburgh, and P. G. Boustead, 'Rainfall at the Jonkershoek Forest Hydrological Research Station', *Annals of the University of Stellenbosch*, 44 Series A (1969), 1–66.

Wicht, C. L., and D. E. Schumann, 'Experimental Investigation of the Effects of Forests on Stream-Discharge', *Paper Presented to the Commonwealth Forestry Conference, Australia and New Zealand, 1957* (Pretoria: Government Printer, Union of South Africa, 1957), 12.

Wicht, M. L., 'Creeping Invasions of the "Green Cancers"', *African Wildlife*, 25 (1971), 11–14.

Winsor, M., 'The Practitioner of Science: Everyone Her Own Historian', *Journal of the History of Biology*, 34 (2001), 229–45.

Wise, R. M., P. J. Dye, and M. B. Gush, 'A Comparison of the Biophysical and Economic Water Use Efficiencies of Indigenous and Introduced Forests in South Africa', *Forest Ecology and Management*, 262 (2011), 906–15.

Witt, H., '"Clothing the Once Bare Brown Hills of Natal": The Origin and Development of Wattle Growing in Natal, 1860–1960', *South African Historical Journal*, 53 (2003), 99–122.

——, 'The Emergence of Privately Grown Industrial Tree Plantations', in S. Dovers, R. Edgecombe, and B. Guest (eds), *South Africa's Environmental History: Cases and Comparisons* (Athens, OH: Ohio University Press, 2002), 90–112.

Worden, N., E. van Heyningen, and V. Bickford-Smith, *Cape Town: Making of a City* (Cape Town: New Africa Books, 2004).

Zhang, L., W. R. Dawes, and G. R. Walker, 'Response of Mean Annual Evapotranspiration to Vegetation Changes at Catchment Scale', *Water Resources Research*, 37 (2001), 701–8.

www.ingramcontent.com/pod-product-compliance
Lightning Source LLC
Chambersburg PA
CBHW061227270326
41928CB00025B/3392